Tribology for Lightweighting

Tribology for Lightweighting

Guest Editors

Montserrat Vilaseca Llosada
Leonardo Pelcastre

MDPI • Basel • Beijing • Wuhan • Barcelona • Belgrade • Novi Sad • Cluj • Manchester

Guest Editors
Montserrat Vilaseca Llosada
Department of Metallic and Ceramic Materials
Eurecat, Technology Centre of Catalonia
Manresa
Spain

Leonardo Pelcastre
Department of Engineering Sciences and Mathematics
Luleå Tekniska Universitet
Luleå
Sweden

Editorial Office
MDPI AG
Grosspeteranlage 5
4052 Basel, Switzerland

This is a reprint of the Special Issue, published open access by the journal *Lubricants* (ISSN 2075-4442), freely accessible at: https://www.mdpi.com/journal/lubricants/special_issues/tribology_lightweighting.

For citation purposes, cite each article independently as indicated on the article page online and as indicated below:

Lastname, A.A.; Lastname, B.B. Article Title. *Journal Name* **Year**, *Volume Number*, Page Range.

ISBN 978-3-7258-7372-2 (Hbk)
ISBN 978-3-7258-7373-9 (PDF)
https://doi.org/10.3390/books978-3-7258-7373-9

Contents

lubricants

MDPI

rticle

nvestigation of the Structural, Mechanical and Tribological roperties of Plasma Electrolytic Hardened hromium-Nickel Steel

auyrzhan Rakhadilov [1,2], Ainur Seitkhanova [3], Zarina Satbayeva [1,4,*], Gulnara Yerbolatova [5], Yulianna Icheva [6] and huldyz Sagdoldina [1]

1 Research Center Surface Engineering and Tribology, Sarsen Amanzholov East-Kazakhstan University, Ust-Kamenogorsk 070000, Kazakhstan; rakhadilovb@mail.ru (B.R.); sagdoldina@mail.ru (Z.S.)
2 PlasmaScience Ltd., Ust-Kamenogorsk 070000, Kazakhstan
3 Department of Physics, Pavlodar Pedagogical University, Pavlodar 140000, Kazakhstan; ainur1179@mail.ru
4 Institute of Composite Materials, Ust-Kamenogorsk 070000, Kazakhstan
5 Department of Physics, D. Serikbayev East-Kazakhstan Technical University, Ust-Kamenogorsk 070000, Kazakhstan; e.gulnara_77@mail.ru
6 Department of Metallurgy and Natural Science Disciplines, Academician K. Satpayev Ekibastuz Engineering and Technical Institute, Ekibastuz 141200, Kazakhstan; isheva1967@mail.ru
* Correspondence: satbaeva.z@mail.ru; Tel.: +7-(777)-300-09-39

Abstract: This paper investigates how electrolytic plasma hardening (PEH) bears upon the changes in the phase structural and tribological properties of steel 0.34C-1Cr-1Ni-1Mo-Fe, which is widely used in manufacturing highly stressed gears. The samples of steel 0.34C-1Cr-1Ni-1Mo-Fe went through the PEH in an electrolyte containing an aqua solution of 20% calcined soda (Na_2CO_3) and 10% carbamide ($(NH_2)_2CO$). The initial steel 0.34C-1Cr-1Ni-1Mo-Fe is stated to have the following structural components: a lamellar pearlite with volume share of 35%, a ferrite-carbide mixture of ~45% and a fragmented ferrite of ~20%; after the PEH it contains lath-lamellar martensite, fine particles of cementite and $M_{23}C_6$ carbide. The durability of steel 0.34C-1Cr-1Ni-1Mo-Fe was found to rise by 3.4 times after the PEH and its microhardness increased in 2.6 times. The curve-tension of the crystal lattice was established to be like plastic ($\chi = \chi_{pl}$) and does not cause the formation of microcracks in the material.

Keywords: electrolytic-plasma hardening (PEH); modified layer; phase composition; microhardness; wear-resisting properties; durability

1. Introduction

The continuous improvement of the operational characteristics of modern machines—along with their basic units and parts, equipment and facilities—is known to be provided by increasing their capacity and performance, which requires intensively improving the performance of these machines and parts [1]. The industry constantly requires the improvement and intensification of the working and operational characteristics of modern equipment (including the equipment, parts and their node), which is provided by increasing their power and productivity. In addition, using modern hardening technologies, the operational characteristics of the parts' working surfaces are improved, leading to an increase in the equipment's service life. The conditions of the working surface layer, where damage is actively developing—and thereby shortens the service life of any equipment's part—provides an assessment of the technical and economic indicators of an equipment's operation. To ensure a high cyclic durability, high wear resistance, and reduced sensitivity to stress concentrators, it is necessary to create a gradient of properties in the hardened section of a part that provides a hard and wear-resistant surface, a viscous but strong core, and compressive stresses in the surface layer [2,3]. This all becomes possible if we apply the method of surface hardening.

The on-site surface hardening of wearing surfaces, combined with several other techniques, is today considered promising, alongside metallurgical techniques and surface hardening, in the conditions of manufacturers to increase the service life of gears. The surface hardening of steel parts is one of the most efficient and forceful ways to increase the service life of the loaded elements of machines and mechanisms, as well as to reduce their material consumption. At the same time, only the most loaded working surface of the part is strengthened, leaving the core intact [4]. At the same time, progress in improving the quality of tempering (hardening) of the working surfaces of parts is associated with the use of concentrated energy sources: electron beams, laser beams and plasma jets. Such methods allow higher performance properties and quality of hardening to be achieved.

High-frequency, gas flame, plasma, electron beam and laser processing are widely used in industry for the surface hardening of gears [5]. Existing strengthening techniques are widely and effectively applied in manufacturing according to their technical and economic indicators [6–8]. Plasma electrolytic hardening (PEH) is one type of plasma surface hardening. The main distinctive features of the PEH method are its lower cost, greater availability of process equipment and consumables, larger area of the hardened zone and a higher cooling rate compared to conventional methods of plasma surface hardening. The idea behind this method is in the thermal phase and structural transformations that take place during the rapid, concentrated heating of the working surface of the part by plasma action, which is followed by rapid cooling due to heat transfer to the part material [9]. The quenching-type structures formed by this method have a high hardness, wear resistance and rupture strength.

By improving the characteristics of the material, PEH allows more efficient structures to be made that decrease the equipment's weight, increase the service life of its components and increase its productivity. An effective method for reducing the weight is to increase the strength of the materials. Unlike the method of increasing stresses by reducing the actual strength reserve related to the risk of a weakening of the part, the reliability in this case does not decrease (i.e., if the value of the strength reserve is maintained). This method works for all parts without exception, whereas the method of increasing stresses is applicable only to design parts.

In view of this fact, the purpose of this paper is to investigate how plasma-electrolytic hardening influences the phase-structural conditions and mechanical and tribological properties of 0.34C-1Cr-1Ni-1Mo-Fe steel.

2. Materials and Methods

Steel 0.34C-1Cr-1Ni-1Mo-Fe was chosen as a subject according to the study's goal. This choice is explained by the fact that this steel is widely used in manufacturing highly stressed gears.

Steel samples underwent plasma electrolytic hardening in a machine consisting of a source of power, a PEH chamber and a computer [10]. Figure 1 illustrates the sequence of the PEH process. The chemical composition of the steel studied is provided by the commercial TU 24-1-12-179-75 characteristics, but the measured chemical composition of the studied steel also include: C: 0.3–0.4%; Si: 0.17–0.37%; Mn: 0.5–0.8%; Ni: 1.3–1.7%; Cr: 1.3–1.7%; Mo: 0.2–0.3%; S: up to 0.035%; P: up to 0.03%.

A sodium-carbonate-based electrolyte was chosen for conducting PEH. Carbamide was added to the electrolyte to prevent decarbonizing. Carbamide was chosen because it is a source of carbon and is cheap and environment-friendly. In addition, sodium carbonate interacts well with carbamide. Therefore, PEH was performed in an electrolyte containing an aqua solution of 20% calcined soda (Na_2CO_3) and 10% carbamide ($(NH_2)_2CO$).

The PEH of steel samples was carried out as follows: the bath (6) was filled with electrolyte; the electrolyte was fed by a pump (4) to the cone-shaped nozzle (3), while the electrolyte jet flowed out through the hole, drained over the edge into the bath (6) and came back into the cone-shaped nozzle. Thus, the electrolyte was constantly in the circulation mode. The electrolyte feed rate (flow rate) was 5–7 L/min. A processed sample was immersed in the electrolyte by means of a holding mechanism so that the processed area of the sample was located at a distance of 2–3 mm from the opening of the cone-shaped nozzle. At the same time, an electrolyte jet was supplied to the processed area through the opening in the cone shaped nozzle located 10 15 mm below the top edge of the dividing wall. An anode (2) was connected to the positive pole of the power supply, and a processed sample—cathode (1)—was connected to its negative pole. A voltage of 320 V was supplied between the electrodes for 2 s to heat the electrolyte to the quenching temperature (800–900 °C); the current density here was 25–30 A/cm^2. Under these stresses, an intensely luminous plasma layer was formed in the near-cathode region and the sample was heated at a velocity of 400–500 °C/s. In such a case, an abnormal arc discharge is formed between the electrodes, resulting in a rapid heating of the in-process part [11].

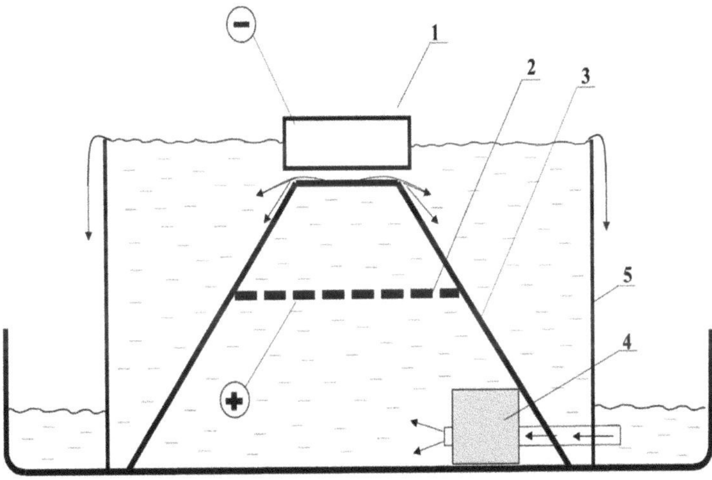

Figure 1. The sequence of the samples' PEH. 1—sample (cathode), 2—anode, 3—conic nozzle, 4—pump, 5—divide wall, 6—bath.

The NEOPHOT-21 optical microscope was used to study general metal structure characteristics. Metallographic cuts of the steel samples were prepared following the techniques described in paper [12]. Morphological and elemental analysis of the samples processed in electrolytic plasma was obtained using a scanning electron microscope JSM-6390LV (JEOL, Japan) with an INCAEnergy add-on device for energy dispersion microanalysis (OXFORD Instruments). X-ray diffraction studies of the steel samples were performed through standard X-ray diffraction analysis methods using X′PertPRO diffractometers. Diffractogram videos were taken using CuKα- radiation (λ = 2.2897 A^0) at a voltage of 40 kV [13]. Fine structure morphology was studied using an EM-125 electron microscope with an accelerating voltage of 125 kV. Working magnification in the column of the microscope was chosen to be from 8000 to 50,000 times.

The microhardness of the steel samples was measured in the PMT-3 device in accordance with GOST 9450-76, under the loads on the indenter P = 1 N and a holding time of 10 s [14]. For the samples, the nanohardness of coatings was determined by the Oliver-Pharr method using the nanoidentation system with the Berkovich indenter at a load of 100 mN/g and a holding time of 5 s. Tribological tests for the sliding friction were performed on a high-temperature tribometer THT-S-BE-0000 using the standard ball–disk technique according to the international standards ASTM G 133-95 and ASTM G 99. A ball with a diameter of 6.0 mm and made of a certified material (Al_2O_3) was used as a counterweight. Tests were performed at a load of 1 N and a linear velocity of 2 cm/s, and with a wear radius of 6 mm and a friction path of 25 m. Tribological characteristics of the modified layer were evaluated by the wear intensity, wear volume and friction coefficient.

Strength characteristics of the steel samples were determined through mechanical tensile testing at room temperature. Mechanical tests were performed in a universal testing machine WDW-5E in accordance with the requirements of GOST 1497-87 for steel 0.34C-1Cr-1Ni-1Mo-Fe. The tests were laid in a uniaxial static stretching of the flat samples up to their rupture with measuring conditional yield strength $\sigma_{0.2}$, fracture strength σ_B and relative elongation to rupture δ.

3. Results and Discussion

3.1. Structure and Morphology

The structural phase conditions of the hardened steel surface layers 0.34C-1Cr-1Ni-1Mo-Fe were investigated before and after PEH. According to the optico-metallographic analysis, the initial structure of steel 0.34C-1Cr-1Ni-1Mo-Fe is similar to the ferrite–pearlite structure (Figure 2a), but there are many more pearlite grains with a cementite network at their borders than there is ferrite. The hardening of 0.34C-1Cr-1Ni-1Mo-Fe surface layers with the PEH technique resulted in the formation of a fine martensite structure (Figure 2b).

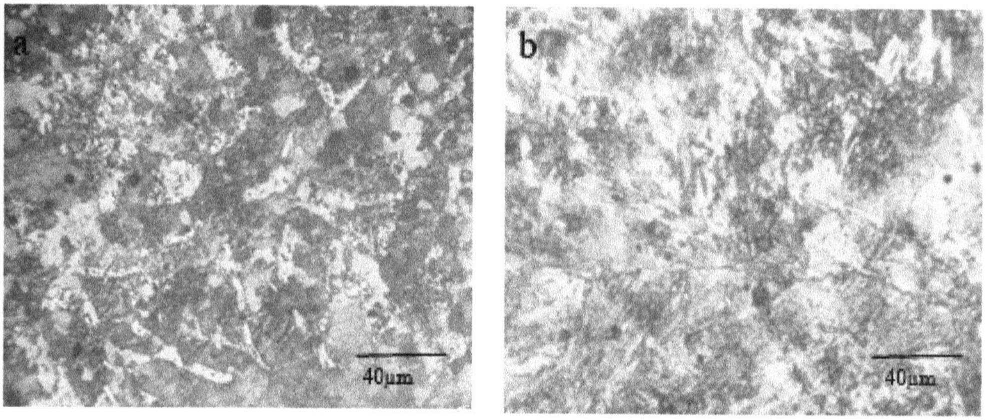

Figure 2. Microstructure of the surface of a 0.34C-1Cr-1Ni-1Mo-Fe sample before and after the PEH. (**a**)—0.34C-1Cr-1Ni-1Mo-Fe initial state, (**b**)—0.34C-1Cr-1Ni-1Mo-Fe after PEH.

Figure 3 shows X-ray diffractograms of 0.34 C-1Cr-1Ni-1Mo-Fe steel before and after the PEH. The X-ray diffraction analysis showed that the initial structure of 0.34 C-1Cr-1Ni-1Mo-Fe steel consisted of an α-phase. After the PEH, there were α-phase reflexes broadening that indicate the formation of martensite. After the PEH, cementite and a small amount of austenite were also formed. The formation of residual austenite was associated with a high cooling rate, and polymorphic transformations did not have enough time to occur. The metallographic analysis showed that there was cementite in the initial state of the steel structure, but the X-ray spectral analysis (XRS) did not detect any cementite in the initial state. This was due to the low content of cementite in the initial state. One of the disadvantages of XRS is the low amplitude of the X-ray scattering of light atoms such as hydrogen, carbon, oxygen, etc., which implies a weak sensitivity to them, especially compared to heavy atoms.

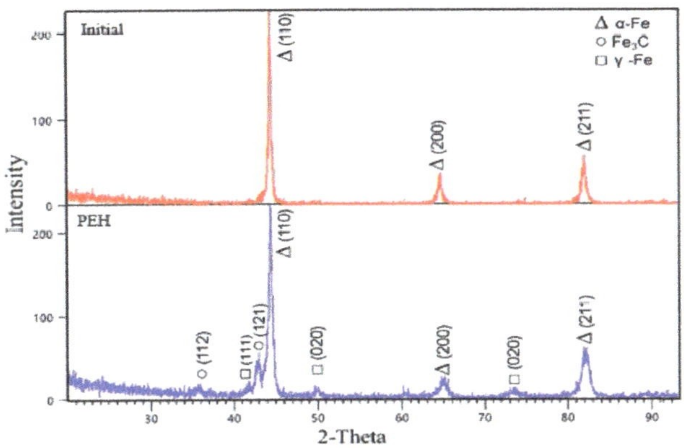

Figure 3. X-ray diffractogram of the 0.34C-1Cr-1Ni-1Mo-Fe steel samples before and after the PEH.

According to numerous studies [15] the formation of a modified surface layer with a phase structural component constituted by fine grained martensite with residual austenite will positively affect the tribological properties of spare parts.

Figure 4 shows fragments of a microstructure obtained using the Scanning electron microscope, a cross section of 0.34C-1Cr-1Ni-1Mo-Fe steel in the course of the PEH, where the zonality of the structure is observed that is typical for surface hardening. The PEH of the steel led to the modification of the sample's surface layer. The layer structure changed as it moved away from the sample's surface. The image of the cross-section microchip shown in Figure 4 was taken using the Scanning electron microscope at a relatively small magnification. It can be seen that a modified surface layer consisting of martensite with an average thickness of ~1.7–2 mm is clearly distinguished on the cross-section of the microchip. Then, moving depthward down the sample we can see a transition layer followed by the layer (zone) of the main material.

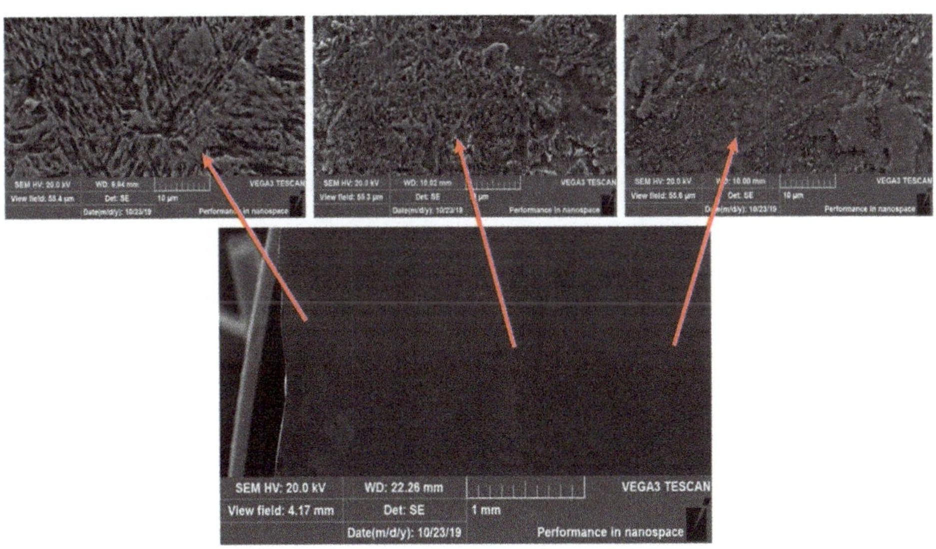

Figure 4. Cross-section of the PEH sample, and photomicrographs from the surface (**left**) to bulk material (**right**).

Figure 5 shows the chemical composition of the modified surface layer through the depth of the steel 0.34C-1Cr-1Ni-1Mo-Fe obtained using the EDS analysis. Based on the results regarding the depth, it can be stated that carbon atoms were collected on the surface of modified layer. That is due to the fact that the PEH process took place in an electrolyte containing aqua solution of 20% sodium carbonate and 10% carbamide where the surface was saturated with carbon.

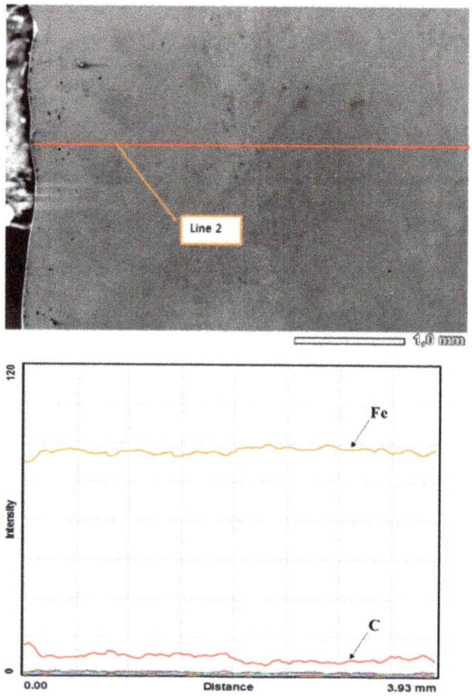

Figure 5. EDS analysis of the modified surface layer through the depth of a 0.34C-1Cr-1Ni-1Mo-Fe steel sample.

The results of the transmission electron microscopy of the initial steel structure confirmed the data from the XRS and metallographic analysis. According to this research and the images of the fine structure obtained using the electron microscopic analysis, the initial state of the 0.34C-1Cr-1Ni-1Mo-Fe steel structure consisted of lamellar pearlite, a ferritocarbide mixture and fragmented ferrite. An electron microscope image of the lamellar pearlite in the 0.34C-1Cr-1Ni-1Mo-Fe steel is shown in Figure 6. The volume fraction of the pearlite in the initial state of the 0.34C-1Cr-1Ni-1Mo-Fe steel was 35%, and the volume fractions of the ideal perlite and fragmented pearlite were 10% and 25%, respectively. The volume fraction of the second component of the structure—the ferritocarbide mixture—was ~45%. The average scalar density of the dislocations inside the fragments of the α-phase was 2.25×10^{10} cm^{-2}. The volume fraction of the third component of the structure—fragmented ferrite—was 20% of the material. The average scalar density of the dislocations inside the fragments of the α-phase was 2.95×10^{10} cm^{-2}.

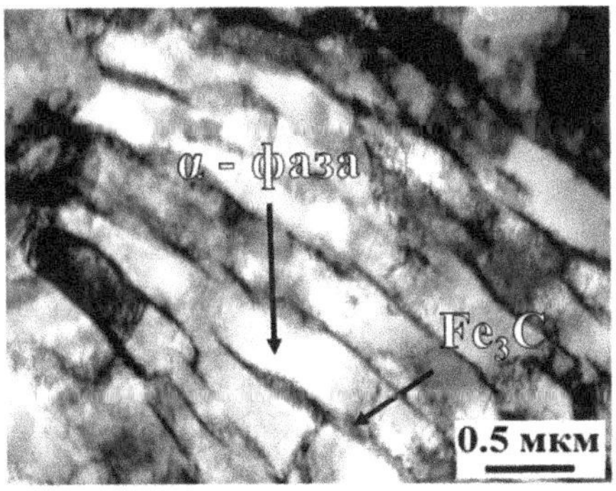

Figure 6. Electron-microimage of the 0.34C-1Cr-1Ni-1Mo-Fe steel in its initial state.

The results of the transmission electron microscopy of the steel after the PEH showed that the structure contained $M_{23}C_6$ carbide along with martensite, cementite and residual austenite. Figure 7 provides an electron microimage of the thin structure of the hardened steel surface layer that shows the allocation of the $M_{23}C_6$ carbide (marked with letter K near the indicated reflexes) in the sublayers of the residual austenite located at the boundaries of the martensitic lamellae. The arrows in Figure 7b indicate the matching directions: (1) $[\ \bar{1}\bar{1}1\ ]_\gamma$ *II* $[\ 0\bar{2}1\ ]_\alpha$, wherein the assumption is performed $(112)_\gamma$ *II* $(012)_\alpha$—the Kurdyumov-Sachs orientation relation; and (2) $[\ \bar{1}10\ ]_\gamma$ *II* $[\ 3\bar{1}0\ ]_K$, wherein the assumption is performed $(112)_\gamma$ *II* $(131)_K$—the cube–cube orientation relation.

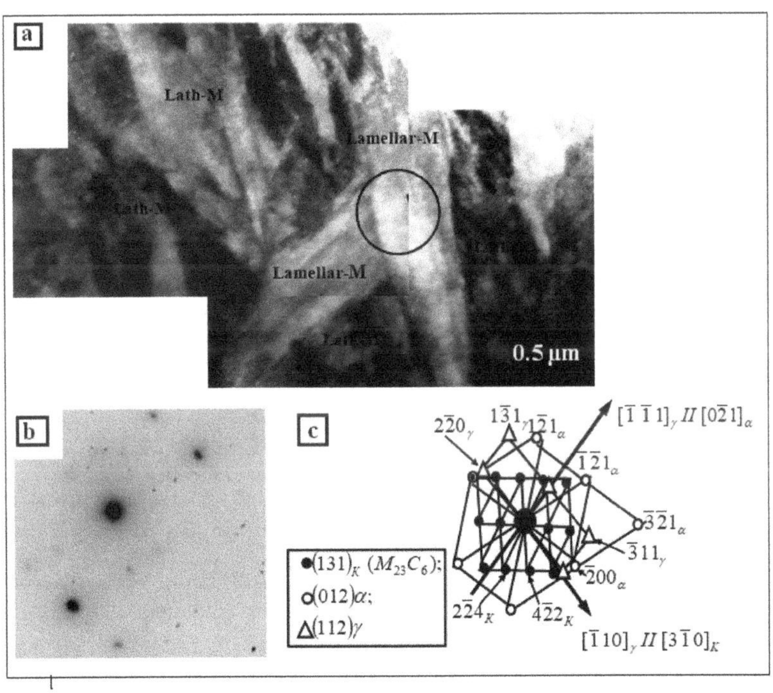

Figure 7. Electron microimage of 0.34C-1Cr-1Ni-1Mo-Fe steel after PEH. (**a**)—light field image; (**b**)—microdiffraction picture taken in the region marked on (**a**) with a circle; (**c**)—its indicated scheme.

During the bending-torsion of a crystal lattice, the amplitude of the internal stresses was calculated. It turned out that σ_d = 285 MPa in the lath martensite, σ_d = 270 MPa in the lamellar martensite and σ_d =280 MPa throughout the material.

One more type of internal stress, called shear stresses σ_L (stresses created by dislocation structure), is known to be present when a defective structure is represented by dislocations in a material. The calculations show that σ_L = 390 MPa in the lath martensite, σ_L = 345 MPa in the lamellar martensite and σ_L = 370 MPa throughout the material.

It can be seen, first, that all quantitative parameters in the lath martensite were higher than in the lamellar martensite. Secondly, the following conditions are met along the material after the surface hardening, as in the initial state: $\rho > \rho_{\pm}$ и$\sigma_L > \sigma_{д}$. This means that the bending-torsion (distortion) of the crystal lattice of the 0.34C-1Cr-1Ni-1Mo-Fe steel after PEH was also purely plastic, which will not lead to the formation of microcracks in the material.

3.2. *Mechanical and Tribological Properties*

The modification of the tribological and mechanical properties of structural steel 0.34C-1Cr-1Ni-1Mo-Fe before and after it has undergone the PEH have been experimentally studied. Figure 8 shows results of the tribological tests of the 0.34C-1Cr-1Ni-1Mo-Fe samples according to the ball-on-disk technique. The wear resistance of the samples was characterized by the intensity of the wear and the volume of the wear on the steel samples 0.34C-1Cr-1Ni-1Mo-Fe before and after the PEH. Figure 8 shows that the processed samples had a low wear rate compared to the initial sample. Figure 8b illustrates that after PEH the samples wore down very little compared to the initial samples, which demonstrated the higher wear resistance of the steel after the PEH.

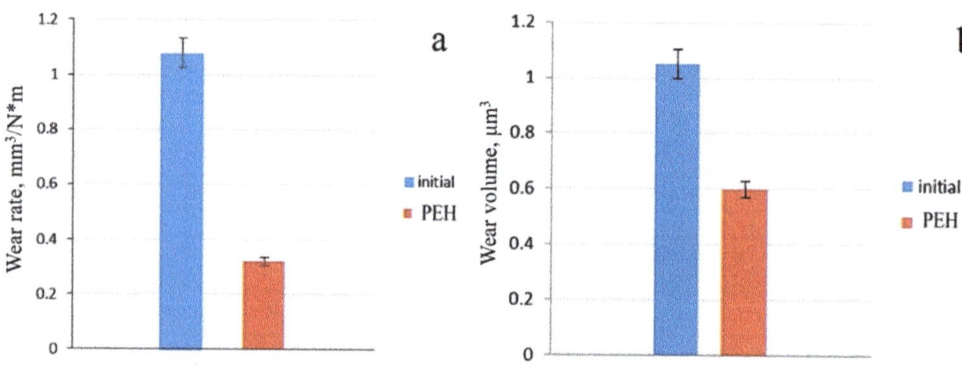

Figure 8. Wear rate (**a**) and wear volume (**b**) of steel 0.34C-1Cr-1Ni-1Mo-Fe before and after the PEH.

For parts of this functional purpose, wear resistance is an important operational property. The level of wear resistance affects both the performance of structures as a whole and the preservation of the geometric sizes of individual parts. The results of tribological tests have shown that the electrolyte-plasma hardening has a significant effect on wear resistance.

According to the results of X-ray phase analysis and TEM studies, an increase in the wear resistance of the surface layer of steel is related to the formation of a modified surface layer with a phase structural component of fine grained martensite with residual austenite.

The intensity of wear under the influence of the tip is calculated based on the volume of material displaced during the test, which was calculated using the following formula:

$$I = \frac{V}{F * l},$$

where: I iswear intensity [mm^3/N*m]; l is the friction path, [m]; F is the nominal pressure [H]; and V is the volume of the wear part [mm^3]. As a result of the calculations, data on the wear intensity for the samples before and after the PEH were obtained.

Images of the wear tracks of the 0.34C-1Cr-1Ni-1Mo-Fe steel samples were taken using a profile meter (Figure 9). When evaluating the samples' wear resistance based on the geometric parameters of the wear tracks we can say that the depth of the sample track after the PEH (Figure 9b) was significantly smaller than that of the initial samples (Figure 9a). The shape of the unevenness generally once again demonstrates improved tribological properties in the testing sample. Figure 9c shows a graph of the behavior of the friction coefficient in the course of the experiment. The measurements showed minor changes in the coefficient of friction. The surface roughness was measured using the Ra parameter using profilometer model 130 on a 7 mm long segment of the sample surface. From the data obtained, the value of the parameter Ra for the initial sample was 0.199 and after processing Ra = 0.446 [16].

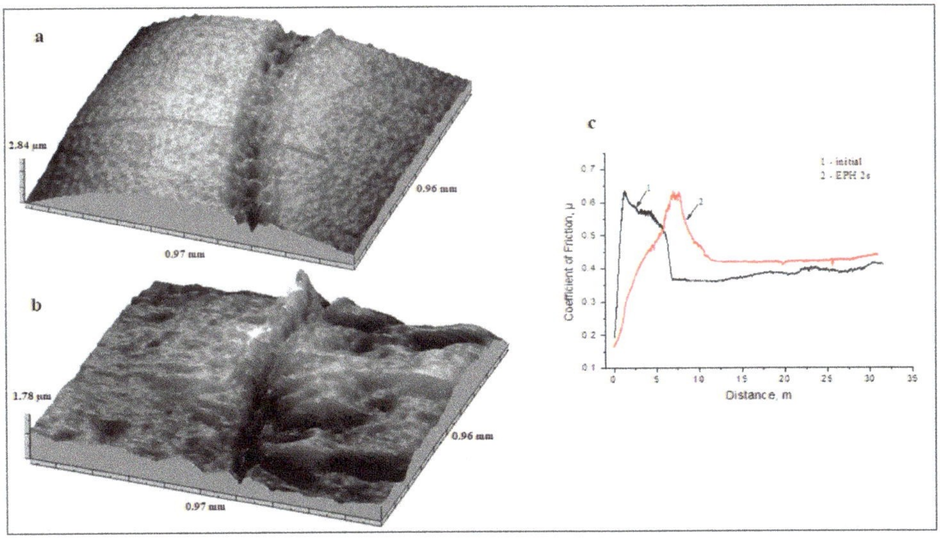

Figure 9. Fragment picture of the wear track and a graph of the behavior of the friction coefficient of the 0.34C-1Cr-1Ni-1Mo-Fe steel samples: (**a**)—untreated sample, (**b**)—sample after the PEH, (**c**)—graph of the change in the coefficient of friction.

Figure 10 shows a graph of the loading dependence on the extension of 0.34C-1Cr-1Ni-1Mo-Fe before and after the PEH. According to the results of the tensile test of 0.34C-1Cr-1Ni-1Mo-Fe steel, the following parameter changes were obtained: the strength limit (σ_B) in the initial state was 743 MPa; after PEH it increased to 835 MPa, i.e., by 1.12 times. There was the same change in the yield strength ($\sigma_{0.2}$); in the initial state, it was 648 MPa, and after the PEH it was 740 MPa, which indicates an increase by 1.14 times in contrast to the initial state. The value of the relative elongation before the break (δ) in the initial state was 10% and after processing it increased to 13%.

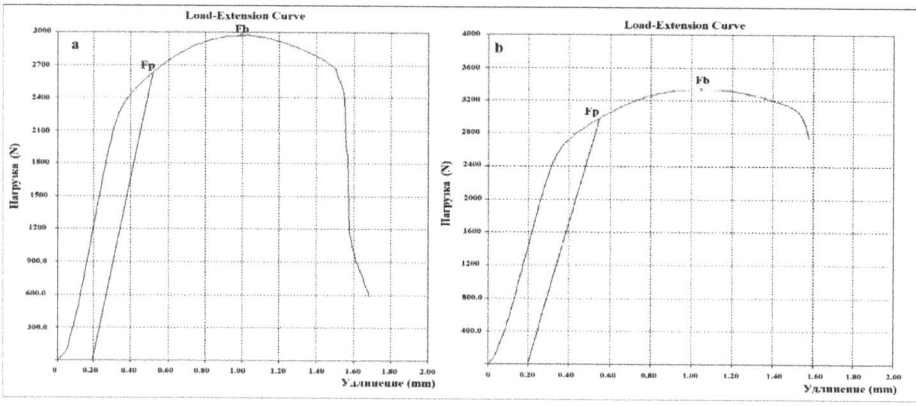

Figure 10. A graph of the loading dependence on the extension of the 0.34C-1Cr-1Ni-1Mo-Fe steel samples before (**a**) and after (**b**) the PEH.

Figure 11 shows a graph of the change in microhardness through the depth of the sample treated by the PEH method. The microhardness data confirmed the formation of a martensitic structure. A significant increase in microhardness is seen near the surface. The transition zone displayed a smooth transition from the hardened layer to the basement, while the microhardness of the transition zone was slightly less than at the basement, and the microhardness of the basement did not change.

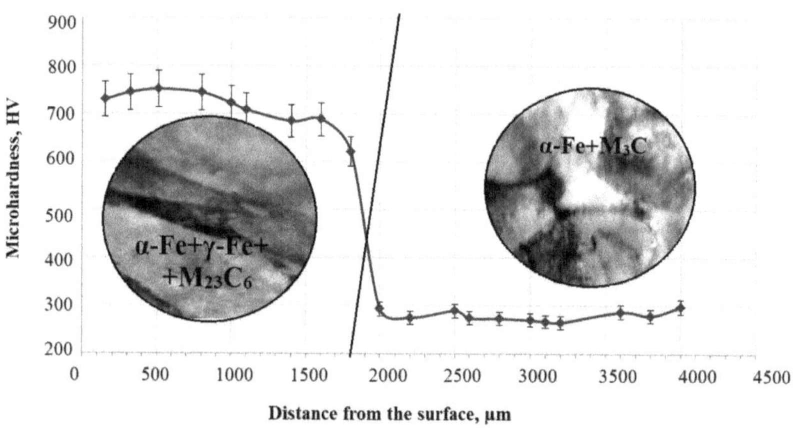

Figure 11. Diagram of the distribution of microhardness through the depth of the steel 0.34C-1Cr-1Ni-1Mo-Fe samples after the PEH.

The results of the nanoindentation of the 0.34C-1Cr-1Ni-1Mo-Fe steel samples before and after the PEH showed that the nanohardness of the sample after PEH—determined by the loading diagram—was on average 8.12 GPa. This was 2.5 times more than the nanohardness of the initial sample.

According to earlier studies [17–19], most ferrite–perlite steels after electrolytic-plasma treatment usually have high wear resistance, microhardness, corrosion resistance and strength characteristics. A correct understanding of the mechanisms of strengthening the surface layer of steel during the PEH allows us to infer the structure of the layer and anticipate the changes that may occur in it depending on the nature of alloying. In this regard, the need to establish the main mechanisms of the electrolyte-plasma hardening to improve the performance of steel parts remains relevant.

Table 1 provides data on the structure and tribological-mechanical characteristics of steel after the PEH. The experimental data clearly illustrate the correlation between the structural and tribological–technical characteristics of the hardened samples.

Table 1. Experimental data on the structure and tribological–mechanical behavior of steel 0.34C-1Cr-1Ni-1Mo-Fe.

Material	Characteristics								
	Phase Composition	H_μ, MPa	f	j, 10^{-4} mm^3/Nm	V, μm^3	σ_B, MPa	$\sigma_{0.2}$, MPa	H, GPa	E, GPa
Initial state	α-phase	2830	0.44	1.08	1.05	743	648	2.88	212.33
After PEH, 2 s	α′-phase, γ-phase, M_3C, $M_{23}C_6$	7420	0.37	0.32	0.6	835	740	8.12	296.25

Note. H_μ—microhardness; f—coefficient of friction; j—wear rate; V—wear volume; σ_B—fracture strength; $\sigma_{0.2}$—tensile yield; N—nanohardness; Y—Young's modulus.

From the generalized data given in Table 1 it can be seen that there was a significant increase in the tribological and mechanical properties of steel after the PEH, which leads to the formation of α′-phase (martensite), γ′-phase, M_3C cementite and $M_{23}C_6$ carbide. Thus, it was found that as a result of the PEH, the surface layer of steel 0.34C-1Cr-1Ni-1Mo-Fe had stronger microhardness and higher wear resistance. The increase in microhardness and wear resistance of 0.34C-1Cr-1Ni-1Mo-Fe after PEH was particularly associated with the formation of martensite, as well as the formation of a defective substructure. A special role was given to fine particles of carbide phases. It is known from the papers [20,21] that the effects of increasing hardness and wear resistance are directly related to the size and number of dispersed inclusions.

4. Conclusions

The following conclusions can be drawn from the analysis of the results obtained:

- Electrolytic-plasma hardening helps to improve the mechanical properties of 0.34C-1Cr-1Ni-1Mo-Fe steel; namely, the microhardness increases by 2.6 times, and the strength limit σ_B and yield strength $\sigma_{0.2}$ increase to 835 MPa and 740 MPa, respectively, compared to those of the initial sample, which were 743 MPa and 648 MPa;
- The PEH improves the tribological properties of steel: the wear rate j of steel 0.34C-1Cr-1Ni-1Mo-Fe is equal to 0.32 mm^3/Nm after the PEH; that is, 3.4 times less than the wear rate of the initial steel;
- The PEH leads to the formation of a modified layer consisting of an α′-phase (packet plate martensite), a γ′-phase, small particles of cementite M_3C and uniformly arranged in the volume of the $M_{23}C_6$ carbide material;
- Strengthening the microhardness and wear resistance of 0.34C-1Cr-1Ni-1Mo-Fe steel is associated with the formation of martensite, as well as the formation of a defective substructure.

Author Contributions: Conceptualization, B.R.; methodology, B.R.; formal analysis, Z.S. (Zhuldyz Sagdoldina), G.Y., A.S., Y.I.; investigation, Z.S. (Zarina Satbayeva) G.Y., A.S., Y.I., writing—original draft preparation, B.R., Z.S. (Zarina Satbayeva), Z.S. (Zhuldyz Sagdoldina); writing—review and editing, B.R., Z.S. All authors have read and agreed to the published version of the manuscript.

Funding: This research was funded by the Ministry of Education of the Republic of Kazakhstan, grant number AP09058547.

Conflicts of Interest: The authors declare no conflict of interest.

References

1. Safonov, E.N. *Plasma Hardening of Machine Parts: Monograph*; NTI (branch) of UFU: Nizhny Tagil, Russia, 2014; p. 116.
2. Bolshakova, M.Y.; Guzanov, B.N.; Migacheva, G.N. *Probabilistic Method for Calculating the Durability of Heavy-Load Gears According to the Wear Criterion*; Theory and technology of metallurgical production: Mezhregion, collection of scientific papers; Kolokoltsev, V.M., Ed.; Publishing house SEI HVE MSTU: Magnitogorsk, Russia, 2010; Volume 10, pp. 193–204.
3. Medelyaev, N.A. Basic regularities of the processes of friction and wear in friction pairs of hydraulic machines. *Vestn. Mashinostroeniya* **2004**, *9*, 42–47.

4. Yeliseev, Y.S.; Krymov, V.V.; Nezhurin, I.P.; Novikov, V.S.; Ryzhov, N.M. *Production of Gear Wheels of Gas Turbine Engines*; Highe school: Moscow, Russia, 2001; p. 495.
5. Kalashnikov, A.S.; Morgunov, Y.A.; Kalashnikov, P.A. *Modern Methods of Processing Gears*; Publishing house Spectr: Moscov Russia, 2012; p. 238.
6. Samotugin, S.S. Plasma micro -and nanostructuring of the surface of tool steels. *Reinf. Technol. Coat.* **2013**, 29–37.
7. Korotkov, V.A. *Plasma Hardening of the Surface*; STI (Branch) of UFU: Nizhny Tagil, Russia, 2012; p. 64.
8. Stavrev, D.S.; Kaputkina, L.M.; Kirov, S.K.; Shamonin, Y.V.; Prokoshkina, V.G. Influence of plasma-arc processing on structura transformations and surface hardening of carbon and alloy steels. *MiTOM* **1996**, 16–19.
9. Stepanova, T.Y. *Technologies of Surface Hardening of Machine Parts: Tutorial*; Ivanovo State Chemical-Technological University Ivanovo, Russia, 2009; p. 64.
10. Zhurerova, L.G.; Rakhadilov, B.K.; Popova, N.A.; Kylyshkanov, M.K.; Buranich, V.V.; Pogrebnjak, A.D. Effect of the PEN/C surface layer modification on the microstructure, mechanical and tribological properties of the 30CrMnSiA mild-carbon steel *J. Mater. Res. Technol.* **2020**, *9*, 291–300. [CrossRef]
11. Suminov, I.V.; Belkin, P.N.; Epelfeld, A.V.; Lyudin, V.; Krit, B.; Borisov, A. *Plasma-Electrolytic Modification of the Surface of Metals and Alloys*; Suminov, I.V., Ed.; Technosphere: Moscow, Russia, 2011; p. 464.
12. *Devices and Methods of Physical Metallurgy*; Weinberg, F. (Ed.) Mir: Moscow, Russia, 1973; Volume 1, p. 427.
13. Gorelik, S.S.; Skakov, Y.A.; Rastorguev, L.N. *X-ray and Electron-Optical Analysis*; MISIS: Moscow, Russia, 2002; p. 360.
14. Grigorevich, V.K. *Hardness and Microhardness of Metal*; Nauka: Moscow, Russia, 1976; p. 230.
15. Zarchi, M.K.; Shariat, M.H.; Dehghan, S.A.; Solhjoo, S. Characterization of nitrocarburized surface layer on AISI 1020 steel by electrolytic plasma processing in an urea electrolyte. *J. Mater. Res. Technol.* **2013**, *2*, 213–220. [CrossRef]
16. Senatore, A.; Risitano, G.; Scappaticci, L.; D'Andrea, D. Investigation of the Tribological Properties of Different Textured Lead Bronze Coatings under Severe Load Conditions. *Lubricants* **2021**, *9*, 34. [CrossRef]
17. Duradji, V.N. Thermochemical treatment of metals with heating in electrolytic plasma. *Technol. Surf. Treat.* **2010**, *6*, 59–61.
18. Rakhadilov, B.K.; Buranich, V.V.; Satbayeva, Z.A.; Sagdoldina, Z.B.; Kozhanova, R.S.; Pogrebnjak, A.D. The cathodic electrolytic plasma hardening of the 20Cr2Ni4A chromium-nickel steel. *J. Mater. Res. Technol.* **2020**, *9*, 6969–6976. [CrossRef]
19. Rakhadilov, B.K.; Satbayeva, Z.; Baizhan, D. Effect of electrolytic-plasma surface strengthening on the structure and properties of steel 40 kHN. In Proceedings of the METAL 2019—28th International Conference on Metallurgy and Materials, Brno, Czech Republic, 22–24 May 2019; pp. 950–955.
20. Qiu, X.; Wei, X.; Xu, X.; Xu, W.; Zhu, M. Dependence of fretting wear resistance on microstructural features of alloyed steels. *Tribol. Int.* **2019**, *137*, 39–45. [CrossRef]
21. Kozlov, E.; Popova, N.; Zhurerova, L.; Nikonenko, E.; Kalashnikov, M.; Skakov, M. Structural and Phase Transformations in 0.3C-1Cr-1Mn-1Si-Fe Steel after Electrolytic Plasma Treatment. In *AIP Conference Proceedings*; AIP Publishing LLC: Melville, NY, USA, 2016; Volume 1783, p. 20112.

lubricants

MDPI

rticle

Mechanical and Tribological Properties of Polytetrafluoroethylene Composites Modified by Carbon Fibers and Zeolite

atyana S. Struchkova, Andrey P. Vasilev *, Aitalina A. Okhlopkova, Sakhayana N. Danilova nd Aleksey G. Alekseev

Department of Chemistry, Institute of Natural Sciences, North-Eastern Federal University, 677000 Yakutsk, Russia; sts_23@mail.ru (T.S.S.); okhlopkova@yandex.ru (A.A.O.); dsn.sakhayana@mail.ru (S.N.D.); alexalekseev.z@gmail.com (A.G.A.)

* Correspondence: gtvap@mail.ru; Tel.: +7-924-8694-792

Abstract: Currently, lightweight and high-strength polymer composites can provide weight savings in the automotive and process equipment industries by replacing metal parts. Polytetrafluoroethylene and polymer composites based on it are used in various tribological applications due to their excellent antifriction properties and thermal stability. This article examines the effect of combined fillers (carbon fibers and zeolite) on the mechanical, tribological properties, and structure of polytetrafluoroethylene. It is shown that the introduction of combined fillers into polytetrafluoroethylene retains the tensile strength and elongation at break at a content of 1–5 wt.% of carbon fibers, the compressive stress increased by 53%, and the yield stress increased by 45% relative to the initial polymer. The wear resistance of polymer composites increased 810-fold compared to the initial polytetrafluoroethylene while maintaining a low coefficient of friction. The structural features of polymer composites are characterized by X-ray diffraction analysis, infrared spectroscopy, and scanning electron microscopy.

Keywords: polytetrafluoroethylene; polymer composite materials; carbon fibers; zeolite; wear; friction

1. Introduction

Polymer composite materials (PCMs), due to their improved performance properties, are widely used as parts in friction units of various equipment. Polytetrafluoroethylene (PTFE) is characterized by thermal stability, chemical resistance, and self-lubricating properties [1]. Due to these properties, PTFE can be used as mechanical components in friction units in the automotive industry. However, low wear resistance and high creep of PTFE in many cases excludes its use in friction units [2]. The disadvantages of PTFE can be reduced by introducing fillers, which effectively increase wear resistance. Typically, fillers for PTFE are dispersed particles (graphite, bronze), carbon and glass fibers, and combinations [3–5]. In recent decades, much research has focused on developing PCMs utilizing a mixture of fillers of different sizes, shapes, and chemical compositions. It is known that carbon fibers (CFs), glass fibers, and aramid fibers are used as fillers to manufacture polymer materials for tribological purposes, together with nanosized and lubricating materials (graphite, MoS_2, PTFE) [5–9].

Li et al. prepared composites of the compositions PTFE/SiO_2 and PTFE/SiO_2/PAO 6, which were obtained at various pressing pressures (from 10 to 50 MPa) [10]. The addition of PAO 6 to PTFE/SiO_2 compositions leads to a significant decrease in the coefficient of friction and an increase in wear resistance compared to PTFE/SiO_2 and the initial PTFE. The improvement in the tribological properties of PCM was explained by the synergistic effect when using PAO 6 (liquid lubricant) and PTFE polymer matrix (solid lubricant). Zhang et al. evaluated the impact of nanosized particles on the tribological characteristics of polyether ether ketone (PEEK) filled with a combination of CF/PTFE/graphite fillers [11,12]. For this, the authors used a commercial composition of the PEEK/CF/PTFE/graphite composite, to which nano-SiO_2 was additionally added. The addition of nano-SiO_2 to the PEEK/CF/PTFE/graphite composite is effective at high contact friction pressures. Song et al. investigated composites based on PTFE filled with CFs with MoS_2 and short glass fibers [13]. The introduction of MoS_2 into PTFE/CF made it possible to improve the tribological properties compared to composites of the composition PTFE/CF/glass fiber. It was shown that the introduction of MoS_2 and glass fiber into the PTFE/CF composition led to a significant improvement in wear resistance. The effect of improving the tribological properties of PTFE/CF/glass fiber/MoS_2 became noticeable at high sliding speeds and pressures.

Previously, the authors of this paper studied the effect of combined fillers consisting of CFs and layered silicates (kaolin and vermiculite) on the tribological and mechanical properties of PTFE [14]. It was shown that the introduction of mechanically activated layered silicates and CFs made it possible to increase the mechanical properties of PCM and significantly improve wear resistance compared to composites containing only fibers. Thus, it is of great interest to introduce CFs with other aluminosilicate particles into PTFE.

Zeolites are crystalline aluminosilicate minerals with microporous structures [15]. They have a high specific surface area, resistance to heating, availability, and low cost, and are widely used to modify polymers [16–19]. Zeolite can have the same exceptional wear-reducing properties with CFs as the above fillers: kaolin and vermiculite. This study will focus on studying the effect of CFs and zeolite on the mechanical and tribological properties of PTFE.

2. Materials and Methods

PTFE (particle size of 46–135 μm and density of 2.16 g/cm^3) powders trademark PN90 were purchased by GaloPolymer, Russia. Chemically modified discrete CF brand «Belum» was obtained from Svetlogorsk Khimvolokno, Belarus. The width and length of the CF were 8–10 μm and 50–500 μm, respectively. The CF were modified by a plasma–chemical treatment in an environment with fluororganic compounds, based on a previous method [20].

Zeolite (Zt) with a chemical composition of $M_{2/n}OAl_2O_3{\cdot}xSiO_2{\cdot}yH_2O$ was used as a mineral filler, sourced from deposits from the Sakha (Yakutia) Republic in Russia. The average specific surface of the Zt used in this study was ~15 m^2/g.

PCM were prepared according to a typical process: dry-mixing the polymer and filler in a paddle mixer, molding under a pressure of 50 MPa at room temperature for 2 min, then sintering the specimen in a programmable muffle furnace (SNOL 180/400) at 375 °C. The mechanical activation of the zeolite was preliminarily carried out in a planetary mill (Activator-2) according to the previous work: the activation time was 2 min at 80 G [14]. PTFE-based composites were prepared, and the detailed compositions were listed in Table 1. The mass content of CF in the composites is 1 and 10 wt.%, respectively.

Table 1. Material compositions of PTFE composites (wt.%).

Abbreviated Form	PTFE	Carbon Fibers	Zeolite
Initial PTFE	100	-	-
PTFE/1CF/Zt	98	1	1
PTFE/3CF/Zt	96	3	1
PTFE/5CF/Zt	94	5	1
PTFE/8CF/Zt	91	8	1
PTFE/10CF/Zt	89	10	1

The mechanical properties, such as the tensile strength, elongation at break, and yield strength, were determined according to the Russian standard GOST 11262-2017 (plastics, tensile test method) using a universal testing machine (Autograph AGS-J, Shimadzu, Kyoto, Japan) at 25 °C, and a strain rate of 100 mm/min. The compressive stress at a deformation of 10% was measured following standard GOST 4651-2014 (plastics, compression test method, ISO 604: 2002, MOD) using the same instrument at a strain rate of 1 mm/min. The densities of the PTFE and PCM samples were determined according to the Russian standard GOST 15139-69 (plastics, methods for the determination of density). This method, also known as hydrostatic weighing (analog ASTM D 792), for determining the bulk density of molded products (e.g., rods, bars, and tubes), and it provides a density measurement accuracy of 0.1%. Distilled water was used as the media.

Tribological properties were evaluated in pin-on-disc tests on a UMT-3 (CETR, Mountain View, CA, USA). The counterbody was a #45 carbon-steel disk with a hardness of 45–50 HRC, and a roughness R_a = 0.06–0.08 μm. The investigated polymer pin samples had a diameter of 10 ± 0.1 mm with 20 ± 1 mm thickness (length), while the wear track centerline on the steel discs was at 20 mm diameter. Nominal pressure is 2 MPa (the normal force was 160 N), an average sliding speed of 0.2 m/s, and a test duration of 3 h. The mean deviation values were calculated from at least three measurements. Coefficient of friction was determined according to the Russian standard GOST 11629-2017 (plastics, method for testing the friction coefficient). The wear rate k (mm^3/N·m) was calculated after wear testing by using Equation (1):

$$k = \frac{\Delta m}{\rho \times F_N \times d}, \quad (1)$$

where F_N, N—normal force; d, m—sliding distance; Δm, g—mass lost during sliding; ρ, g/mm^3—density of specimens.

The crystal structure of the PTFE and its PCM was determine using X-ray powder diffractometry (XRD; ARL X'Tra, Thermo Fisher Scientific, Ecublens, Switzerland) with a CuK$_\alpha$ (λ = 0.154 nm) radiation source. The degree of crystallinity was estimated from the ratio of the areas of reflections corresponding to amorphous and crystalline regions using Equation (2):

$$\alpha = \frac{I_C}{I_C + I_A}, \quad (2)$$

where α, %—degree of crystallinity; I_C—area under the crystalline peaks; $I_C + I_A$—area under all peaks. Win XRD software (v. 2.0-6, Thermo Fisher, Ecublens, Switzerland) was used for data analysis.

The worn surface and supramolecular structures of the PTFE and PCM samples were observed by scanning electron microscopy (SEM; JSM-7800F, JEOL, Tokyo, Japan). The specimens for supramolecular structure studies were obtained by cold chipping in liquid nitrogen.

Fourier transform infrared spectroscopy (FT-IR; Varian 7000, Varian, Palo Alto, CA, USA) was used to record IR spectra with an attenuated total reflection (ATR) attachment over the range of 550–4000 cm^{-1}. IR spectra were obtained before and after friction tests.

3. Results and Discussion

The plot of tensile strength and elongation at break of PTFE and PTFE/CF/Zt in terms of the content of CFs is presented in Figure 1.

As can be seen from Figure 1, the introduction of combined fillers in PTFE in an amount of 1–5 wt.% of CFs leads to the preservation of the tensile strength and elongation at break of PCM at the level of the initial polymer. The mechanical properties of polymer composites are related to the structure of the material; in the case of polymers filled with fibers, the significant role is played by the interfacial interaction at the fiber–matrix interface [21]. It is known that modified CFs of the Belum brand are characterized by strong interfacial adhesion to PTFE [22]. However, a CF content of 8–10 wt.% in PCM leads to a decrease in tensile strength up to 25% and elongation at break up to 30%, possibly associated with the formation of defective areas with a high content of fillers.

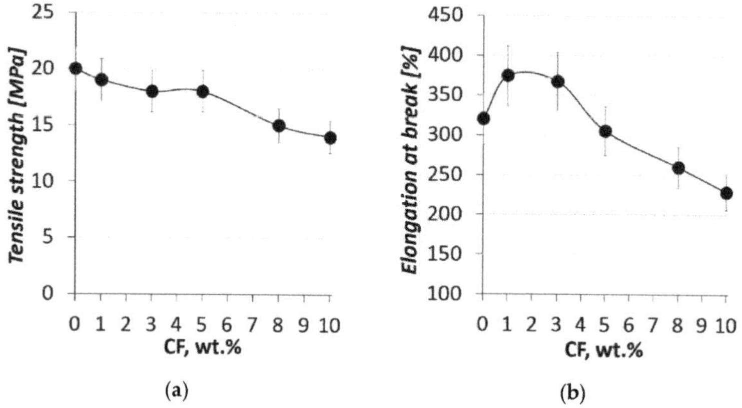

Figure 1. Mechanical characteristics of PTFE and PCM: (**a**) tensile strength; (**b**) elongation at break.

The compressive strength (10%) and the yield strength of PTFE and PTFE/CF/Zt depending on the content of CFs is shown in Figure 2.

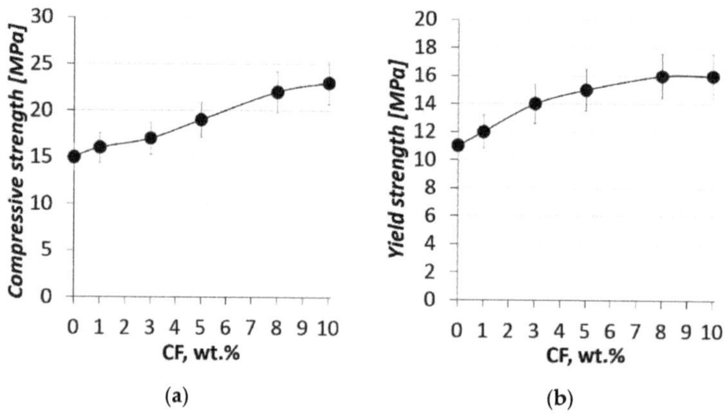

Figure 2. Mechanical characteristics of PTFE and PCM: (**a**) compressive stress at 10 % of relative deformation; (**b**) yield strength.

Figure 2 shows that the compressive value strength of PCM monotonically increases with an increase in the CF content. It can be seen that the compressive strength of PCM increased by 7–53% compared to the initial PTFE at a filler with increasing content of 1 to 10 wt.%. The yield point of the composites increased by 45% relative to the initial PTFE at a CF content of 8–10 wt.%. In composites, a significant increase in compressive strength and yield strength indicates the reinforcing effect of fillers. Thus, the developed materials have improved compression characteristics and yield strength, which is necessary for materials for tribological purposes.

Structural studies of PTFE and PCM were carried out using an X-ray diffractometer to determine the structural parameters of the developed materials. The X-ray diffraction patterns of PTFE and PCM are shown in the Figure 3.

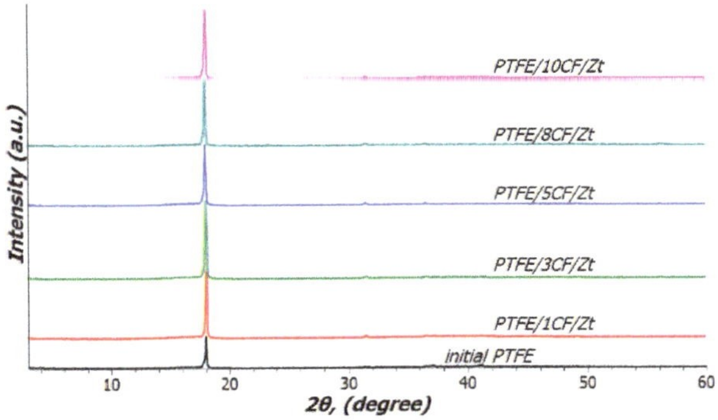

Figure 3. X-ray diffraction patterns of PTFE and PCM samples.

As can be seen from Figure 3, in the initial PTFE and polymer composites, an intense diffraction peak was revealed in the region 2θ = 18° (100), which corresponds to the crystalline phase of PTFE [23]. No other peaks were found in all X-ray diffraction patterns of PCM. It was found that the intensity of the peak at 2θ = 18° PCM increased with an increase in the content of fillers. An increase in the intensity of the PCM peak is consistent with an increase in the degree of crystallinity, as shown in Table 2.

Table 2. The degree of crystallinity and density of PTFE and PCM.

Sample	Degree of Crystallinity (%)	Density (g/cm^3)
Initial PTFE	55	2.16 ± 0.01
PTFE/1CF/Zt	56	2.17 ± 0.01
PTFE/3CF/Zt	63	2.16 ± 0.01
PTFE/5CF/Zt	60	2.16 ± 0.01
PTFE/8CF/Zt	62	2.11 ± 0.01
PTFE/10CF/Zt	61	2.08 ± 0.01

The degree of crystallinity is determined, from the diffraction pattern, by calculating the relative values of the integrated intensity of the crystal reflections according to Equation (2). As can be seen from Table 2, the degree of crystallinity of PTFE increases with the introduction of fillers. PTFE/CF/Zt has a degree of crystallinity according to X-ray diffraction analysis that is higher than the initial PTFE by 1–8%. Such a change in the degree of crystallinity of PCM indicates that the fillers become crystallization centers.

The density of PCMs with a content of 1–5 wt.% CF remains at the level of the initial PTFE. With an increase in the fiber content by 8–10 wt.% CF in the polymer matrix, the density of the composites decreases in comparison with the initial polymer, reaching a minimum at 10 wt% CF. The decrease in PTFE density upon the introduction of CFs is possibly associated with a "loosening" of the supramolecular structure PCM [14]. In addition, the contribution is made by the lower density of the initial carbon fibers (~1.45 g/cm^3), which is significantly lower in comparison with the density of the PTFE polymer matrix (~2.16 g/cm^3).

The results of a study of the supramolecular structure of the initial PTFE and PCM by scanning electron microscopy (SEM) can be seen in Figure 4.

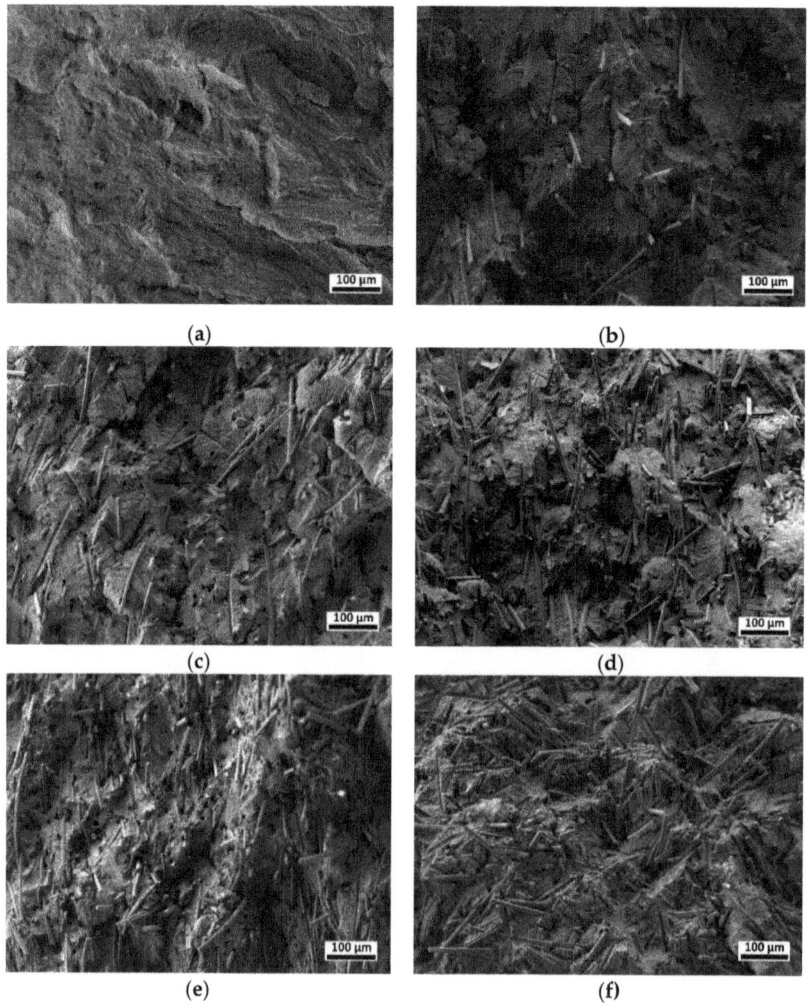

Figure 4. SEM image of the supramolecular structure of PCM (×150): (**a**) initial PTFE; (**b**) PTFE + 1 wt.% CF + 1 wt.% Zt; (**c**) PTFE +3 wt.% CF + 1 wt.% Zt; (**d**) PTFE + 5 wt.% CF + 1 wt.% Zt; (**e**) PTFE + 8 wt.% CF + 1 wt.%; (**f**) PTFE + 10 wt.% CF + 1 wt.% Zt.

As seen from Figure 4a, the supramolecular structure of the initial PTFE is characterized by the formation of a lamellae structure. In all composites, CFs were randomly oriented and relatively uniformly distributed in the supramolecular structure of the initial polymer (Figure 4b–f). The SEM images of PCM show that the amount of fibers visually increases with an increase in the CFs content from 1 to 10 wt.%. Due to this, the yield strength and compressive strength values increase by 7–53% and 8–45% compared to the initial PTFE. The zeolite content is not visible in the supramolecular structure, associated with its low content (1 wt.%) and high dispersion mechanically activated filler.

The tribological properties of PTFE are attributed to the relatively easy sliding between the lamellae or fibrils, resulting in a low coefficient of friction but also a high of wear rate [24]. Fillers are known to introduced into PTFE reduce the subsurface crack propagated and part of the normal load is transferred to the solid filler [25]. In addition, wear of the PTFE-based polymer composite formation a transfer film on the surface of the steel counterbody and on the friction surface polymer composites, which contributes significantly to wear reduction. Thus, the tribofilm formed on the surface of the composite and on the counterbody protects them from wear. The surface of the composites was investigated in this work since there is practically no wear of the counterbody, since the disc is made of sufficiently strong grade 45 steel. The tribological test results for PTFE and PTFE/CF/Zt as a function of fiber content are shown in Figure 5.

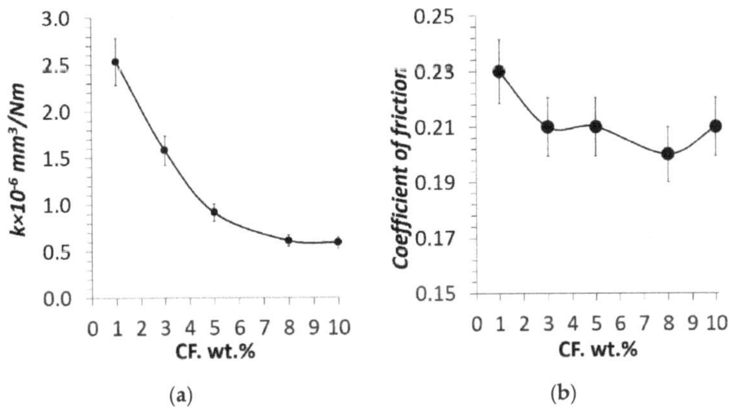

Figure 5. Results of wear rate (**a**) and coefficient of friction (**b**) of PCM.

As can be seen from Figure 5a, the addition of combined fillers to PTFE increased the wear resistance by order of magnitude compared to the original polymer. In this work, the wear rate for the initial PTFE was ~4.79 × 10^{-4} mm^3/N·m, and the friction coefficient was ~0.22. The results of comparing the PTFE/CF/Zt composites with the initial PTFE showed a decrease in the wear rate by a factor of ~190–810. Moreover, the wear rate of PTFE/CF/Zt depends on the CF content, so with an increase in the fiber content, the wear resistance increased from 2.53 × 10^{-6} to 0.59 × 10^{-6} mm^3/N·m. Composites containing 8–10 wt.% of CF have the highest wear resistance than composites containing 1–5 wt.% of CF.

The friction coefficient of PTFE/CF/Zt, regardless of the CF content, practically does not change over the entire concentration range of values, as shown in Figure 5b. It can be seen that the friction coefficient of PTFE/CF/Zt is comparable with the results of the initial PTFE and is in a relatively narrow range of values, i.e., 0.19–0.23. An earlier work [14] showed that the introduction of CFs with kaolin and vermiculite into PTFE increased the friction coefficient by 20–36% relative to the initial PTFE. In [26], PTFE filled with ultrafine kaolin particulates was investigated. The friction of PTFE filled with 5 wt.% kaolin increased the coefficient of friction of the composite relative to the initial polymer. In the work [27], the effect of vermiculite on the mechanical and tribological properties of PTFE was investigated. Shown that the introduction of vermiculite increases the coefficient of friction PCM to compare the initial PTFE. At the same time, PTFE filled with zeolite (5 wt.%), where the friction coefficient of the composite was at the level of the initial polymer, was studied in [19]. Based on this, it is evident that the PTFE/CF friction coefficient with the additional introduction of natural aluminosilicate depends on the structure of the used filler. Thus, the developed PTFE/CF/Zt materials are characterized by high wear resistance and a low coefficient of friction.

The surface of the PCM before and after friction was investigated by infrared (IR) spectroscopy. The IR spectra of PCM before and after friction can be seen in Figure 6.

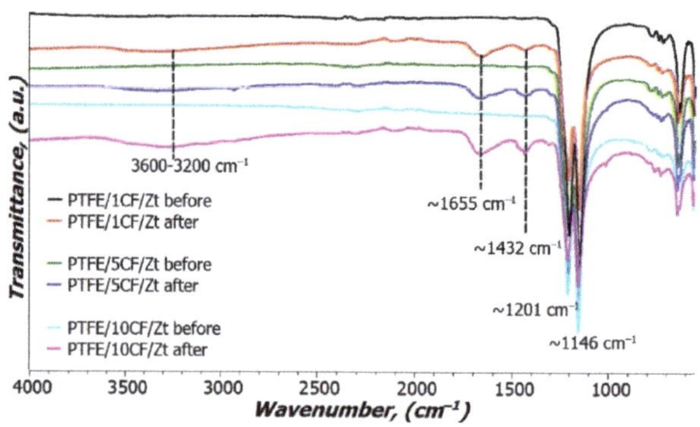

Figure 6. IR spectra before and after wear of the PTFE/CF/Zt.

As seen in Figure 6, the results of analysis of the obtained IR spectra showed the presence of a region of characteristic absorption of PTFE; new peaks were also revealed on the friction surfaces of PCM. Intense peaks at ~1201 and ~1146 cm^{-1} are characteristic of PTFE (-CF_2 asymmetric and symmetric stretching bands) [28]. On the worn surface of the composites, new absorption peaks were revealed in the range of ~1432 cm^{-1}, ~1655 cm^{-1}, and 3200–3600 cm^{-1}. From literature data, it is known that the peaks ~1432 cm^{-1} and ~1655 cm^{-1} are attributed to metal chelates of salts of perfluorinated carboxylic acids $M_2(R_f\text{-}COO^-)$ [29]. These peaks are found on the friction surface of the counterbody and on the worn surface of the composites [28]. The broad peak at 3600–3200 cm^{-1} is attributed to the associated hydroxyl groups –OH. The new absorption peaks recorded in the IR spectra of PCM after friction are associated with the rupture of the polymer chain at C–C bonds. Further, these broken bonds form radicals that can react with oxygen and ambient moisture to form end groups of carboxylic acids [30]. Thus, the friction of the PTFE/CF/Zt composite against a steel counterbody leads to tribochemical reactions.

Figure 7 shown the results of studying the worn surface of PCM depending on the composition of fillers by the SEM method.

As shown in Figure 7a, the worn surface of the initial PTFE is characterized by tiny grooves and is relatively smooth along the friction direction. Such a friction surface of unfilled PTFE is typical and was observed in the work [31]. The formation of such morphology of the PTFE worn surface is due to the structure of the polymer [25]. In the case of the introduction of combined fillers (CF/Zt) into PTFE, fibers visible on the friction surface protect the polymer surface layer from fracture during friction. Figure 7 shows that grooves are visible in PCM at 1 wt.% CF, which is characterized by the increased wear, while in other composites, such a morphology on the friction surface is not observed. Thus, in all composites, CFs on the friction surface were uniformly distributed and randomly oriented; the number of CFs visible on the worn surfaces increased with increasing CF content in the PCM, which is consistent with the results of the wear rate.

One of the theories explaining the decrease in wear with the introduction of microsized fillers in PTFE is associated with the interruption of the propagation of subsurface cracks by filler particles [32]. However, the additional introduction of layered silicates into PTFE/CF leads to a significant increase in wear resistance compared to PTFE filled only with fibers, which is explained by the formation of secondary reinforced structures in the form of tribofilms between fibers on the worn surface [14]. Such tribofilms were found on the PTFE/CF/Zt worn surface at high magnification of SEM images (Figure 8).

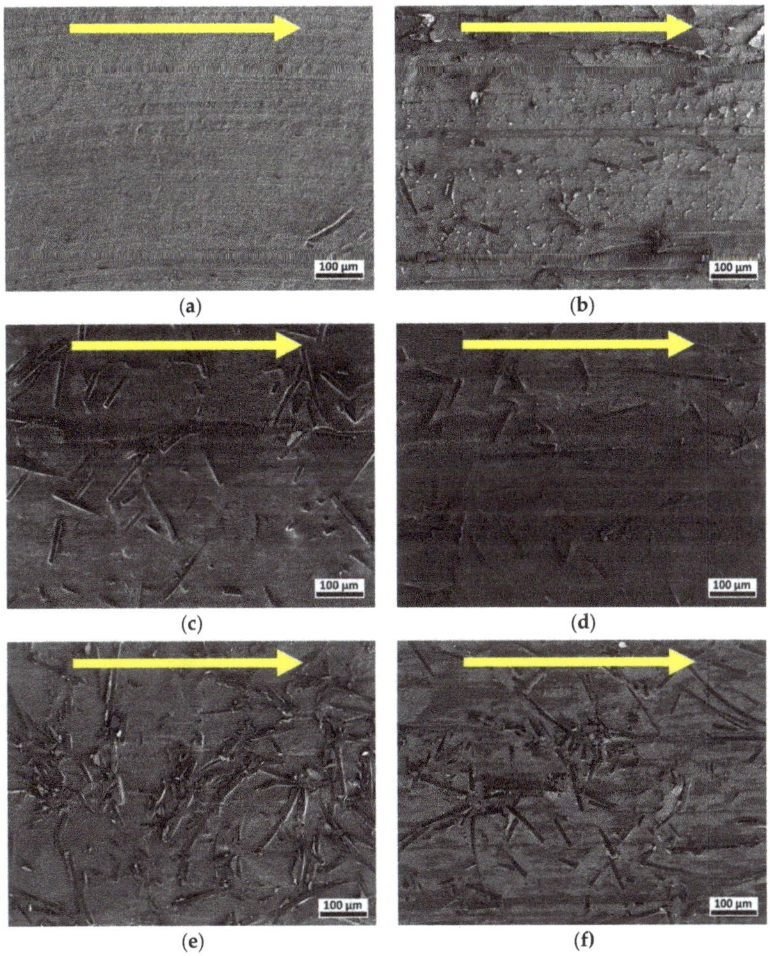

Figure 7. SEM image of worn surface of PCM (×150): (**a**) initial PTFE; (**b**) PTFE/1CF/Zt; (**c**) PTFE/3CF/Zt; (**d**) PTFE/5CF/Zt; (**e**) PTFE/8CF/Zt; (**f**) PTFE/10CF/Zt.

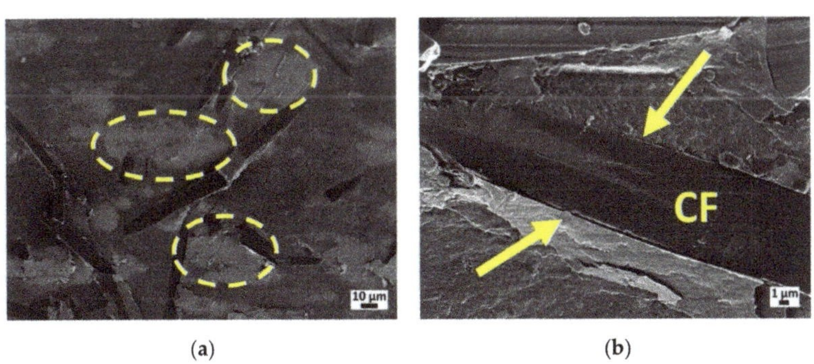

Figure 8. SEM image of worn surface of PCM: (**a**) at magnification ×500; (**b**) at magnification ×3000.

It is known that under friction, PTFE/CF occurs over a steel counterbody, the CF becomes a protruding on the worn surface that protects the polymer material from wear, and in addition, the fibers wear and can be pulled out of the polymer matrix [33]. Tribofilms are formed on the worn surface of the PCM between the fibers, as shown in Figure 8a. The formation of such tribofilms protects the fibers from destruction, thereby significantly reducing material wear (Figure 8b). It should be noted that tribofilms on the PCM friction surface are formed when mechanically activated aluminosilicates are introduced, which can also be evidenced by the occurrence of tribochemical reactions during friction by IR spectroscopy [14,33]. In addition, many works showed that the appearance of carboxyl groups on the friction surface of PCM and the metal counterbody coincided with a low wear rate due to the formation of a strong transfer film on the metal counterbody [24,34,35].

The wear rate and the coefficient of friction of PTFE/CF/Zt composites compared with PTFE/CF/Kl and PTFE/CF/Vl (referred to as Kl—kaolin, Vl—vermiculite) from [14] are shown in Figure 9. For comparison of tribological characteristics, polymer composites based on PTFE filled with 8–10 wt.% CF were selected.

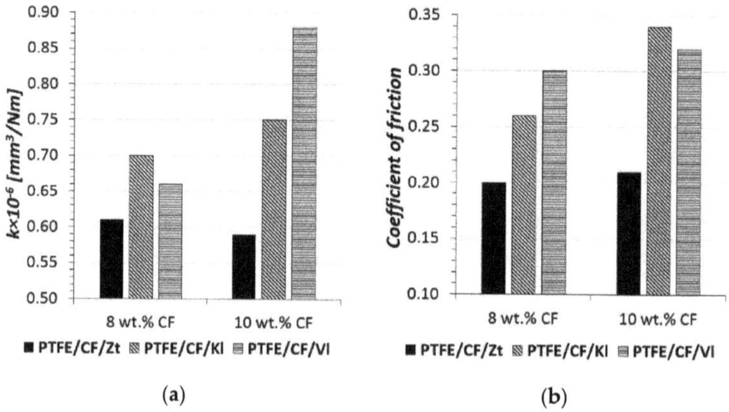

Figure 9. Compare of wear rate (**a**) and coefficient of friction (**b**) of PCM.

As shown in Figure 9, PTFE/CF/Zt composite is characterized by high wear resistance and a low coefficient of friction in comparison with polymer composites of the compositions PTFE/CF/Kl and PTFE/CF/Vl. The wear rate of PCM with zeolite at a CF content of 8 wt.% is higher by 8–14%, at a CF content of 10 wt.% by 27–49% compared to composites with PTFE/CF/Kl and PTFE/CF/Vl. The coefficient of friction of PCM with zeolite is lower by 30–50% and 52–61% compared to composites containing Kl and Vl at a content of 8–10 wt.%, respectively. Perhaps one of the explanations for the higher wear resistance of PTFE/CF/Zt relative to PTFE/CF/Kl and PTFE/CF/Vl is the low coefficient of friction of composites with zeolite. Obviously, such a difference in tribological characteristics between the composites is associated with the type of used aluminosilicate.

4. Conclusions

In this work, the effect of CFs with zeolite on the mechanical, tribological properties, and structure of PTFE was investigated. The developed materials with a CF content of 1–5 wt.% retained their deformation and strength properties at the level of the initial polymer. The compressive stress of PCM increased by 7–53%, and the yield point by 30% relative to the initial polymer. It was found that with an increase in the content of fillers, the degree of crystallinity increased, and the density decreased in comparison with unfilled PTFE. Combining fillers (CF/Zt) into PTFE reduced the wear rate by 810 times relative to the initial polymer. Tribochemical reactions were shown by IR spectroscopy. SEM established the formation of secondary structures in the form of tribofilms on the friction surface, which, together with CFs, protect the surface layer of the material from destruction during friction. The wear resistance of the composite material PTFE/CF/Zt was effectively improved, and the coefficient of friction was low compared to PTFE/CF/Kl and PTFE/CF/Vl.

Author Contributions: Conceptualization, T.S.S. and A.P.V.; methodology, A.A.O.; software, A.G.A.; validation, T.S.S. and A.P.V.; formal analysis, T.S.S. and A.P.V.; investigation, A.P.V.; resources, S.N.D.; data curation, A.G.A.; writing—original draft preparation, A.P.V.; writing—review and editing, T.S.S. and A.P.V.; visualization, A.P.V.; supervision, T.S.S.; project administration, A.A.O.; funding acquisition, T.S.S. All authors have read and agreed to the published version of the manuscript.

Funding: This research was funded by the Ministry of Science and Higher Education of the Russian Federation (grant numbers [FSRG-2020-0017]).

Data Availability Statement: The data presented in this study are available on request from the corresponding author.

Conflicts of Interest: The authors declare no conflict of interest.

References

1. Mazur, K.; Gądek-Moszczak, A.; Liber-Kneć, A.; Kuciel, S. Mechanical Behavior and Morphological Study of Polytetrafluoroethylene (PTFE) Composites under Static and Cyclic Loading Condition. *Materials* **2021**, *14*, 1712. [CrossRef] [PubMed]
2. Zhang, Y.; Kou, K.; Zhang, S.; Wu, G. Effect of Fiber Diameter on Thermal Properties of Short-Glass-Fiber-Reinforced PTFE-Based Composites. *J. Mater. Sci. Mater. Electron.* **2020**, *31*, 10715–10723. [CrossRef]
3. Tanaka, K.; Kawakami, S. Effect of Various Fillers on the Friction and Wear of Polytetrafluoroethylene-Based Composites. *Wear* **1982**, *79*, 221–234. [CrossRef]
4. Khedkar, J.; Negulescu, I.; Meletis, E.I. Sliding Wear Behavior of PTFE Composites. *Wear* **2002**, *252*, 361–369. [CrossRef]
5. Conte, M.; Igartua, A. Study of PTFE Composites Tribological Behavior. *Wear* **2012**, *296*, 568–574. [CrossRef]
6. Friedrich, K. Polymer Composites for Tribological Applications. *Adv. Ind. Eng. Polym. Res.* **2018**, *1*, 3–39. [CrossRef]
7. Chang, L.; Friedrich, K. Enhancement Effect of Nanoparticles on the Sliding Wear of Short Fiber-Reinforced Polymer Composites: A Critical Discussion of Wear Mechanisms. *Tribol. Int.* **2010**, *43*, 2355–2364. [CrossRef]
8. Zhang, K.; Ji, X.; Mi, Y.; Gao, L.; Wang, T. Effects of Carbon Fibers with Different Particle Sizes on the Physical Properties of MoS 2 -Filled PTFE Composites. *Philos. Mag. Lett.* **2021**, *101*, 277–286. [CrossRef]
9. Zhang, F.; Zhang, J.; Zhu, Y.; Wang, X.; Jin, Y. Microstructure and Properties of Polytetrafluoroethylene Composites Modified by Carbon Materials and Aramid Fibers. *Coatings* **2020**, *10*, 1103. [CrossRef]
10. Li, X.; Ling, Y.; Zhang, G.; Yin, Y.; Dai, Y.; Zhang, C.; Luo, J. Preparation and Tribological Properties of Solid-Liquid Synergetic Self-Lubricating PTFE/SiO2/PAO6 Composites. *Compos. Part B Eng.* **2020**, *196*, 108133. [CrossRef]
11. Zhang, G.; Chang, L.; Schlarb, A.K. The Roles of Nano-SiO2 Particles on the Tribological Behavior of Short Carbon Fiber Reinforced PEEK. *Compos. Sci. Technol.* **2009**, *69*, 1029–1035. [CrossRef]
12. Zhang, G.; Schehl, M.; Burkhart, T. *Effect of Low-Loading Nanoparticles on the Tribological Property of Short Carbon Fiber (SCF)/PTFE/Graphite Reinforced PEEK*; University of Kaiserslautern: Kaiserslautern, Germany, 2016.
13. Song, F.; Wang, Q.; Wang, T. Effects of Glass Fiber and Molybdenum Disulfide on Tribological Behaviors and PV Limit of Chopped Carbon Fiber Reinforced Polytetrafluoroethylene Composites. *Tribol. Int.* **2016**, *104*, 392–401. [CrossRef]
14. Vasilev, A.P.; Struchkova, T.S.; Nikiforov, L.A.; Okhlopkova, A.A.; Grakovich, P.N.; Shim, E.L.; Cho, J.-H. Mechanical and Tribological Properties of Polytetrafluoroethylene Composites with Carbon Fiber and Layered Silicate Fillers. *Molecules* **2019**, *24*, 224. [CrossRef] [PubMed]
15. Stocker, K.; Ellersdorfer, M.; Lehner, M.; Raith, J.G. Characterization and Utilization of Natural Zeolites in Technical Applications. *Berg. Huettenmaenn. Mon.* **2017**, *162*, 142–147. [CrossRef]
16. Król, M. Natural vs. Synthetic Zeolites. *Crystals* **2020**, *10*, 622. [CrossRef]
17. Haque, M.M.; Hasan, M. TiO_2 and Zeolite Nanopowder Enhanced Mechanical Properties of Hybrid Polymer Composites. *J. Thermoplast. Compos. Mater.* **2021**, *34*, 382–395. [CrossRef]
18. Petrova, P.N.; Gogoleva, O.V. Development of tribotechnical materials based on mixtures of fluoroplasts. *Dream* **2018**, *2018*, 184–190. [CrossRef]
19. Gogoleva, O.V.; Okhlopkova, A.A.; Petrova, P.N. Development of Self-Lubricating Antifriction Materials Based on Polytetrafluoroethylene and Modified Zeolites. *J. Frict. Wear* **2014**, *35*, 383–388. [CrossRef]
20. Shelestova, V.A.; Zhandarov, S.F.; Danchenko, S.G.; Grakovich, P.N. Surface Modification of Carbon Fiber by Fluoropolymer in a Low-Temperature Plasma. *Inorg. Mater. Appl. Res.* **2015**, *6*, 219–224. [CrossRef]
21. Shesan, O.J.; Stephen, A.C.; Chioma, A.G.; Neerish, R.; Rotimi, S.E. Fiber-Matrix Relationship for Composites Preparation. In *Renewable and Sustainable Composites*; Pereira, A.B., Fernandes, F.A.O., Eds.; IntechOpen: London, UK, 2019; ISBN 978-1-78984-215-9.

22. Grakovich, P.N.; Shelestova, V.A.; Shumskaja, V.J.; Shylko, I.S.; Gucev, D.M.; Bashlakova, A.L.; Celuev, M.J. Influence of the Type c Carbon Fiber Filler on the Physical–Mechanical and Tribological Properties of PTFE Composites. *J. Frict. Wear* **2019**, *40*, 11–16. [CrossRef]
23. Yuan, Y.; Yu, D.; Yin, Y.; Tang, B.; Li, E.; Zhang, S. Effect of Sintering Temperature on the Crystallization Behavior and Propertie of Silica Filled PTFE Composites. *J. Mater. Sci. Mater. Electron.* **2016**, *27*, 13288–13293. [CrossRef]
24. Ye, J.; Burris, D.; Xie, T. A Review of Transfer Films and Their Role in Ultra-Low-Wear Sliding of Polymers. *Lubricants* **2016**, *4*, 4. [CrossRef]
25. Blanchet, T.A.; Kennedy, F.E. Sliding Wear Mechanism of Polytetrafluoroethylene (PTFE) and PTFE Composites. *Wear* **1992**, *15*3, 229–243. [CrossRef]
26. Xiang, D.; Gu, C. A Study on the Friction and Wear Behavior of PTFE Filled with Ultra-Fine Kaolin Particulates. *Mater. Lett.* **2006**, *60*, 689–692. [CrossRef]
27. Lazareva, N.N.; Kirillina, D.Y.; Sleptsova, S.A.; Okhlopkova, A.A.; Kapitonova, Y.V. *Development of Polymer-Silicate Composites Based on Polytetrafluoroethylene and Natural Vermiculite with FORUM Addition*; American Institute of Physics Conference Series Ekaterinburg, Russia, 2020; p. 050015.
28. Piwowarczyk, J.; Jędrzejewski, R.; Moszyński, D.; Kwiatkowski, K.; Niemczyk, A.; Baranowska, J. XPS and FTIR Studies of Polytetrafluoroethylene Thin Films Obtained by Physical Methods. *Polymers* **2019**, *11*, 1629. [CrossRef]
29. Pitenis, A.A.; Harris, K.L.; Junk, C.P.; Blackman, G.S.; Sawyer, W.G.; Krick, B.A. Ultralow Wear PTFE and Alumina Composites: It Is All About Tribochemistry. *Tribol. Lett.* **2015**, *57*, 4. [CrossRef]
30. Campbell, K.L.; Sidebottom, M.A.; Atkinson, C.C.; Babuska, T.F.; Kolanovic, C.A.; Boulden, B.J.; Junk, C.P.; Krick, B.A. Ultralow Wear PTFE-Based Polymer Composites—The Role of Water and Tribochemistry. *Macromolecules* **2019**, *52*, 5268–5277. [CrossRef]
31. Shen, M.; Li, B.; Zhang, Z.; Zhao, L.; Xiong, G. Abrasive Wear Behavior of PTFE for Seal Applications under Abrasive-Atmosphere Sliding Condition. *Friction* **2020**, *8*, 755–767. [CrossRef]
32. Tanaka, K.; Uchiyama, Y.; Toyooka, S. The Mechanism of Wear of Polytetrafluoroethylene. *Wear* **1973**, *23*, 153–172. [CrossRef]
33. Okhlopkova, A.A.; Vasilev, A.P.; Struchkova, T.S.; Alekseev, A.G.; Grakovich, P.N. Study of the Influence of Carbon Fibers and Bentonite on the Tribological Properties of PTFE. *J. Frict. Wear* **2020**, *41*, 571–577. [CrossRef]
34. Krick, B.A.; Pitenis, A.A.; Harris, K.L.; Junk, C.P.; Sawyer, W.G.; Brown, S.C.; Rosenfeld, H.D.; Kasprzak, D.J.; Johnson, R.S.; Chan, C.D.; et al. Ultralow Wear Fluoropolymer Composites: Nanoscale Functionality from Microscale Fillers. *Tribol. Int.* **2016**, *95*, 245–255. [CrossRef]
35. Harris, K.L.; Pitenis, A.A.; Sawyer, W.G.; Krick, B.A.; Blackman, G.S.; Kasprzak, D.J.; Junk, C.P. PTFE Tribology and the Role of Mechanochemistry in the Development of Protective Surface Films. *Macromolecules* **2015**, *48*, 3739–3745. [CrossRef]

lubricants

MDPI

rticle

Carbon Nanotori Reinforced Lubricants in Plastic Deformation Processes

ose Jaime Taha-Tijerina [1,2,*], Juan Manuel Martínez [2], Daniel Euresti [1] and Patsy Yessenia Arquieta-Guillén [3]

1 Engineering Technology Department, University of Texas Rio Grande Valley, Brownsville, TX 78520, USA; daniel.euresti01@utrgv.edu
2 Departamento de Ingeniería, Universidad de Monterrey, Av. Ignacio Morones Prieto 4500 Pte., San Pedro Garza García 66238, Mexico; juan.martineza@udem.edu
3 Facultad de Ciencias Físico-Matemáticas, Universidad Autónoma de Nuevo León, San Nicolás de los Garza 66455, Mexico; patsy.arquietagl@uanl.edu.mx
* Correspondence: jose.taha@utrgv.edu

Abstract: This research presents the effects of carbon nanotori structures (CNst) dispersed as reinforcement for metal-working and metal-forming lubricants. Synthetic (SL) and deep drawing (DD) nanolubricants were prepared following a two-step method at 0.01 wt.%, 0.05 wt.%, and 0.10 wt.% filler fractions. Slight increases in viscosity (<6%) for nanolubricants were observed as filler fraction was increased through various measured temperatures. Tribological behavior of nanolubricants displayed superb improvements under antiwear and extreme pressure conditions. The load carrying capacity (p_{oz}) increased by 16% and 22% at merely 0.01 wt.% CNst reinforcement and up to 73% and 107% at 0.10 wt.% filler fraction for SL and DD nanolubricants, respectively, compared to conventional materials. Additionally, at 0.10 wt.% wear scar evaluations showed a highest benefit of 16% and 24%, for SL and DD nanolubricants, respectively. This enhancement is attributed to diverse mechanisms such as rolling/sliding and load bearing effects, tribofilm formation, and CNst tribosintering behavior (at high pressures) onto metallic surfaces due to nanostructures size and morphology and their interlayer relationship among conventional lubricants.

Keywords: friction; wear; nanolubricant; carbon nanotori; plastic deformation

1. Introduction

Sustainable and environmentally friendly materials are being developed and incorporated to diverse industrial metal-mechanic manufacturing processes such as grinding, milling, turning, stamping, punching, and hydroforming, among others [1–8]. For lightweight purposes, thinner materials with higher mechanical properties are used in fields such as automotive and agriculture, among others. These materials require higher processing loads, needing a more robust performance of lubrication due to higher friction and wear among tooling and base substrates.

The incorporation of diverse nanostructures to cutting fluids and lubricants for plastic deformation processes has demonstrated great benefits, particularly for improving the efficiency, quality, and production in industrial operations. Through the comparison of tribological properties, researchers have been able to determine effective methods of improving lubricants under specific conditions such as antiwear (AW) and extreme pressures (EP). According to diverse studies, the most relevant advantages and improvements of using nanostructures as additives are: coefficient of friction (COF) and wear reduction, hence, decreasing machining operations cutting forces as well as energy consumption [9,10]. Another critical aspect is the improvement on the surface finishing of the produced components, which has a great impact on quality and post-processing. Tooling and machinery components are also benefited by using nanofluids and nanolubricants, such as also acting as thermal transport of the generated heat by metal-to-metal contact [11–14].

The reinforcing role of nanostructures in decreasing friction, wear, and improving plastic deformation manufacturing processes [14–16] attracts attention and substantial interest for its micro-rolling effect, self-repairability, polishing effect, and tribolayer formation [10,17]. These nanostructures, due to their intrinsic physico-chemical characteristics, convey desired performance characteristics to the conventional materials [16]. Due to the size and morphology of nanostructures, they possess high surface energies; thus, these are susceptible to agglomerate and sediment. The filler fraction of these nanostructures also contributes to this phenomenon, which is very important for machinery and devices.

Lubrication mechanisms using carbonaceous structures such as single wall/multiwall carbon nanotubes (SWCNTs/MWCNTs), graphene nanoplatelets (GNPs), carbon fibers (CFs), nanodiamonds (NDs), carbon nano-onions, and carbon nanotori structures (CNst) have attracted significant attention as potential lubrication reinforcement for conventional fluids and lubricants [18–23] due to their good corrosion behavior, superb mechanical properties, high thermal conductivity, and excellent lubricating properties. Widely studied carbon nanotubes (CNTs) have demonstrated superb enhancements in tribological behavior when incorporated as additives or reinforced solid material of diverse systems [24], showing enhancement with diverse mechanisms such as tribofilm formation, rolling/sliding, and load bearing effects [7,21,25].

According to Jansson et al. [26] carbon nano-allotropes structures may contribute to the tribochemical layer creation through their chemical reactivity, rather than through their surface functional groups [26]. Conventional fluids and lubricants depend on their intrinsic characteristics such as chemical properties and structural parameters, among others. Ali et al. observed outstanding improvements in wear and friction reduction for diverse fluids and lubricants by adding carbon and graphene nanostructures [27].

Tribological effects of graphene nanoflakes-blended within engine oil were studied by Rasheed et al. [28]. At merely 0.01 wt.% graphene within conventional engine oil, COF was reduced by 21%. It was also observed that graphene nanostructures also reduce the ring wear over extended hour engine evaluation. Bhaumik et al. [29] observed that adding MWCNTs enhanced the AW characteristics and load-carrying capacity of conventional mineral oil. Wear reduction of 80% was reported for the nanolubricant at 0.05 wt.%. MWCNTs were incorporated as additives for synthetic and mineral oils for metal-working processes. At 0.1 wt.% reinforcement, an enhancement of 21% and 36%, respectively, in the load-carrying capacity was observed [30]. Kumar et al. investigated [31] MWCNTs and SWCNTs as additives in SAE10W40 motor oil. Using these reinforcements, worn scar diameter was reduced by 67% and 38%, respectively. The average COF was also decreased by 48% and 56%, respectively. One important aspect is that nanolubricants viscosity raised by increasing the CNTs filler fraction. Kumar et al. concluded that the AW mechanism is due to the accumulation of SWCNTs and MWCNTs on the material's surface, reducing shearing stress and improving the tribological performance. Mohamed et al. [32] demonstrated that 1.0 wt.% of CNTs incorporated to lithium grease contributed to improvements in EP performance and load-carrying capacity by 52%, and COF reduction by 82%. Other research on commercial greases reinforced with MWCNTs at various filler fractions showed an increase of 10% at 0.05 wt.% [33]. According to Ni et al. [34], this improvement may be due to the flattening behavior of the nanostructures by compression at high pressures, making them move between surfaces similar to a "tank belt". Hong et al. [35] investigated the behavior of CNTs-based greases within PAO (polyalphaolefin) oil with incorporation of 1.0 wt.% MoS_2 and other commercial greases (lithium, calcium, etc.). A COF reduction of 28% was observed for the CNTs-based grease with MoS_2, compared to conventional grease.

Another critical aspect of lubrication is the material's viscosity. The lubricant viscosity is one parameter for optimizing drawing processes, for instance. Viscosity is related to temperature; a system with high temperature reduces the protecting lubricant layer thickness and generates higher pressures. In diverse studies, dynamic viscosity effects with the addition of nanofillers were observed. In general, viscosity of nanofluids and nanolubricants increases with the addition of solid nanostructures; however, the flow characteristics, such as rheology, depend on diverse parameters and operating circumstances [36–39]. Sadeghinezhad et al. [10] observed that viscosity performance of GNP-based nanofluids depends on temperature and reinforcement filler fraction. In their study, GNP concentrations vary from 0.025 wt.% to 0.1 wt.%. Viscosity increments were 9–38% compared with distilled water, from room temperature to 60 °C. Similar research by Mehrali et al. [40] demonstrated that viscosity behavior of GNP nanofluids was reduced with filler fraction and temperature increase. Similarly, Moghaddam et al. [41] found that viscosity decreases with shear rate and temperature increments. Particularly, viscosity of GNP nanofluid at 0.1 wt.% was increased by 44%, when compared with conventional material. Afrand et al. [42] studied the effects of hybrid MWCNT- SiO_2 nanostructures dispersed in concentrations from 0 to 1 vol.% within SAE40 lubricant. Kinematic viscosity was evaluated over a 25 °C to 60 °C temperature range. In general, viscosity increases with a filler fraction raise, and decrease with temperature increase. The maximum observed improvement in viscosity was 37% for the 1.0 vol.% filler fraction at 60 °C. According to Dhanola et al. [43], the reason for the high viscosity of canola-reinforced lubricants is attributed to the incorporation of small amounts of nanoadditives, which attracted each other due to Van der Waals bonding. Furthermore, agglomeration of additives was observed due to the high concentration of reinforcement, leading to an increase in viscosity [43].

A toroidal form of carbon nanotubes, or toroidal "crop circles" as some authors name them, were experimentally observed by Liu et al. [44] while investigating laser-grown single-wall carbon nanotube (SWCNT) materials. Carbon nanotori structures are novel configurations of carbon-based materials considered as prospect objects for many investigations, with similar mechanical properties and characteristics as CNTs. These structures have been emerging as a sustainable option for improving industrial manufacturing processes. Most recently, nanotori structures have been utilized in the evaluation of the thermal dissipation of oil-based and synthetic lubricants [45]. These structures have been found to have excellent properties for reinforcing the enhancing thermal transport behavior of common lubricants. This has led to findings that show carbon nanotori structures significantly improve the thermal transport properties of lubricants [46].

Through the comparison of tribological properties, researchers have been able to determine effective methods of improving lubricants under EP and AW conditions. The primary method is preparing lubricants with additives such as carbon nanotori structures. Examining these properties before and after using different additives can help organizations in several ways such as increasing tool life, reducing energy consumption, and decreasing surface damage and COF. The rest of this paper is dedicated to an in-depth review of topics in tribology, focusing on the previous topics mentioned and how they can benefit individual organizations and the industry.

Due to the superb properties of carbon nanotori (CNst) (high mechanical properties such as elastic modulus ~1 TPa and tensile strengths [19]), their characteristics, and likeness to CNTs, these nanostructures are a suitable alternative to reinforce fluids and lubricants for plastic deformation manufacturing processes. In our work, deep drawing (DD) and synthetic (SL) nanolubricants at various concentrations (by weight) of CNst are evaluated for their effect in tribological performance and viscosity behavior over a temperature range of up to 70 °C (343 K).

2. Materials and Methods

Nanolubricants Preparation

Two common lubricants for metal-working and metal-forming processes (stamping, drawing, and punching, among others) (FUCHS Industry) were investigated (see Table 1). The synthetic lubricant (SL) possesses high lubricity properties which reduces the formation of cracks or severe surface damage (galling). The DD lubricant is a pre-emulsified automotive deep drawing compound. The SL and DD medium viscosity lubricants provide a thin protective and effective layer within the metal sheets and tooling to reduce wear and dissipate the generated heat due to friction. Nanolubricants prepared with carbon nanotori homogenously dispersed structures were obtained by a two-step method in three concentrations: 0.01, 0.05, and 0.10 wt.%, according to the procedure by Kharissova [47].

Table 1. Material Characteristics.

Materials	Properties		
Base lubricant	**Density (15 °C)**	**Kinetic Viscosity (mm^2/s)**	**Water: oil ratio**
Deep-drawing (DD)	0.967 g/cm^3	2.02 @ 24 °C; 0.86 @ 70 °C	4:1
Synthetic (SL)	1.033 g/cm^3	63.3 @ 24 °C; 40.7 @ 70 °C	6:1
Nanoparticles	**Size**	**Polydispersity index**	**Zeta potential**
Carbon Nanotori structures (CNst)	Torus diameter ~130 nm Tube diameter ~40 nm	0.248	−26.5 mV

MWNTs with 35–50 nm in diameter were functionalized with carboxylic acid groups. The next step was to dissolve them in a solution of DiW, potassium permanganate, nitric acid, hydrochloric acid, and sulfuric acid. The solution was maintained in an ice bath which was magnetically stirred for 60 h. Afterwards, the solution was centrifuged and the supernatant was collected to obtain carbon nanotori from other carbon-based materials. Henceforth, nanotori structures were homogeneously incorporated within SL and DD lubricants at various concentrations.

The zeta-potential of the SL and DD nanolubricants at 0.10 wt.% CNst is found to be ~22 mV and 18 mV, indicating the stability of the nanolubricants.

3. Experimental Details

3.1. Tribological Evaluations

A four-ball tribotester was employed to evaluate the load-carrying capacity of nanolubricants under EP conditions and COF. Tests were conducted according to the ITeE-PIB evaluation methodology to evaluate materials under scuffing criteria [48]. This methodology was chosen since it is more sensitive to EP additives [49], and it is also less material and time consuming.

The evaluation involves the application of a load over a top ball, which acts against a set of three stationary bottom balls on a cylindrical container (oil cup) (Figure 1), which are fully covered with the sample nanolubricant while the upper ball rotates at 500 rpms, at room temperature (~24 °C). The applied load increased progressively up to 7200 N in a period of 18 s. For each set of evaluations, four to six evaluations are performed. Each test consisted of pouring 12.5 mL of nanolubricant over the bottom stationary balls. Ball material is an AISI 52,100 steel grade, 12.7 mm in diameter, with surface roughness average (R_a) of 0.29 μm and 63 HRc. If a seizure occurred, the loading would stop, which means that the protective lubricant film on the balls was damaged or destroyed. This point is where the torque reached 10 N m. Here, the load corresponds to the seizure load (P_{oz}). If the 10 N m frictional torque is not achieved, P_{oz} is considered as the maximum load of 7200 N. After the evaluation is performed, the wear scar diameter (WSD) is measured for the three bottom balls with the aid of an optical microscope. Average values are then employed to determine the load-carrying capacity (p_{oz}) of the nanolubricants according to the following equation:

$$p_{oz} = 0.52 \frac{P_{oz}}{\mathrm{WSD}^2} \left[\mathrm{N/mm^2}\right] \quad (1)$$

Furthermore, the nanolubricant with the better tribological characteristics will be obtained with the greater load-carrying capacity. This evaluation helps to determine the EP behavior of nanolubricants, the load, and time at which the loss of lubricant film and wear occur.

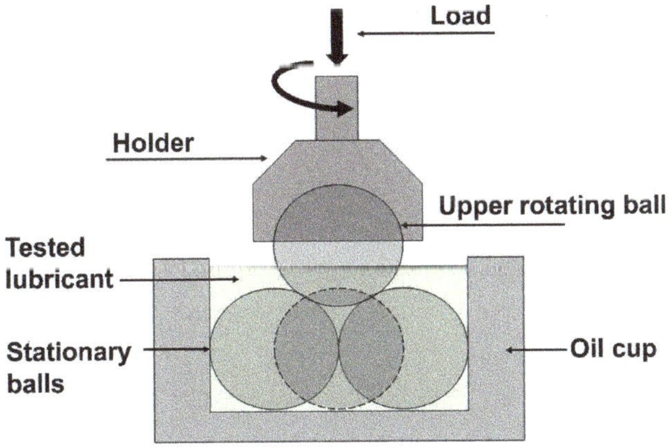

Figure 1. Schematic of four-ball testing setup.

3.2. *Viscosity Evaluations*

Rheological performance of nanolubricants has remarkable importance due to the relationship among shear rates, viscosity, pressure drop, and other characteristics. For instance, viscosity is associated with temperature, pressure, and tribofilm form [25]. Higher viscosity indicates higher flow resistance. Hence, as lubricant viscosity increases, a thicker tribofilm is created [50]. It controls the sealing effect of lubricants and the oil consumption rate. It also determines the ease of machine operation under varying temperature conditions, particularly in cold environments.

Therefore, in our research, the viscosity of nanolubricants is investigated. Shear viscosity evaluations were measured with an ARES rheometer (TA Instruments) with a Couette holder configuration. Temperature-dependent evaluations were obtained at room temperature (~300 K), 30 °C (303 K), 40 °C (313 K), 50 °C (323 K), 60 °C (333 K), and 70 °C (343 K) as well as varying filler fraction as reinforcement of the metal-working and metal-forming lubricants.

4. Results and Discussion

4.1. *Tribological Performance*

Improvements in tribological behavior with the incorporation of CNst within SL and DD lubricant systems have been obtained. Figure 2 depicts the scanning electron micrographs (SEM) of worn steel balls for SL and DD nanolubricants at the evaluated concentrations. For SL material, a reduction of 16% in the WSD was shown for the 0.10 wt.% nanolubricant. Furthermore, DD lubricants reflected a 24% improvement at the same filler fraction of CNst. This behavior for both lubricants is mainly associated with the size and morphology of the incorporated carbon nanotori structures, whose sliding-rolling effect acted as a promotor for reducing wear and minimizing COF. Viscosity of base lubricants also played a paramount factor, together with its relationship to applied pressure and temperature, which are responsible for the formed tribolayer thickness and, subsequently, the effect on the tribological properties.

As filler fraction increases, a favorable benefit is reflected in the SL nanolubricants. For instance, at merely 0.01 wt.% an enhancement of 16% is observed, achieving a maximum increment of 73% at 0.10 wt.%. DD nanolubricants displayed a similar effect on load-carrying capacity. Figure 3 shows the effect of the incorporation of CNst at various concentrations. In this case, load carrying capacity was higher than observed for SL lubricants, which is attributed to the affinity of carbon nanotori structures to deep drawing lubricant. Therefore, their dispersion is easy without using additives or surfactants. For instance, at 0.01 wt.% a reduction of 22% was observed, increasing to 53% at 0.05 wt.% and reaching a maximum improvement of 107% at 0.10 wt.%.

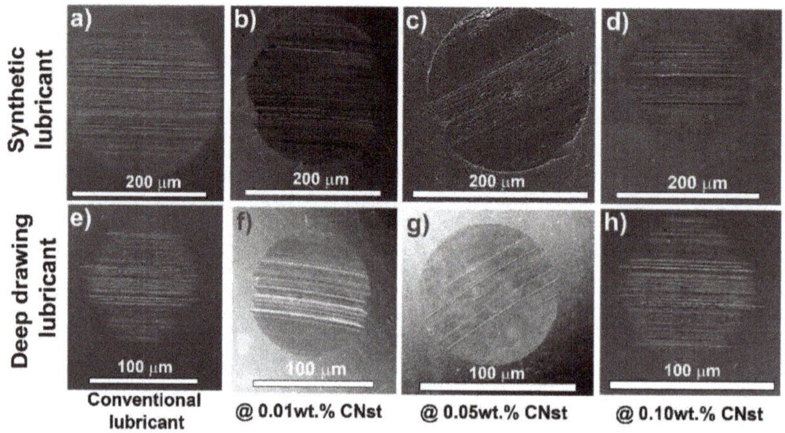

Figure 2. SEM images of evaluated steel balls wear scars for SL (**a**–**d**) and DD (**e**–**h**) nanolubricants.

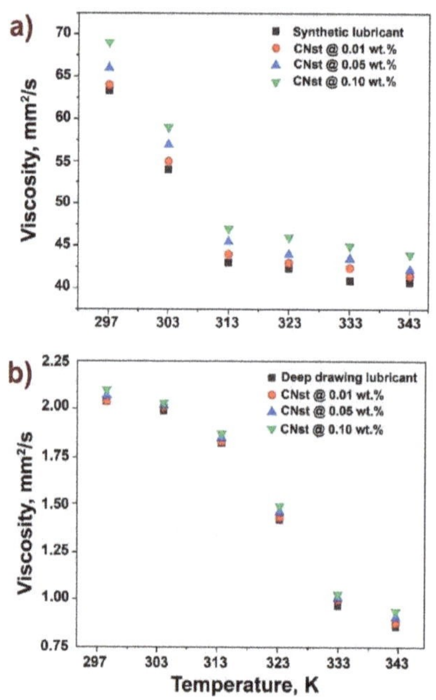

Figure 3. Temperature-dependence evaluation for viscosity behavior. (**a**) SL lubricants; (**b**) DD lubricants.

Table 2 shows the enhancements during four-ball evaluations by carbon nanotori structures reinforcing SL and DD lubricants under EP conditions. Seizure load (P_{oz}), pressure loss limit (p_{oz}), and WSD are integrated into this table, with enhancement in parentheses. The solid nanostructures are pushed into the metal contacts, forming a protective film, as other authors have also seen and called the tribo-sintering effect [51–53]. The observed improvements can be attributed to this tribo-sintering effect of the carbon nanotori structures onto worn areas, filling valleys or vacancies and the shearing mechanism of captured nanostructures at the surfaces interface, hence providing smoothness and reducing the frictional forces. Another critical characteristic can be associated with the substrate hardness and the tribo-layer formation, as other authors have also described [7,45,47,52–54].

Table 2. Enhancements by CNst reinforcement to SL and DD lubricants under EP conditions.

Lubricant	Seizure Load, P_{oz} (N)	Load-Carrying Capacity, p_{oz} (N/mm^2) (Enhancement)	WSD (mm) (Reduction)
Synthetic lubricant (SL)	4950	705	1.93
@ 0.01 wt.% CNst	5435	826 (16%)	1.85 (4%)
@ 0.05 wt.% CNst	5760	1007 (42%)	1.72 (10%)
@ 0.10 wt.% CNst	6150	1218 (73%)	1.62 (16%)
Deep drawing lubricant (DD)	6100	3300	0.98
@ 0.01 wt.% CNst	6400	4018 (22%)	0.91 (7%)
@ 0.05 wt.% CNst	6700	5055 (53%)	0.83 (15%)
@ 0.10 wt.% CNst	7200	6840 (107%)	0.74 (24%)

As previously explained, during the four-ball evaluations an extreme pressure is applied. A tribo-membrane is formed, resulting from a chemical response among the nanolubricants and the working surfaces. Due to the carbon nanotori morphology, higher loads could be supported, maintaining the lubrication layer between metal components. The load-bearing effect provided by the CNst-nanolubricants due to the high elastic modules is promoted at EP conditions.

4.2. Viscosity Evaluations

Temperature-dependent measurements for shear viscosity behavior for various filler fractions of nanolubricants are shown in Figure 3. Viscosity slightly increased with reinforcement of carbon nanotori structures to both lubricants, as expected. Moreover, for SL lubricants (Figure 3a), viscosity is reduced with temperature raise (from 63 mm^2/s at room temperature to 40 mm^2/s at 70 °C), while a slight increment in viscosity due to the incorporation of carbon nanotori (<6%) was observed.

Similar to reported research [37,55–58], an incremental viscosity behavior trend is seen as filler fraction is increased, as well as a decrease in viscosity with temperature raise. Figure 3b shows the viscosity behavior for DD nanolubricants over the evaluated temperatures. In this case, due to the initial low viscosity of the DD lubricant, viscosity increments due to the incorporation of reinforcing nanostructures is minimal, at less than 4%.

5. Conclusions

Two common lubricants for metal-working and metal-forming processes were investigated. Carbon nanotori structures (CNst) at diverse concentrations reinforced SL and DD lubricants. CNst demonstrated good affinity and improved tribological behavior under EP conditions, providing a sliding-rolling effect that promotes the wear reduction. Tribological performance was greatly benefited; the load carrying capacity was observed to be enhanced by 16% and 22% at merely 0.01 wt.% reinforcement and up to 73% and 107% at 0.10 wt.% for SL and DD nanolubricants, respectively, compared to pure lubricants. Wear scar evaluations were performed; the highest improvement was found at 0.10 wt.% reinforcement of 16% and 24%, for SL and DD, respectively. This improvement may be due to diverse mechanisms such as rolling effect and their tribo-sintering behavior (at high pressures) onto metal surfaces and their spacer effect due to the size and morphology of nanostructures as well as the interlayer relationship with conventional lubricants. Further investigation is needed to fully show these effects. Viscosity behavior was analyzed by comparing nanolubricants and pure lubricants. The incorporation of nanostructures slightly increased (<6% at 0.10 wt.%) the viscosity of conventional lubricants, as expected.

Carbon nanotori as reinforcing material for lubricants in industrial applications such as in metal-mechanic plastic deformation processes is an emerging alternative. Furthermore, predictive simulations and methodologies for tribological performance is an opportunity field in which additional development could be addressed.

Author Contributions: J.J.T.-T. contributed to the conceptualization, project administration, methodology, literature research, measuring campaign, data analysis, data interpretation, validation, formal analysis, resources, investigation, figures, study design, supervision, and writing—original. P.Y.A.-G. contributed to the methodology, resources, validation, formal analysis, investigation, and writing—original. J.M.M. contributed to the measuring campaign, data interpretation, and investigation. D.E. contributed to the investigation, literature research, and formal analysis. All authors have read and agreed to the published version of the manuscript.

Funding: This research received no external funding.

Institutional Review Board Statement: Not applicable.

Informed Consent Statement: Not applicable.

Data Availability Statement: Not applicable.

Acknowledgments: Authors acknowledge Metalsa and Universidad de Monterrey for their support given in this research.

Conflicts of Interest: The authors declare no conflict of interest.

References

1. Taha-Tijerina, J.J.; Calderón, R.; Rodríguez, B. Optimization and nanoreinforcements of lubricant Concentration for steel shee forming process. *Int. J. Mod. Manuf. Technol.* **2021**, *13*, 137–142. [CrossRef]
2. Chai, Y.; Yusup, S.; Chok, V. A Review on Nanoparticle Addition in Base Fluid for Improvement of Biodegradable Ester-Base Drilling Fluid Properties. *Chem. Eng. Trans.* **2015**, *45*, 1447–1452.
3. Rout, I.S.; Sahoo, M. Optimization of Drilling Parameters Using Nanofluid MinimumQuantity Lubrication. *Int. J. Emerg. Technc Adv. Eng.* **2016**, *6*, 27–34.
4. Minh, D.T.; The, L.T.; Bao, N.T. Performance of Al_2O_3 nanofluids in minimum quantity lubrication in hard milling of $60Si_2M$ steel using cemented carbide tools. *Adv. Mech. Eng.* **2017**, *9*, 1–9. [CrossRef]
5. Zareh-Desari, B.; Davoodi, B. Assessing the lubrication performance of vegetable oil-based nano-lubricants for environmentall conscious metal forming processes. *J. Clean. Prod.* **2016**, *135*, 1198–1209. [CrossRef]
6. Zhang, Y.; Li, H.N.; Li, C.; Huang, C.; Ali, H.M.; Xu, X. Nano-enhanced biolubricant in sustainable manufacturing: Fron processability to mechanisms. *Friction* **2021**, *10*, 803–841. [CrossRef]
7. Morshed, A.; Wu, H.; Jiang, Z. A Comprehensive Review of Water-Based Nanolubricants. *Lubricants* **2021**, *9*, 89. [CrossRef]
8. Babu, M.N.; Anandan, V.; Babu, M.D.; Muthukrishnan, N. Turning SKD 11 Steel Using Silver Nanofluids with Minimum Quantit Lubrication. *Int. J. Manuf. Mater.* **2021**, *11*, 74–95. [CrossRef]
9. Chinchanikar, S.; Kore, S.S.; Hujare, P. A review on nanofluids in minimum quantity lubrication machining. *J. Manuf. Proces* **2021**, *68*, 56–70. [CrossRef]
10. Tang, H.; Sun, J.; He, J.; Wu, P. Research Progress of Interface Conditions and Tribological Reactions: A Review. *J. Ind. Eng. Chen* **2021**, *94*, 105–121. [CrossRef]
11. Sharma, A.K.; Tiwari, A.K.; Dixit, A.R. Progress of Nanofluid Application in Machining: A Review. *Mater. Manuf. Process.* **201** *30*, 813–828. [CrossRef]
12. Ben Said, L.; Kolsi, L.; Ghachem, K.; Almeshaal, M.; Maatki, C. Application of nanofluids as cutting fluids in machining operation A brief review. *Appl. Nanosci.* **2021**, 1–32. [CrossRef]
13. Kishawy, H.A.; Hegab, H.; Deiab, I.; Eltaggaz, A. Sustainability assessment during machining Ti-6Al-4V with nano-additive based minimum quantity lubrication. *J. Manuf. Mater. Process.* **2019**, *3*, 61. [CrossRef]
14. Kadirgama, K. A comprehensive review on the application of nanofluids in the machining process. *Int. J. Adv. Manuf. Techno* **2021**, *115*, 2669–2681. [CrossRef]
15. Reverberi, A.P.; D'Addona, D.M.; Bruzzone, A.A.G.; Teti, R.; Fabiano, B. Nanotechnology in machining processes: Recen advances. *Procedia CIRP* **2019**, *79*, 3–8. [CrossRef]
16. Guo, D.; Xie, G.; Luo, J. Mechanical properties of nanoparticles: Basics and applications. *J. Phys. D Appl. Phys.* **2014**, *47*, 13001 [CrossRef]
17. Bai, M.J.; Liu, J.L.; Qi, Z.N.; He, J.; Wei, J.J.; Miao, J.Y. Research progress in nanofluids with graphene addition. *J. Mater. Eng.* **202** *48*, 46–59.
18. Delgado, J.L.; Herranz, M.Á.; Martín, N. The nano-forms of carbon. *J. Mater. Chem.* **2008**, *18*, 1417–1426. [CrossRef]

19. Liu, L.; Zhao, J. Toroidal and Coiled Carbon Nanotubes. In *Syntheses and Applications of Carbon Nanotubes and Their Composites*; Suzuki, S., Ed.; IntechOpen: London, UK, 2013.
20. Sarapat, P.; Hill, J.; Baowan, D. A Review of Geometry, Construction and Modelling for Carbon Nanotori. *Appl. Sci.* **2019**, *9*, 2301. [CrossRef]
21. Jammoria, N.S.; Ul Haq, M.I.; Raina, A. Carbon-Related Materials for Tribological Application. In *Advances in Sustainability Science and Technology, Proceedings of Fourth International Conference on Inventive Material Science Applications, Coimbatore, India, 4–15 May 2021*; Springer: Singapore, 2022; pp. 469–483.
22. Shafi, W.K.; Charoo, M.S. An overall review on the tribological, thermal and rheological properties of nanolubricants. *Tribol.-Mater. Surf. Interfaces* **2020**, *15*, 20–54. [CrossRef]
23. Opia, A.C.; Kameil, A.H.M.; Syahrullail, S.; Johnson, C.A.N.; Izmi, M.I.; Mamah, S.C. Tribological behavior of organic formulated anti-wear additive under high frequency reciprocating rig and unidirectional orientations: Particles transport behavior and film formation mechanism. *Tribol. Int.* **2022**, *167*, 107415. [CrossRef]
24. Chen, Y.; Yang, K.; Lin, H.; Zhang, F.; Xiong, B.; Zhang, H. Important contributions of multidimensional nanoadditives on the tribofilms: From formation mechanism to tribological behaviors. *Compos. Part B Eng.* **2022**, *234*, 109732. [CrossRef]
25. Jason, Y.J.J.; How, H.G.; Teoh, Y.H.; Chuah, H.G. A Study on the Tribological Performance of Nanolubricants. *Processes* **2020**, *8*, 1372. [CrossRef]
26. Jansson, N. Carbon Nanostructures as Lubricant Additives. Master's Thesis, Norwegian University of Science and Technology, Trondheim, Norway, 2021.
27. Ali, I.; Basheer, A.A.; Kucherova, A.; Memetov, N.; Pasko, T.; Ovchinnikov, K. Advances in carbon nanomaterials as lubricants modifiers. *J. Mol. Liq.* **2019**, *279*, 251–266. [CrossRef]
28. Rasheed, A.K.; Khalid, M.; Javeed, A.; Rashmi, W.; Gupta, T.C.S.M.; Chan, A. Heat transfer and tribological performance of graphene nanolubricant in an internal combustion engine. *Tribol. Int.* **2016**, *103*, 504–515. [CrossRef]
29. Bhaumik, S.; Prabhu, S.; Singh, K.J. Analysis of Tribological behavior of carbon nanotube based industrial mineral gear oil 250 cSt viscosity. *Adv. Tribol.* **2014**, *2014*, 341365. [CrossRef]
30. Peña-Parás, L.; Taha-Tijerina, J.; García, A.; Maldonado, D.; González, J.A.; Molina, D. Antiwear and Extreme Pressure Properties of Nanofluids for Industrial Applications. *Tribol. Trans.* **2014**, *57*, 1072–1076. [CrossRef]
31. Kumar, N.; Goyal, P. Experimental study of Carbon Nanotubes to enhance Tribological Characteristics of Lubricating Engine Oil SAE 10W40. *IOP Conf. Ser. Mater. Sci. Eng.* **2022**, *1225*, 012052. [CrossRef]
32. Mohamed, A.; Osman, T.A.; Khattab, A.; Zaki, M. Tribological behavior of carbon nanotubes as an additive on lithium grease. *J. Tribol.* **2015**, *137*, 011801. [CrossRef]
33. Peña-Parás, L.; Taha-Tijerina, J.; García, A.; Maldonado, D.; Nájera, A.; Cantú, P. Thermal transport and tribological properties of nanogreases for metal-mechanic applications. *Wear* **2015**, *332–333*, 1322–1326. [CrossRef]
34. Ni, B.; Sinnott, S.B. Tribological properties of carbon nanotube bundles predicted from atomistic simulations. *Surf. Sci.* **2001**, *487*, 87–96. [CrossRef]
35. Senatore, A.; Hong, H.; D'urso, V.; Younes, H. Tribological Behavior of Novel CNTs-Based Lubricant Grease in Steady-State and Fretting Sliding Conditions. *Lubricants* **2021**, *9*, 107. [CrossRef]
36. Mishra, P.C.; Mukherjee, S.; Nayak, S.K.; Panda, A. A brief review on viscosity of nanofluids. *Int. Nano Lett.* **2014**, *4*, 109–120. [CrossRef]
37. Hamze, S.; Cabaleiro, D.; Estellé, P. Graphene-based nanofluids: A comprehensive review about rheological behavior and dynamic viscosity. *J. Mol. Liq.* **2021**, *325*, 115207. [CrossRef]
38. Koca, H.D.; Doganay, S.; Turgut, A.; Tavman, I.H.; Saidur, R.; Mahbubul, I.M. Effect of particle size on the viscosity of nanofluids: A review. *Renew. Sustain. Energy Rev.* **2018**, *82*, 1664–1674. [CrossRef]
39. Rubbi, F.; Das, L.; Habib, K.; Aslfattahi, N.; Saidur, R.; Alam, S.U. A comprehensive review on advances of oil-based nanofluids for concentrating solar thermal collector application. *J. Mol. Liq.* **2021**, *338*, 116771. [CrossRef]
40. Sadeghinezhad, E.; Mehrali, M.; Tahan, L.S.; Mehrali, M.; Kazi, S.N.; Oon, C.S. Experimental investigation of convective heat transfer using graphene nanoplatelet based nanofluids under turbulent flow conditions. *Ind. Eng. Chem. Res.* **2014**, *53*, 12455–12465. [CrossRef]
41. Moghaddam, M.B.; Goharshadi, E.K.; Entezari, M.H.; Nancarrow, P. Preparation, characterization, and rheological properties of graphene-glycerol nanofluids. *Chem. Eng. J.* **2013**, *231*, 365–372. [CrossRef]
42. Afrand, M.; Nazari Najafabadi, K.; Akbari, M. Effects of temperature and solid volume fraction on viscosity of SiO_2-MWCNTs/SAE40 hybrid nanofluid as a coolant and lubricant in heat engines. *Appl. Therm. Eng.* **2016**, *102*, 45–54. [CrossRef]
43. Dhanola, A.; Garg, H.C. Dispersion stability and rheology study of canola oil containing TiO_2 nanoadditives for tribological applications: An experimental approach. *Proc. Inst. Mech. Eng. Part J J. Eng. Tribol.* **2020**, *235*, 1765–1781. [CrossRef]
44. Liu, J.; Dai, H.; Hafner, J.H.; Colbert, D.T.; Smalley, R.E. Fullerene "crop circles". *Nature* **1997**, *385*, 780–781. [CrossRef]
45. Taha-Tijerina, J.; Aviña, K.; Martínez, J.M.; Arquieta-Guillén, P.Y.; González-Escobedo, M. Carbon Nanotori Structures for Thermal Transport Applications on Lubricants. *Nanomaterials* **2021**, *11*, 1158. [CrossRef] [PubMed]
46. Liu, L.; Guo, G.Y.; Jayanthi, C.S.; Wu, S.Y. Colossal Paramagnetic Moments in Metallic Carbon Nanotori. *Phys. Rev. Lett.* **2002**, *88*, 217206. [CrossRef] [PubMed]

47. Kharissova, O.V.; Garza-Castañón, M.; Kharisov, B.I. Inorganic nanorings and nanotori: State of the art. *J. Mater. Res.* **2019**, 3 3998–4010. [CrossRef]
48. Michalczewski, R.; Piekoszewski, W.; Tuszynski, W.; Szczerek, M.; Wulczynski, J. The new methods for scuffing and pittin investigation of coated materials for heavy loaded lubricated elements. In *Tribology—Lubricants and Lubrication*; Kuo, C.-H., Ec InTech: Rijeka, Croatia, 2011; pp. 305–320.
49. Szczerek, M.; Tuszynski, W. A method for testing lubricants under conditions of scuffing. Part I. Presentation of the metho *Tribotest* **2002**, *8*, 273–284. [CrossRef]
50. Rudnick, L.R. *Synthetics, Mineral Oils, and Bio-Based Lubricants: Chemistry and Technology*; CRC Press: Boca Raton, FL, USA, 202 p. 1194.
51. Tóth, Á.D.; Szabó, Á.I.; Leskó, M.Z.; Rohde-Brandenburger, J.; Kuti, R. Tribological Properties of the Nanoscale Spherical Y_2O Particles as Lubricant Additives in Automotive Application. *Lubricants* **2022**, *10*, 28. [CrossRef]
52. Hernández-Battez, A.; González, R.; Viesca, J.L.; Fernández, J.E.; Díaz-Fernández, J.M.; Machado, A. CuO, ZrO_2 and ZnC nanoparticles as antiwear additive in oil lubricants. *Wear* **2008**, *265*, 422–428. [CrossRef]
53. Kato, H.; Komai, K. Tribofilm formation and mild wear by tribo-sintering of nanometer-sized oxide particles on rubbing stee surfaces. *Wear* **2007**, *262*, 36–41. [CrossRef]
54. Rabaso, P.; Ville, F.; Dassenoy, F.; Diaby, M.; Afanasiev, P.; Cavoret, J. Boundary lubrication: Influence of the size and structure o inorganic fullerene-like MoS_2 nanoparticles on friction and wear reduction. *Wear* **2014**, *320*, 161–178. [CrossRef]
55. Ali, N.; Bahman, A.M.; Aljuwayhel, N.F.; Ebrahim, S.A.; Mukherjee, S.; Alsayegh, A. Carbon-based nanofluids and their advances towards heat transfer applications—A review. *Nanomaterials* **2021**, *11*, 1628. [CrossRef]
56. Ibrahim, M.; Saeed, T.; Chu, Y.M.; Ali, H.M.; Cheraghian, G.; Kalbasi, R. Comprehensive study concerned graphene nano-sheets dispersed in ethylene glycol: Experimental study and theoretical prediction of thermal conductivity. *Powder Technol.* **2021**, *386* 51–59. [CrossRef]
57. Rasheed, A.K.; Khalid, M.; Rashmi, W.; Gupta, T.C.S.M.; Chan, A. Graphene based nanofluids and nanolubricants–Review of recent developments. *Renew. Sustain. Energy Rev.* **2016**, *63*, 346–362. [CrossRef]
58. Kamel, B.M.; Tirth, V.; Algahtani, A.; Shiba, M.S.; Mobasher, A.; Hashish, H.A. Optimization of the Rheological Properties and Tribological Performance of SAE 5W-30 Base Oil with Added MWCNTs. *Lubricants* **2021**, *9*, 94. [CrossRef]

lubricants

MDPI

rticle

nteraction of Co_3O_4 Nanocube with Graphene and Reduced Graphene Oxide: Adhesion and Quantum Capacitance

'ladislav Shunaev [1,*] and Olga Glukhova [1,2]

[1] Department of Physics, Saratov State University, 410012 Saratov, Russia; glukhovaoe@sgu.ru
[2] Institute for Bionic Technologies and Engineering, Sechenov University, 119991 Moscow, Russia
* Correspondence: shunaevvv@sgu.ru; Tel.: +7-8452-514562

Abstract: The composites on the base of Co_3O_4 and graphene are in demand in the field of portable, flexible energy storage devices due to their small size, lightweight, big specific capacitance, good cycle stability and appropriate capacitance retention. The synthesis of this material always starts from the treatment of graphene oxide, so as a result, experimenters receive Co_3O_4 nanocubes incorporated into reduced graphene oxide indicates the presence of different oxygen-containing groups in the compound. This fact may limit the advantages of the considered material. Our theoretical quantum chemical calculations show that the process of Co_3O_4 incorporation between reduced graphene oxide layers is more energetically favorable in comparison to pure graphene. However, the win in the quantum capacitance in the case of pure graphene is in the range of 300–500 F/g in dependence on the applied voltage. The obtained result may indicate the need for modification of the current methods of graphene/Co_3O_4 synthesis to improve its application in supercapacitors and lithium-ion batteries.

Keywords: graphene; Co_3O_4; lightweight supercapacitors; DFTB; heat of formation; adhesion; quantum capacitance

1. Introduction

Lightweight and flexible supercapacitors (SCs) with the planar and fiber-shaped structures are essential elements of portable and wearable devices—one of the most rapidly developing technologies right now [1,2]. Transition metal oxides are considered the most promising materials for chemical power sources such as supercapacitors (SCs) and lithium-ion batteries (LIBs) [3,4]. Among them, Co_3O_4 attracts increased attention due to its high theoretical capacitance (3560 F/g), low cost, ease of synthesis, large surface area and the best compromise between initial capacity and capacity retention [5–7]. However, its application is limited by extremely low electrical conductivity and drastic volume change during Li+ ion insertion/extraction leading to poor cyclability. The integration of Co_3O_4 with superconductive and ultimately strong graphene (G) solves this problem. The electrodes on the base of Co_3O_4/G demonstrate high-specific capacitance, fast electron transfer and long-cycle life [8–16]. Besides that, graphene and its derivatives have already been considered as an element base for fully flexible, lightweight, high-performance all-solid-state SCs. [17,18].

Li et al. reported on a prepared nanosheet assembly route-layered G/Co_3O_4 hybrid electrode that exhibited a high specific capacitance of 331 F/g at a current density of 5 A/g and good cycle stability with 122% capacitance retention after 5000 continuous charge–discharge cycles [8]. The Co_3O_4/G/Co_3O_4 sandwich architecture obtained by sonochemical synthesis demonstrated high specific capacitance of 276.6 F/g at a current density of 0.5 A/g and good cycling stability with 92.4% capacitance retention after 5000 cycles at 5 A/g [9]. Synthesized via a hydrothermal method and constant current electropolymerization, the sandwich-structured Co_3O_4@C@PPy electrode delivered a higher areal capacitance of 2.71 F/cm^2 at 10 mA/cm^2 (1663 F/g at 6.1 A/g) [10]. Synthesized by a facile hydrothermal process and subsequent thermal treatment, the sandwich-like Co3O4/G nanocomposites maintained a specific capacity of 639.8 mAh/g at a current density of 0.5 C after 50 cycles with a capacity retention rate of 71%. Cong et al. received a hierarchal porous structure that represented ultrasmall Co_3O_4 anchored into ionic liquid (IL)-modified graphene oxide (GO), ultrasmall/Co_3O_4-GO-IL electrodes were delivered by a remarkably simple hydrothermal and annealing process [11]. The obtained structure demonstrated electrochemical properties with an excellent initial capacity of 1304 mAh/g at a current density of 0.5 C and durable long-term cycling stability of capacity retention of 98.4% even after 500 cycles. Yang et al. reported a novel one-step laser irradiation method that synthesized 2D graphene sheets decorated by numerous ultrafine Co_3O_4 nanoparticles. This structure reached 978.1 F/g at the current densities of 1 A/g and showed 93.7% capacitance retention of 916.5 F/g at current densities up to 10 A/g [12].

Several attempts were performed to discover the energy and electron properties of Co_3O_4/G nanocomposite by density functional theory (DFT) methods. Feng et al. studied the coupling effect at the interface of graphene. It was found that the graphene sheet bent on the Co_3O_4 surface due to the strong electron coupling at the interface, and a Co-C bond with a length of 1.897 Å was formed [19]. Beatty et al. reported that the strong interaction of G with Co_3O_4 might be explained by the charge transfer while the binding energy was equal to 0.193 eV/atom [20]. Cong confirmed the significant charge transfer from G to the top surface of Co_3O_4 but noted that the minimal distance between G and Co_3O_4 was equal to 3.4 Å [11]. Odedairo reported that the shortest equilibrium distances between the graphene layer and the Co_3O_4 nanosheet were calculated to be 2.6 Å [21]. However, all these theoretical works considered the interaction of Co_3O_4 particles with pure graphene, while in all real experiments, the synthesis of GO was the essential initial stage [8–16]. In fact, the Co_3O_4/G nanocomposite represented the compound of Co_3O_4 and reduced graphene oxide (rGO); that is why the presence of oxygen atoms at the graphene surface should be taken into account. Besides that, recent studies have demonstrated the presence of in-plane oxygen in graphene [22,23]. In particular, the so-called "oxygen pair" and "graphitic oxygen" are two configurations where oxygen directly substitutes carbon atoms in the graphene lattice and constitute ~50% of the total relative counts. Substitutional oxygen in the graphene lattice could modify the adhesion parameters between Co_3O_4 and graphene, as well as the overall calculated capacitance of the nanocomposite, so graphene with in-plane oxygen should also be considered in this study.

This paper aims to theoretically find out the regularities of Co_3O_4 nanocubes binding to the surface of pure graphene, graphene with substitutional oxygen and graphene oxide with the following obtaining of Co_3O_4/rGO nanocomposite by the self-consistent charge density-functional tight-binding (SCC DFTB) method. The most favorable hybrid materials for lightweight SCs will be found from the view of energetic and capacitive parameters.

2. Methods

The optimization of the atomic structure, as well as the calculation of zone structure, was performed by the SCC DFTB method [24]. Besides the traditional DFTB terms, the Lennard-Jones potential [25] was introduced to evaluate the interaction of non-bonded atoms. All calculations were performed at the real temperature of 0 K; the SCC tolerance was 1×10^{-5}. The Brillion zone was sampled according to the Monkhorst-Pack scheme with parameters $8 \times 8 \times 1$ for 2D films and $8 \times 8 \times 8$ for 3D composites.

The quantum capacitance (QC) C_Q was first introduced as a part of the total capacitance C_T that required extra energy for filling a quantum well with electrons due to the Pauli principle [26]:

$$\frac{1}{C_T} = \frac{1}{C_G} + \frac{1}{C_Q} \quad (1)$$

For low-dimensional structures, QC was comparable to geometric capacitance C_G and played a key role in processes of charge and discharge. In our research, C_Q was calculated by the formula [27,28]:

$$C_Q = \frac{1}{mV}\int_0^V eD(E_F - eV')dV' \tag{2}$$

where m is the mass structure, V is the applied voltage that corresponds to shift in the Fermi level E_F, D is the area under the plot of density of state (DOS) in the considered range and e is the elementary charge.

3. Results

3.1. Adhesion

In the first stage, we attached Co_3O_4 nanocubes from the Fd3m space group [29] that was observed in [30] to 160 atoms of pure graphene. The most favorable structure was received when the atomic cell was optimized in the periodic box with lattice vectors Lx = 19.78 Å and Ly = 21.41 Å. Energy calculations showed that the complex was stabilized by 7.37 eV:

$$\Delta H\ (G \rightarrow G/Co_3O_4) = E(G/Co_3O_4) - E(G) - E(Co_3O_4) = -7.37\ \text{eV} \tag{3}$$

We observed the formation of the covalent bond between Co and C atoms with the length of 2.07 Å, which was a repeat of the results of Beatty et al. [20]. After binding, the Co_3O_4 nanocube reported the charge of 1.16 e to the graphene. We noticed that the Co_3O_4 particle maintained its cubic form in contrast to Fe_3O_4 particles that demonstrated partial loss of crystal after attachment to the graphene layer [31].

The atomic structure of the graphene with substitutional graphitic oxygen (G*) is shown in Figure S1. Incorporating into the graphene carcass, the oxygen atom formed two C–O bonds with a length of 1.47 Å and one C-O bond with a length of 1.48 Å. The C–C length of carbon atoms bonded to oxygen decreases to 1.40 Å. After the attachment of the Co_3O_4 nanocube, the C–O bond with a length of 1.48 Å was broken, and the C atom was bonded to the Co atom with a length of 2.09 Å (Figures 1b and S2). Another C–Co bond of 2.07 Å was observed, as in the case of the G/Co_3O_4 nanocomposite. The heat of the G*/Co3O4 formation was lower than for G/Co_3O_4 due to the presence of two chemical bonds and was equal to:

$$\Delta H\ (G^* \rightarrow G^*/Co_3O_4) = E(G^*/Co_3O_4) - E(G^*) - E(Co_3O_4) = -9.69\ \text{eV} \tag{4}$$

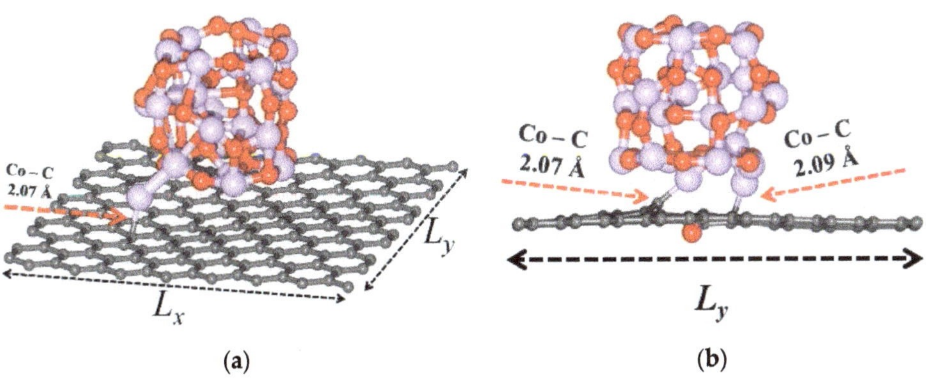

Figure 1. The atomic structure of (**a**) the G/Co_3O_4 nanocomposite and (**b**) the G^*/Co_3O_4 nanocomposite. The translation vectors are Lx = 19.78 Å and Ly = 21.41 Å.

As was mentioned in the Introduction, in the process of G/Co_3O_4 nanocomposite synthesis, Co_3O_4 nanoparticles are firstly disposed of at the GO surface, and then the oxygen-containing groups are etched. The scheme of this process modeling is shown in Figure 2. In the first stage, the graphene particle was oxidized with the maximum possible energy epoxy functional groups. The number of oxygen atoms was equal to 17. Thus the received mass ratio in obtained GO was m(G):m(O) = 14.2%. Note that after optimization, the lattice vectors were enlarged and were equal to Lx = 19.90 Å, Ly = 21.62 Å. The Co–C bond that was observed in the case of the G/Co_3O_4 nanocomposite was absent. Therefore, the main type of interaction between the rGO and the Co_3O_4 nanocube was determined by van der Waals forces. The process was exothermic, and the heat of formation was:

$$\Delta H\,(G \rightarrow GO) = E(GO) - E(G) - 17 \times E(O) = -51.14\ \text{eV}. \quad (5)$$

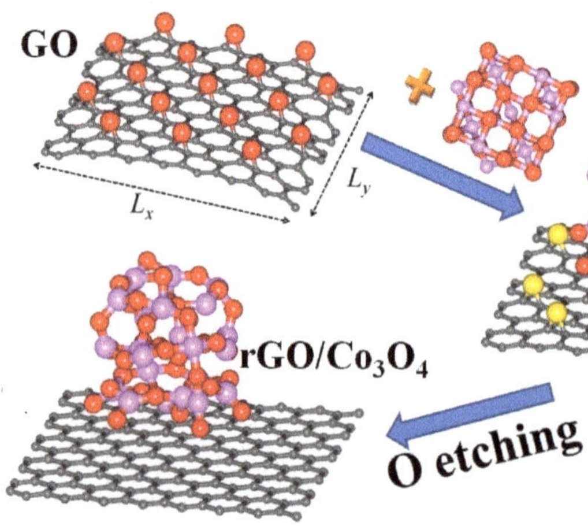

Figure 2. The process of the rGO/Co_3O_4 atomic supercell building. In the first stage, the Co_3O_4 supercell was attached to the surface of the GO cell, then the O atoms that did not bind to the Co_3O_4 nanocube (marked by yellow) were removed.

Then, the Co_3O_4 nanocube was attached to the obtained GO. After optimization, seven oxygen atoms from the GO surface formed covalent bonds with the Co_3O_4 nanocube and the energy of:

$$\Delta H\,(GO \rightarrow GO/Co_3O_4) = E(GO/Co_3O_4) - E(GO) - E(Co_3O_4) = -52.49\ \text{eV} \quad (6)$$

was released. After that, we removed 10 O-atoms from the GO surface that did not bind to the nanocube (these atoms are marked by yellow in Figure 2). This process required additional energy of:

$$\Delta H\,(GO/Co_3O_4 \rightarrow rGO/Co_3O_4) = E(rGO/Co_3O_4) - E(GO/Co_3O_4) + 10 \times E(O) = 32.71\ \text{eV} \quad (7)$$

that indicated an endothermic process. Finally, we calculated the heat of formation of rGO cell from GO:

$$\Delta H\,(GO \rightarrow rGO/Co_3O_4) = \Delta H\,(GO \rightarrow GO/Co_3O_4) + \Delta H\,(GO/Co_3O_4 \rightarrow rGO/Co_3O_4) = -19.78\ \text{eV}. \quad (8)$$

Therefore it is seen that such a process is also exothermic, but it is much more favorable than the formation of the rGO/Co_3O_4 nanocomposite ($\Delta H\,(G \rightarrow G/Co_3O_4) = -7.37$ eV). That is why in most experiments, the presence of GO is essential for the synthesis of the G/Co_3O_4 nanocomposite.

The same procedure was performed to obtain the energy properties of layered nanocomposites. For this goal, the additional vector of translation (Lz) was added to the periodic box (Figure 3). The optimal Lz in the case of G/Co_3O_4 was 11.9 Å, in the case of G^*/Co_3O_4 it was 11.8 Å, and in the case of rGO/Co_3O_4 it was 14.5 Å. Note that during incorporation, the Co3o4 nanocube interacted with two graphene oxide layers, so it formed chemical bonds with 15 O atoms. It is notable that in the case of 3D G*/Co_3O_4, the rotation of the nanocube was observed. The incorporation of Co_3O_4 between G, G* and rGO released −2.76, −6.46 and −7.13 eV, correspondingly. Therefore, it can be concluded that the processes of 3D G/Co_3O_4, G*/Co_3O_4 and rGO/Co_3O_4 and nanocomposite synthesis are less favorable than in the case of 2D nanofilms.

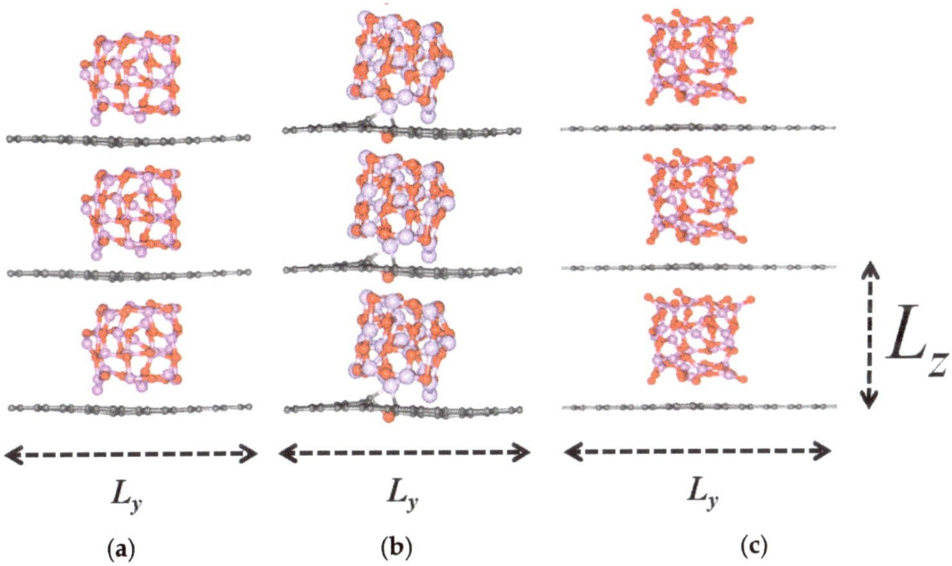

Figure 3. The expanded atomic supercells of 3D nanocomposites (**a**) G/Co_3O_4 with translations vectors Ly = 21.41 Å, Lz = 11.90 Å; (**b**) G*/Co_3O_4 with translations vectors Ly = 21.41 Å, Lz = 11.80 Å; (**c**) rGO/Co_3O_4 with translations vectors Ly = 21.62 Å, Lz = 14.50 Å.

3.2. Capacitive and Electronic Properties

The plots of QC for the considered structures are shown in Figure 4. The incorporation of in-plane oxygen into graphene carcasses significantly changed the QC near 0 V and created the peak at 0.2 V. The peak follows from the DOS of the structure in which the peak at −4.55 eV is mainly contributed by carbon's p-orbital and slightly by oxygen's p-orbitals (Figures S3 and S4). As it was predicted, the addition of the Co_3O_4 nanocube significantly (≈30–40 times) raised the level of graphene's QC, especially at U = 0 V (Table 1). The large peaks of QC were observed near −1 V, which indicated the major contribution of QC to the total capacitance at this value during the charge cycle. We also note that the QC for GO/Co_3O_4 and rGO/Co_3O_4 composites are almost similar, especially at the level of 0 V. However, the QC of G/Co_3O_4 surpasses the QC of GO/Co_3O_4 and rGO/Co_3O_4 at ≈190 F/g in the case of 2D films (Figure 4a) and at ≈210 F/g in the case of 3D structures. At −1 V, the differences in QC reached ≈300 F/g in the case of 2D films and ≈500 F/g in the case of 3D structures. The QC plot of G*/Co_3O_4 almost repeated the QC of G/ Co_3O_4 except for the range 0 < U < 0.6 V, where the QC of Co_3O_4 is still higher.

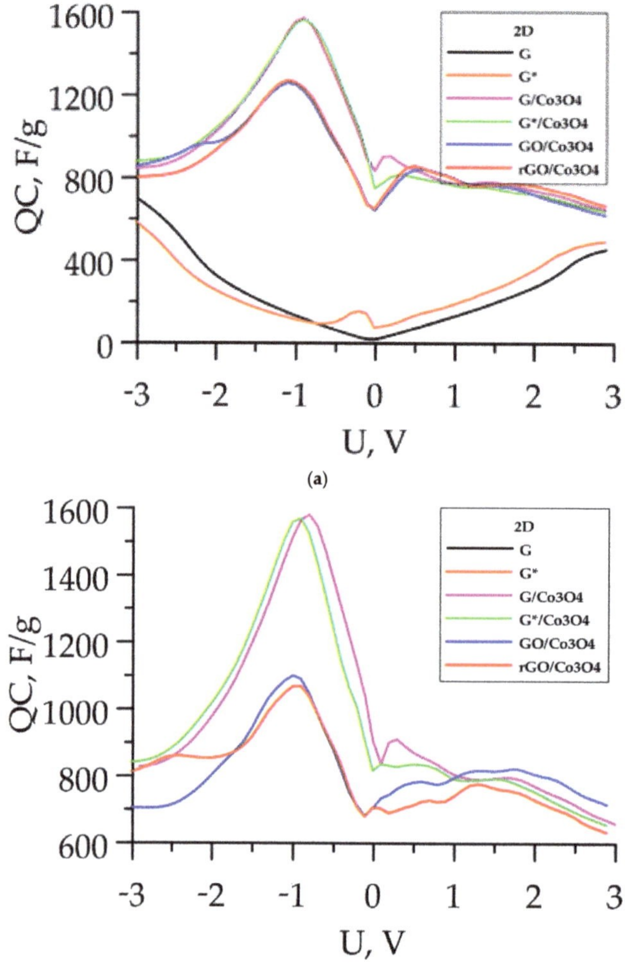

Figure 4. The QC for G, G*, G/Co_3O_4, G*/Co_3O_4, GO/Co_3O_4 and rGO/Co_3O_4: (**a**) 2D films; (**b**) 3D composites.

Table 1. Fermi level, charge transfer and the value of QC at the U = 0 V for pure graphene and 2D and 3D composites: G/Co_3O_4, GO/Co_3O_4, rGO/Co_3O_4.

Structure	Fermi Level, eV	Charge, e	QC (0 V), F/g
G	−4.67		20.78
G*	−4.41		111.5
2D G/Co_3O_4	−3.50	1.15	837.55
2D G*/Co_3O_4	−3.57	1.32	832.50
2D GO/Co_3O_4	−3.57	1.38	644.25
2D rGO/Co_3O_4	−3.48	0.88	652.17
3D G/Co_3O_4	−3.41	1.73	880.34
3D G*/Co_3O_4	−3.40	2.25	862.81
3D GO/Co_3O_4	−3.65	1.99	692.13
3D rGO/Co_3O_4	−3.26	1.13	691.13

Table 1 also lists the Fermi level of the considered structures. The substitution of the C atom by the O atom shifted the Fermi level of pure graphene from −4.67 to −4.41 eV. The addition of the Co_3O_4 nanocube raises the Fermi level to −3.50 eV for 2D G/Co_3O_4, to −3.57 eV for 2D G*/Co_3O_4 and 2D G*/Co_3O_4 and to −3.48 for 2D rGO/Co_3O_4. Further, it can be seen that in the case of the 3D structure, the Fermi level was less, which indicated the enhancement of its conductive properties in comparison to 2D films. Another interesting fact that confirms the advantage of sandwich-like structures of SCs application is the amount of charge that flowed to graphene layers. In the case of 2D G and rGO particles, this charge was equal to 1.15 and 0.88 e, correspondingly, while in the case of 3D structures, the amount of reported charge enlarges to 1.99 and 1.13 e. The most noticeable charge transfer was observed in the case of 3D G*/Co_3O_4 and was equal to 2.25 e. It can be explained by the formation of the two chemical bonds between graphene and the nanocube.

4. Conclusions and Discussion

The attempt to compare adhesion and electronic properties of G/Co_3O_4, G*/Co_3O_4 and rGO/Co_3O_4 nanocomposites was performed for the first time by the SCC DFTB method. Our calculation showed that the formation of the rGO/Co_3O_4 started from the attachment of Co_3O_4 nanocubes to the GO surface with consequent etching of oxygen atoms is more favorable than the straight decoration of G and G* (graphene with substitutional oxygen) with Co_3O_4 nanoparticles both for 2D films and 3D composites. The addition of Co_3O_4 nanoparticles to G shifts the Fermi level to 1.10–1.41 eV, which indicates the enhancement of the conductive properties. The maximum charge transfer that was given by Co_3O_4 nanocubes was observed in the case of 3D G*/Co_3O_4 and was equal to 2.25 e. The dependence of QC on voltage showed a significant improvement of capacitance in the case of G and G* in comparison to rGO. Independent of the applied voltage, this difference varies in the range of 300–500 F/g. In the Introduction, it was mentioned that G/Co_3O_4 is obtained from GO that leads to the retention of oxygen atoms and to a decrease in total capacitance. Therefore, one of the main challenges in this field is connected to the straight synthesis of G/Co_3O_4 from pure G. One of the possible methods is the two-step method in which graphene was initially deposited on a Ni foam with Co_3O_4 reported by Zhang [32]. Another variant to obtain materials with high capacity is connected with G*/Co_3O_4 material, the capacitive properties of which are a little less than G/Co_3O_4. The high QC, along with its small size, may make the G/Co_3O_4 nanocomposite an important element of lightweight and flexible supercapacitors for future portable and wearable devices.

Supplementary Materials: The following supporting information can be downloaded at: https://www.mdpi.com/article/10.3390/lubricants10050079/s1, Figure S1: The atomic structure of graphene with substitutional graphitic oxygen (G*). The oxygen is located in one plane with a graphene carcass. There are two C–O bond lengths of 1.47 Å (blue line), and one C–O bond length of 1.48 Å. (green line). The C–C bond length near O atoms decreases to 1.40 Å (orange); Figure S2: The atomic structure of graphene with substitutional graphitic oxygen after attachment to the Co_3O_4 nanocube. It is seen that one of the C–O bonds was broken; Figure S3: The density of state (DOS) for pure graphene (G) and graphene with substitutional graphitic oxygen (G*). Dashed lines mark the Fermi level; Figure S4: The total density of state (DOS) for graphene with substitutional graphitic oxygen (G*) and partial DOS (pDOS) for p-orbitals of carbon and oxygen atoms. Dashed line marks the Fermi level.

Author Contributions: Conceptualization, V.S. and O.G.; methodology, V.S.; software, O.G.; validation, V.S. and O.G.; formal analysis, O.G.; investigation, V.S.; resources, O.G.; data curation, O.G.; writing—original draft preparation, V.S.; writing—review and editing, V.S. and O.G.; visualization, V.S.; supervision, O.G.; project administration, O.G.; funding acquisition, O.G. All authors have read and agreed to the published version of the manuscript.

Funding: This research was funded by the Ministry of Science and Higher Education of the Russian Federation (project no. FSRR-2020-0004). V. V. Shunaev wishes to thank Russian President scholarship, project no. SP-3976.2021.1.

Institutional Review Board Statement: Not applicable.

Informed Consent Statement: Not applicable.

Data Availability Statement: The data presented in this study are available on request from the corresponding author.

Conflicts of Interest: The authors declare no conflict of interest.

References

1. An, B.; Shin, J.; Kim, S.-Y.; Kim, J.; Ji, S.; Park, J.; Lee, Y.; Jang, J.; Park, Y.-G.; Cho, E.; et al. Smart sensor systems for wearab electronic devices. *Polymers* **2017**, *9*, 303. [CrossRef] [PubMed]
2. Xu, C.; Song, Y.; Han, M.; Zhang, H. Portable and wearable self-powered systems based on emerging energy harvesting technolog *Microsyst. Nanoeng.* **2021**, *7*, 25. [CrossRef] [PubMed]
3. Zhu, X. Recent advances of transition metal oxides and chalcogenides in pseudo-capacitors and hybrid capacitors: A review o structures, synthetic strategies, and mechanism studies. *J. Energy Storage* **2022**, *49*, 104148. [CrossRef]
4. Ye, H.; Zheng, G.; Yang, X.; Zhang, D.; Zhang, Y.; Yan, S.; You, L.; Hou, S.; Huanga, Z. Application of different carbon-base transition metal oxide composite materials in lithium-ion batteries. *J. Electroanal. Chem.* **2021**, *898*, 115652. [CrossRef]
5. Hou, C.; Wang, B.; Murugadoss, V.; Vupputuri, S.; Chao, Y.; Guo, Z.; Wang, C.; Du, W. Recent Advances in Co_3O_4 as Anod Materials for High-Performance Lithium-Ion Batteries. *Eng. Sci.* **2020**, *11*, 19–30. [CrossRef]
6. Meher, S.K.; Rao, G.R. Ultralayered Co_3O_4 for High-Performance Supercapacitor Applications. *J. Phys. Chem. C* **2011**, *115* 15646–15654. [CrossRef]
7. Shi, Y.; Pan, X.; Li, B.; Zhao, M.; Pang, H. Co_3O_4 and its composites for high-performance Li-ion batteries. *Chem. Eng. J.* **2018**, *34* 427–446. [CrossRef]
8. Li, Q.; Yang, Q.; Yan, Z.; Kang, L.; Lei, Z.; Yang, Z.; Liu, Z. Electrocapacitive performance of graphene/Co_3O_4 hybrid material prepared by a nanosheet assembly route. *Electrochim. Acta* **2014**, *119*, 184–191. [CrossRef]
9. Han, X.; Huang, Z.; He, C.; Zhang, Q.; Zhang, X.; Yang, Y. Sonochemical synthesis of Co_3O_4/graphene/Co_3O_4 sandwich architecture for high-performance supercapacitors. *J. Appl. Electrochem.* **2019**, *49*, 1133–1142. [CrossRef]
10. Guo, D.; Zhang, M.; Chen, Z.; Liu, X.-X. The construction of a sandwich structured Co_3O_4@C@PPy electrode for improving pseudocapacitive storage. *RSC Adv.* **2018**, *8*, 33374. [CrossRef]
11. Cong, L.; Zhang, S.; Zhu, H.; Chen, W.; Huang, X.; Xing, Y.; Xia, J.; Yang, P.; Lu, X. Structure-design and theoretical-calculation for ultrasmall Co_3O_4 anchored into ionic liquid modified graphene as anode of flexible lithium-ion batteries. *Nano Res.* **2022**, *15*, 2104–2111. [CrossRef]
12. Yang, S.; Liu, Y.; Hao, Y.; Yang, X.; Goddard, W.A.; Zhang, X.L.; Cao, B. Oxygen-Vacancy Abundant Ultrafine Co_3O_4/Graphene Composites for High-Rate Supercapacitor Electrodes. *Adv. Sci.* **2018**, *5*, 1700659. [CrossRef] [PubMed]
13. Wang, S.; Wang, R.; Chang, J.; Hu, N.; Xu, C. Self-supporting Co_3O_4/Graphene Hybrid Films as Binder-free Anode Materials for Lithium Ion Batteries. *Sci. Rep.* **2018**, *8*, 3182. [CrossRef] [PubMed]
14. Li, Y.; Jia, M.; Xu, L.; Gao, J.; Zhang, F.; Jin, X.-J. Graphene and activated carbon-wrapped and Co_3O_4-intercalated 3D sandwich nanostructure hybrid for high-performance supercapacitance. *N. J. Chem.* **2018**, *42*, 10733. [CrossRef]
15. Jing, M.; Zhou, M.; Li, G.; Chen, Z.; Xu, W.; Chen, X.; Hou, Z. Graphene-Embedded Co_3O_4 Rose-Spheres for Enhanced Performance in Lithium Ion Batteries. *ACS Appl. Mater. Interfaces* **2017**, *9*, 9662–9668. [CrossRef] [PubMed]
16. Lin, G.; Jiang, Y.; He, C.; Huang, Z.; Zhanga, X.; Yang, Y. In Situ encapsulation of Co_3O_4 polyhedra in graphene sheets for high-capacitance supercapacitors. *Dalton Trans.* **2019**, *48*, 5773. [CrossRef] [PubMed]
17. Ramadoss, A.; Yoon, K.-Y.; Kwak, M.-J.; Kim, S.-I.; Ryu, S.-T.; Janga, J.-H. Fully flexible, lightweight, high performance all-solid-state supercapacitor based on 3-Dimensional-graphene/graphite-paper. *J. Power Sources* **2017**, *337*, 159–165. [CrossRef]
18. Chee, W.K.; Lim, H.N.; Zainal, Z.; Huang, N.M.; Harrison, I.; Andou, Y. Flexible Graphene-Based Supercapacitors: A Review. *J. Phys. Chem. C* **2016**, *120*, 4153–4172. [CrossRef]
19. Feng, Q.; Zeng, Y.; Xu, P.; Lin, S.; Feng, C.; Li, X.; Wang, J. Tuning the electrical conductivity of amorphous carbon/reduced graphene oxide wrapped-Co_3O_4 ternary nanofibers for highly sensitive chemical sensors. *J. Mater. Chem. A* **2019**, *7*, 27522. [CrossRef]
20. Beatty, J.; Cheng, T.; Cao, Y.; Sky Driver, M.; Goddard, W.A.; Kelber, J.A. Nucleation of Graphene Layers on Magnetic Oxides: Co_3O_4(111) and Cr_2O_3(0001) from Theory and Experiment. *J. Phys. Chem. Lett.* **2017**, *8*, 188–192. [CrossRef]
21. Odedairo, T.; Yan, X.; Ma, J.; Jiao, Y.; Yao, X.; Du, A.; Zhu, Z. Nanosheets Co_3O_4 Interleaved with Graphene for Highly Efficient Oxygen Reduction. *ACS Appl. Mater Interfaces* **2015**, *7*, 21373–21380. [CrossRef] [PubMed]
22. Hofer, C.; Skákalová, V.; Görlich, T.; Tripathi, M.; Mittelberger, A.; Mangler, C.; Reza, M.; Monazam, A.; Susi, T.; Kotakoski, J.; et al. Direct imaging of light-element impurities in graphene reveals triple-coordinated oxygen. *Nat. Commun.* **2019**, *10*, 4570. [CrossRef] [PubMed]
23. Mackenzie, D.; Galbiati, M.; de Cerio, X.; Sahalianov, I.; Radchenko, T.; Sun, J.; Peña, D.; Gammelgaard, L.; Jessen, B.; Thomsen, J. Unraveling the electronic properties of graphene with substitutional oxygen. *2D Mater.* **2021**, *8*, 045035. [CrossRef]
24. Elstner, M.; Porezag, D.; Jungnickel, G.; Elsner, J.; Haugk, M.; Frauenheim, T. Self-consistent-charge density-functional tight-binding method for simulations of complex materials properties. *Phys. Rev. B* **1998**, *58*, 7260–7268. [CrossRef]
25. Lennard-Jones, J.E. On the determination of molecular fields.—II. From the equation of state of a gas. *Proc. R. Soc.* **1924**, *106*, 463.

6. Luriy, S. Quantum capacitance devices. *Appl. Phys. Lett.* **1988**, *52*, 501–503. [CrossRef]
7. Shunaev, V.V.; Ushakov, A.V.; Glukhova, O.E. Increase of γ-Fe_2O_3/CNT composite quantum capacitance by structural design for performance optimization of electrode materials. *Int. J. Quantum Chem.* **2020**, *120*, e26165. [CrossRef]
8. Xu, Q.; Yang, G.; Fan, X.; Zheng, W. Improving the Quantum Capacitance of Graphene-Based Supercapacitors by the Doping and Co-Doping: First-Principles Calculations. *ACSOmega* **2019**, *4*, 13209–13217. [CrossRef]
9. Persson, K. Materials Data on Co_3O_4 (SG:227) by Materials Project. 2016. Available online: https://doi.org/10.17188/1193429 (accessed on 23 March 2022).
30. Mu, J.-C.; Wang, E.-Q.; Zhang, Y.-L.; Zhang, L.-P. Sandwich-Like Co_3O_4/Graphene Nanocomposites as Anode Material for Lithium Ion Batteries. *J. Nanosci. Nanotechnol.* **2019**, *19*, 7819–7825. [CrossRef]
31. Shunaev, V.V.; Glukhova, O.E. Graphene/Fe_3O_4 Nanocomposite as a Promising Material for Chemical Current Sources: A Theoretical Study. *Membranes* **2021**, *11*, 642. [CrossRef]
32. Zhang, J.; Li, P.; Wang, Z.; Qiao, J.; Rooney, D.; Sun, W.; Sun, K. Three-dimensional graphene–Co_3O_4 cathodes for rechargeable Li–O_2 batteries. *J. Mater. Chem. A* **2015**, *3*, 1504. [CrossRef]

lubricants

MDPI

Article

Super-Hard DLC Coatings as an Alternative to Polycrystalline Diamond for Cutting Tools: Predictive Analysis of Aluminium Alloy Surface Quality

Giselle Ramírez [1], Jose M. Gonzalez Castro [2], Jordi Orrit-Prat [1], Raül Bonet [1], Nuria Cuadrado [1], Montserrat Vilaseca [1], Lluís Carreras [3] and Jaume Caro [1,*]

1 Unit of Metallic and Ceramic Materials, Eurecat, Centre Tecnològic de Catalunya, Plaça de la Ciència 2, 08243 Manresa, Spain; giselle.ramirez@eurecat.org (G.R.); jordi.orrit@eurecat.org (J.O.-P.); raul.bonet@eurecat.org (R.B.); nuria.cuadrado@eurecat.org (N.C.); montserrat.vilaseca@eurecat.org (M.V.)
2 Unit of Applied Artificial Intelligence, Eurecat, Centre Tecnològic de Catalunya, Av. Universitat Autònoma, 23, 08290 Cerdanyola del Vallès, Spain; jose.gonzalez@eurecat.org
3 Tratamientos Térmicos Carreras S.A., Passatge de l'Enginyer Playà, 38-42, 08205 Sabadell, Spain; lluiscg@grupttc.com
* Correspondence: jaume.caro@eurecat.org; Tel.: +34-938-777-373

Abstract: The use of lightweight materials, such as ultra-high-strength aluminium alloys, is in high demand in the automotive and aerospace industries where weight savings are critical. The tool materials used for high-speed cutting of these aluminium alloys are subjected to severe conditions that promote premature failure of cutting tools. The application of polycrystalline diamond (PCD) coatings provides cutting tools with increased mechanical and thermal fatigue resistance and improved tribological performance. Despite these good properties, their high cost remains a major limitation in this sector. Super-hard Diamond-Like Carbon (DLC) coatings offer a technologically and economically feasible alternative to PCD-coated tools for cutting and machining non-ferrous materials. In this paper, the machining performance of coated and un-coated hard metal inserts in the turning of 7075 aluminium alloy has been explored. The surface quality of machined parts, the cutting tool wear resistance and the vibrations generated during turning of un-coated, PCD and super-hard thin DLC coatings on tungsten carbide inserts were compared. The results obtained demonstrate that DLC coatings are a potentially interesting alternative to PCD coatings for machining ultra-high-strength aluminium alloys, where surface component finish is a key factor.

Keywords: DLC coatings; PCD coatings; cathodic arc deposition; tool wear; tool vibrations; aluminium alloys

1. Introduction

Lightweight manufacturing in the automotive industry is one of the topics at the forefront of research activities for two main reasons: the reduction of emissions and the positive effect on road safety [1,2]. In this regard, ultra-high-strength aluminium alloys are in high demand in the automotive and aerospace industries, where weight savings are critical. The alloys of the 7000 series, which are primarily mixed with zinc and small amounts of magnesium and copper, are hardenable and very resistant. The most popular alloys from this group, 7050 and 7075, are already widely used in aviation and aerospace, e.g., for aircraft wings, and are currently starting to be used in the automotive industry [3]. On the other hand, tool materials used for high-speed cutting of ultra-high-strength aluminium alloys are subjected to severe conditions that promote premature failure of cutting tools.

The constant evolution of these materials and their applications requires the development of tools that can withstand such severe working conditions while optimizing the performance, durability and costs of cutting and machining processes. These requirements can be achieved through the application of hard coatings. Compared to an unprotected tool, a coated tool exhibits a higher resistance to mechanical and thermal fatigue, a lower coefficient of friction, less interaction between the tool and the machined material, and an improved wear resistance over a wide temperature range. In this respect, the hardest materials known and used in industry are diamond and cubic boron nitride (cBN). Both can be obtained in polycrystalline form (PCD, PcBN) and applied as a coating. While the former is more suitable for machining non-ferrous materials (because carbon reacts chemically with iron), the latter applies to iron-containing alloys [4].

There are two types of polycrystalline diamond coatings used in the tool industry: those obtained by the High-Pressure High-Temperature method (HPHT), and those applied by Chemical Vapor Deposition (CVD). It should be noted that both types of coatings meet the most stringent requirements for cutting and machining materials for the aeronautical sector. These include high hardness (50–100 GPa), high abrasion resistance, high thermal conductivity (800 W/mK), high temperature resistance, low coefficient of friction (between 0.05 and 0.2), low coefficient of expansion, chemical inertness and modulable electrical conductivity [5]. Despite these good properties, their high cost remains a major limitation in this sector.

Super-hard Diamond-Like Carbon (DLC) coatings offer a technologically and economically feasible alternative to PCD-coated tools for cutting and machining non-ferrous materials. In the last two decades, DLC coatings have received a great deal of interest due to their exceptional properties, such as high hardness, low friction, chemical inertness, corrosion protection, biocompatibility, optical transparency in the IR spectral range and tuneable electrical resistivity [6,7]. Currently, DLC coatings are industrially implemented in many engineering applications where excellent tribological properties are required.

The tribomechanical properties of DLC coatings depend on the hydrogen content, the ratio of carbon sp^2 to sp^3 bonds and doping with metallic or non-metallic elements. Within this large family of coatings, hydrogen-free tetrahedral amorphous carbon (ta-C) coatings with a high proportion of sp^3 bonds exhibit very high hardness (50 to 80 GPa) and very low coefficient of friction values (0.05–0.25). However, the main drawback of ta-C coatings is the high level of compressive stress (leading to poor adhesion with the substrate), which limits their thickness and requires the application of adhesion and stress relief interlayers. On the other hand, ta-C coatings can only be obtained by means of cathodic arc Physical Vapor Deposition (PVD) technique.

In comparative terms, while PCD coatings have a thermal resistance of up to 700 °C, DLC coatings start to degrade from 400 °C. In addition, DLC coatings are limited to a few microns in thickness, whereas PCD coatings can reach thicknesses of several tens of microns, which increases their durability in machining operations.

In this paper, the machining performance of coated and un-coated hard metal inserts in the turning of 7075 aluminium alloy is explored. The surface quality of machined parts, the wear resistance and the vibrations generated during turning of un-coated, polycrystalline diamond CVD-coated and super-hard ta-C PVD-coated tungsten carbide inserts are compared.

2. Materials and Methods

Hydrogen-free ta-C coatings were deposited on commercial tungsten carbide inserts (Canela Tools, SPUN 120308E PM25) using the cathodic vacuum arc deposition system shown in Figure 1 (Eurecat-Tratamientos Térmicos Carreras S.A.). In order to reduce the incorporation of detrimental microparticles in the growing films, the system is provided with a DC filtered cathodic arc source (Plasma Technology Limited, Hong Kong, China), where the generated arc plasma is guided into the vacuum chamber by an electromagnetic field applied to the curved duct.

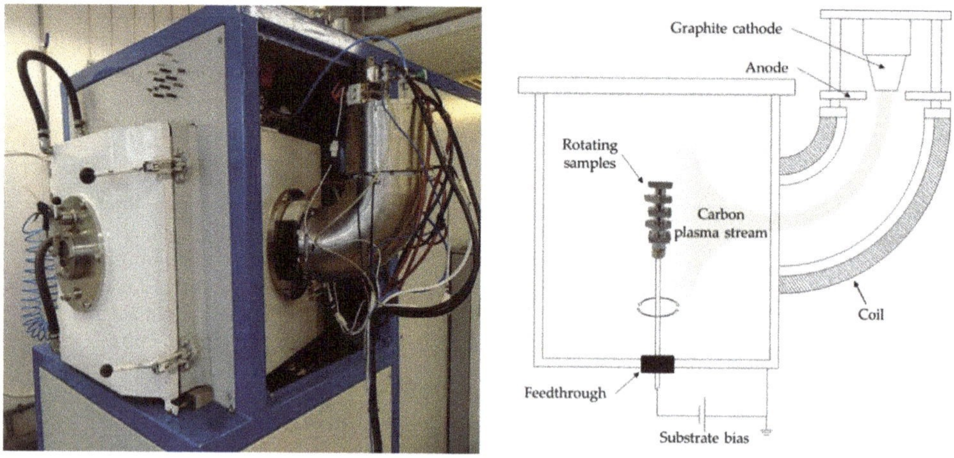

Figure 1. DC filtered cathodic vacuum arc PVD system.

Up to five inserts were positioned in the center of the chamber (30 cm away from the duct exit) and were continuously rotated at 2.75 rpm as shown in Figure 1. The cathodic arc source was supplied by a pure graphite cathode (99.997%). The temperature was measured using a floating K-type thermocouple placed 27 cm from the inserts. Table 1 summarizes the experimental parameters of the different stages used for the preparation of ta-C coatings. The base pressure in the vacuum chamber was 2.6×10^{-5} mbar. Prior to the carbon coating deposition, an argon ion etching was applied to the samples. Subsequently, a carbon bonding layer was applied by gradually decreasing the bias voltage from 1000 to 100 V over 2 min. Finally, the top ta-C coating was deposited at 50 V. The carbon arc current was maintained at 40 A. The final thickness obtained was about 0.40 μm.

Table 1. Experimental parameters of the different deposition stages used for the preparation ta-C coatings.

Stage	T (°C)	P (mbar)	Ar (sccm)	I_{arc} (A)	t (min)	V_{bias} (V)
Ion etching	20	1.3×10^{-2}	128	-	30	1000
Carbon adhesion layer	21	4.8×10^{-5}	20	40	2	1000 to 100
ta-C coating	31	4.8×10^{-5}	20	40	75	50

It is interesting to note that unlike other DLC coatings grown on metallic or hard metal substrates, where adhesion layers based on nitrides and/or carbides of transition metals are commonly deposited, in this work, a carbon adhesion layer was applied, which simplifies the coating process.

Commercial PCD coatings with a thickness of about 9 μm and a hardness of 10,000 HV were deposited by hot-filament CVD on the same tungsten carbide inserts.

The surface morphology of the coatings was studied by Field Emission Scanning Electron Microscope (FESEM) using a Carl Zeiss Neon 40 system (Zeiss, Germany) equipped with an Energy Dispersive X-ray Spectroscopy probe (EDX).

The Daimler-Benz Rockwell-C adhesion test was used to qualitatively assess the coating adhesion. The hardness (H) of the coatings was measured using a NanoIndenter XP (MTS, Eden Prairie, MN, USA) system fitted with a Berkovich diamond tip. The tip was calibrated using a fused silica sample following the Oliver and Pharr method [8]. The evaluation of H was conducted as a function of depth using the continuous stiffness measurement (CSM) operation mode.

The surface quality of the machined components and the hard metal inserts was assessed by measuring three roughness parameters: the arithmetical mean height for line (Ra) and surface (Sa), and the developed interfacial area ratio (Sdr). These values were measured by means of confocal laser scanning microscopy (Sensofar Plμ 2300, Terrassa, Spain). Data analysis was performed using MountainsMap 5.1 software (Digital Surf, Besançon, France) according to ISO 4287 and ISO 25178.

The turning tests were carried out on a HAAS ST-10Y CNC lathe machine, under dry conditions, cutting speed of V_c = 440 m/min, feed of f = 0.26 mm/rev and with a constant cutting depth of a_p = 2 mm corresponding to the finishing conditions. The tests were carried out using 7075 aluminium alloy cylindrical samples of 250 mm in length and a diameter of 70 mm. Two samples were tested for each insert condition.

The wear of the inserts was measured in terms of adhered material. For this purpose, 3D images of cutting edge of pre- and post-tested inserts were acquired using infinite focus microscopy (Alicona InfiniteFocus SL, Graz, Austria). In addition, elemental mapping was used to compile specific chemical composition data across the rake face of the worn inserts that aimed to discern the integrity of the coating and the origin of the adhered material.

Finally, the analysis of tool vibrations produced during turning were recorded using a Piezo Star triaxial accelerometer (Kistler type 8766A, Winterthur, Switzerland) with a sampling frequency of 10 kHz. The purpose of this test was to investigate the effects of coating material and roughness on tool vibration and their correlation with the machined surface components. Thus, all the cutting parameters such as cutting speed, feed rate, depth of cut, tool and work piece length were kept constant during the test. The signal was stored on a time series database with separated tables for each experiment. The data were segmented to consider only periods where the selected tool was working, to obtain a region of interest to be analyzed. The first and last working periods were selected to compare the initial and final states of the cutting edges. Once the different time series segments were extracted, it was possible to perform a set of analyses to compare the data between each tool and cutting edge condition.

3. Results and Discussion

Figure 2a shows the cross-section SEM image of deposited ta-C coating, which is characterized by a dense microstructure and the presence of some surface defects that can be attributed to the small fraction of carbon microdroplets that, through multiple collisions, have escaped from the magnetic filter. It is worth noting that no carbon microparticles have been observed either at the substrate interface or within the coating. The microparticles generated by the arc discharge on the graphite cathode that have reached the substrate have been segregated towards the surface during the growth of the coating. The same phenomenon was already observed in a previous work related to the growth of silver-doped DLC coatings using a similar experimental setup [9].

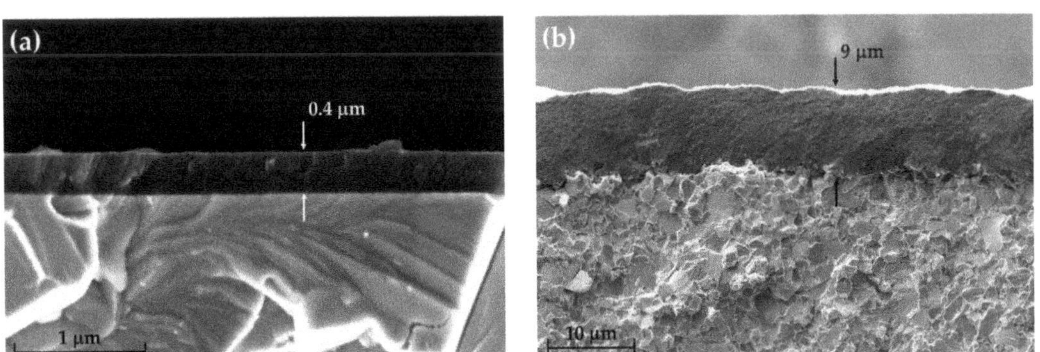

Figure 2. Cross-section SEM images of: (**a**) ta-C coating; (**b**) PCD coating.

The cross-section of the PCD coating is shown in Figure 2b. Comparatively, the PCD coating is much thicker and rougher than the ta-C coating. The roughness parameters of un-coated, ta-C-coated and PCD-coated inserts are shown in Table 2. As can be seen, while the deposition of the ta-C coating doubles the surface roughness of the substrate, the PCD coating increases it by two orders of magnitude due to its cauliflower-like surface morphology (not shown).

Table 2. Roughness parameters of un-coated, ta-C-coated and PCD-coated inserts.

Sample	Ra (μm)	Sa (μm)	Sdr (%)
Un-coated	0.004 ± 0.001	0.005 ± 0.001	0.004 ± 0.003
ta-C coating	0.010 ± 0.001	0.011 ± 0.001	0.017 ± 0.003
PCD coating	0.341 ± 0.010	0.386 ± 0.006	3.85 ± 0.34

Figure 3a shows the variation of hardness as a function of the tip penetration depth obtained by nanoindentation on ta-C coating. The curve shows a maximum hardness of 68.5 ± 6.6 GPa, which can be roughly attributed to the intrinsic hardness value of the coating. In addition to this extremely high hardness, one of the most interesting results achieved in this work is the excellent adhesion shown by the ta-C coating without the application of any bonding layer based on nitrides and/or carbides of transition metals. Figure 3b displays an optical microscopy image corresponding to the residual imprint generated by the Rockwell-C adhesion test, where no delamination has been generated.

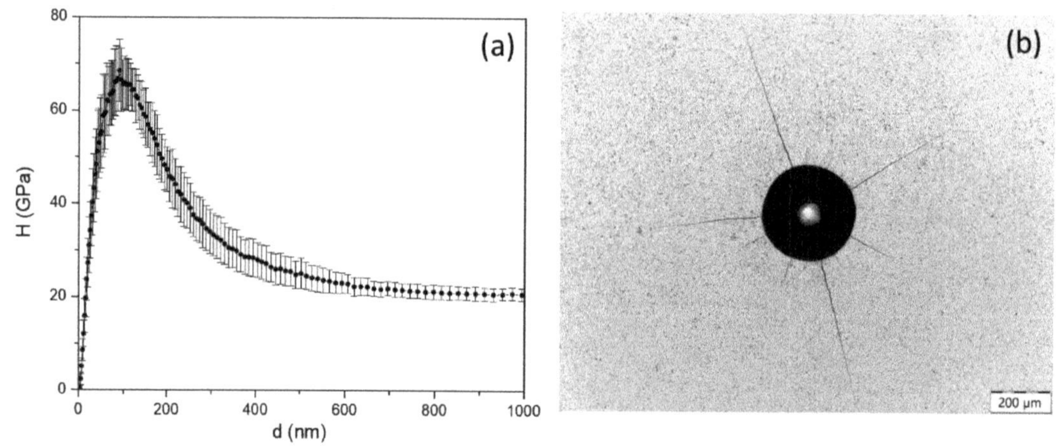

Figure 3. (**a**) Hardness as a function of the tip penetration depth, and (**b**) Rockwell-C adhesion test performed on ta-C coating.

Concerning the surface quality, Figure 4 shows optical microscopy images corresponding to the turning finish obtained using each of the hard metal inserts considered in this study. As can be seen, even with the naked eye, the best surface finish of the machined parts was obtained using ta-C coated inserts. This observation is confirmed from the measurements of the roughness parameters obtained from the machined parts, as shown in Figure 5. After a reduction of the workpiece diameter from 70 mm to 40 mm (15 passes), the values of Ra, Sa and Sdr are significantly lower for the inserts coated with ta-C, while the worst results are obtained with the inserts coated with PCD. It should be noted that the percentage of developed interfacial area (Sdr) is the value that shows the greatest decrease in the case of ta-C coated inserts compared to un-coated and PCD coated inserts.

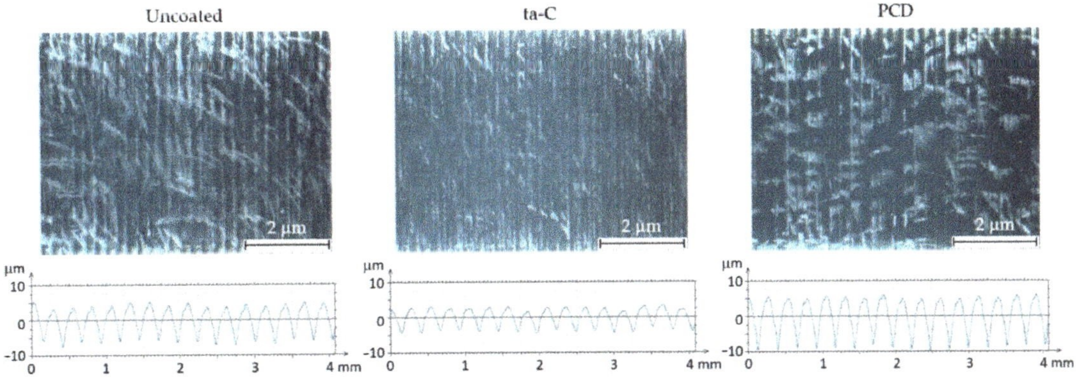

Figure 4. Optical microscopy images and height profiles corresponding to the turning finish obtained using un-coated, ta-C-coated and PCD-coated inserts.

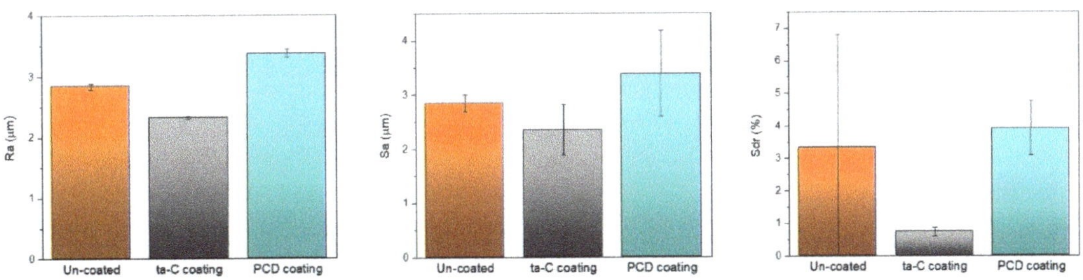

Figure 5. Roughness parameters Ra, Sa and Sdr of machined parts using un-coated, ta-C-coated and PCD-coated inserts.

The machining conditions used in these experiments were quite aggressive for the cutting tool, since tests were performed under dry conditions. Despite not being useful to industry due to the reduction in tool life that this would entail, for this research, the use of dry conditions was useful to obtain information on tool wear resistance. Figure 6 shows SEM images of the cutting edge of the tested inserts taken at the rake face, and their corresponding mapping element distribution and total atomic content of the adhered aluminium as determined by EDX. From these figures, it can be seen that the machining test conditions do not induce abrasive wear or severe chipping damage at the cutting edge. In addition, the test conditions reproduced the same wear area for the three inserts. However, the distribution and total atomic content of the adhered aluminium show that the un-coated insert presented a high adhesion of aluminium, homogeneously distributed over the entire contact area. On the other hand, in the ta-C-coated insert, although there is an area where the coating has partially delaminated, the content of adhered aluminium is considerably lower, and is basically concentrated in the transition between the delaminated zone and the coated area, where excessive cutting temperatures and pressures tend to increase. Finally, in the PCD-coated insert the coating remains intact (probably due to its greater thickness), showing a low aluminium content in the crater area. Despite the low atomic content of aluminium in the crater area, the element mapping shows aluminium particles mechanically adhered to the PCD coating irregularities (cauliflower-like morphology). These surface asperities induce a higher mechanical sticking which is present along the rake surface. This phenomenon has also been observed in coated tools for forming aluminium alloys [10].

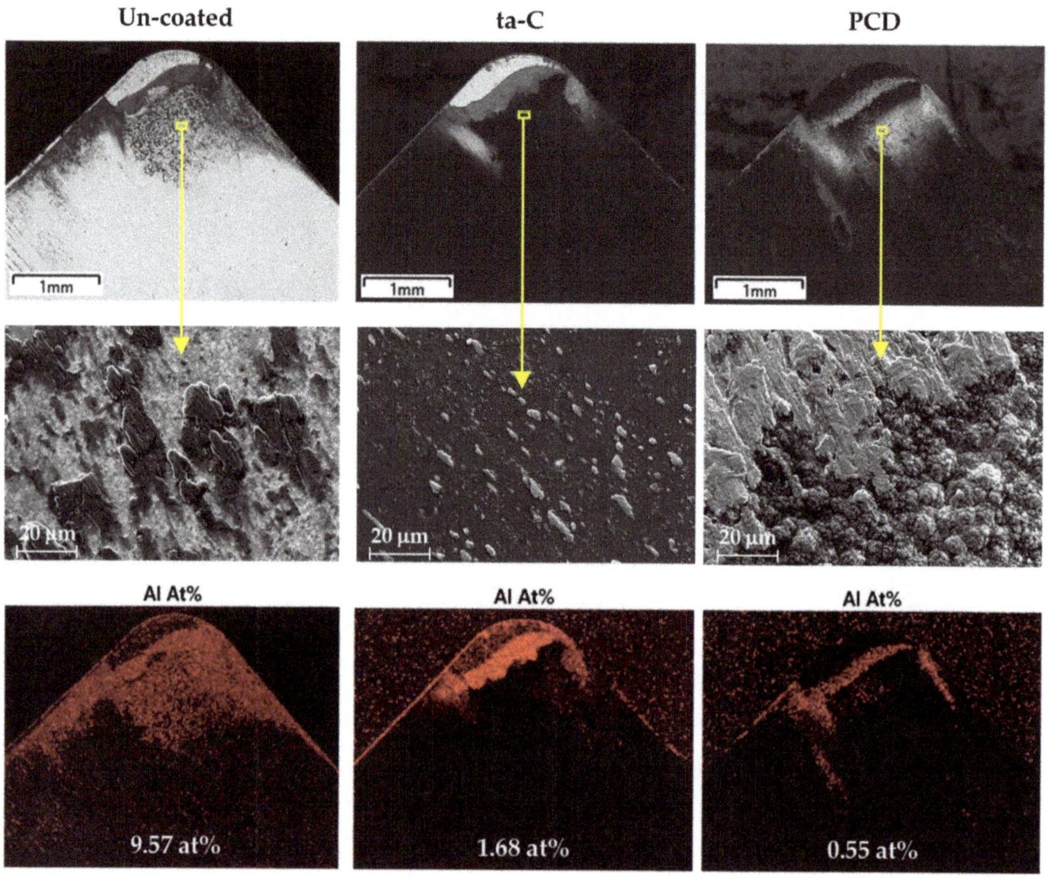

Figure 6. SEM images of the cutting edge of the tested inserts taken at the rake face, and their corresponding mapping distribution of total atomic content of the adhered aluminium.

The high roughness of parts machined with PCD inserts could be attributed to the high irregular topography of the coating displayed at the cutting surface of the tool, as shown previously in Table 2. This phenomenon is more evident in the 3D images obtained at the cutting edge of the worn inserts and their respective cross-section profiles (Figure 7). These images were acquired automatically by comparing the reference geometry (insert in the initial state) with the worn tool (Figure 7a). Form deviations are clearly visible by advanced color visualization scale, where the violet color indicates the reference geometry, the bright cyan color indicates the adhered material with a thickness between 10 and 20 µm, and dark blue shows the adhered material between 5 and 10 µm. The 3D images show that un-coated and ta-C inserts presented more adhered material in the rake face, as previously observed in SEM images, whereas in the flank face the maximum worn material was around 5 µm in depth, as indicated by the magenta-colored area. In contrast, the most affected area in PCD inserts was the flank face, with a large amount of adhered material, in some cases reaching a thickness of up to 50 µm.

Bi-dimensional profiles were extracted from the cutting edges radius (Figure 7b), as indicated by the dash-line circle. Once again, the un-coated insert presents the highest material adhesion with particles up to 20 µm in size, while the radius edge of ta-C and PCD-coated inserts show adhered particles with a maximum size of 5 µm. However, PCD coated tool shows an irregular cutting edge radius due to the high roughness of the coating. Thomas et al. [11] reported that the cutting-edge radius is one of the variables affecting the surface roughness of the machined component, interpreting the results obtained in terms of tool vibration. Nevertheless, the influence of tool surface roughness at the cutting edge was not mentioned.

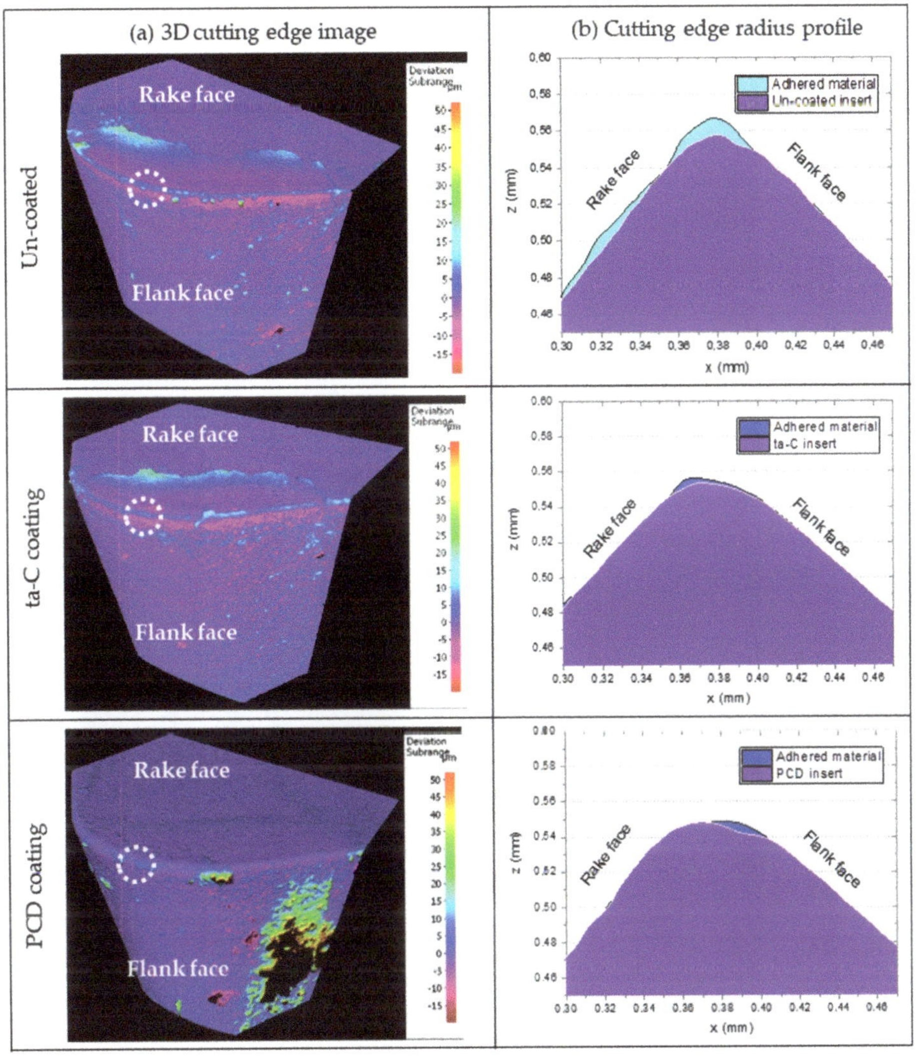

Figure 7. 3D images obtained at the cutting edge of the worn inserts and their respective cross-section profiles.

The effect of tool vibrations on surface quality of machining components during the high-speed cutting process has been extensively analyzed, but these studies mainly focus on tool vibrations caused by variations in cutting parameters, geometrical cutting factors, dynamic factors, part geometries, lubricants, component materials and tools' geometries [12,13]. However, the influence of the tool surface material and the roughness has not been widely reported. In previous works, authors evaluated the cutting tool vibrations for CNC turning machines using a predictive model to identify tool wearing that can affect surface integrity quality of the manufactured component [14]. These preliminary results were orientated towards implementing a predictive maintenance methodology in inserts with the same material but with different grades of abrasive wear.

In this study, the tool vibrations were analyzed as a function of insert surface roughness and material. For this purpose, the tools were monitored using an accelerometer sensor placed in the insert holder, and all cutting parameters were kept constant. A comparative between initial and final tool state based on the vibrations generated with the accelerometer on the cutting axis can be seen in Figure 8. The frequency spectrum for each tool signal was computed using the Power Spectral Density (PSD), which measures the spectral energy distribution over frequency, for each initial and final state. The increment on the vibration was obtained by computing the ratio between the final and initial PSD.

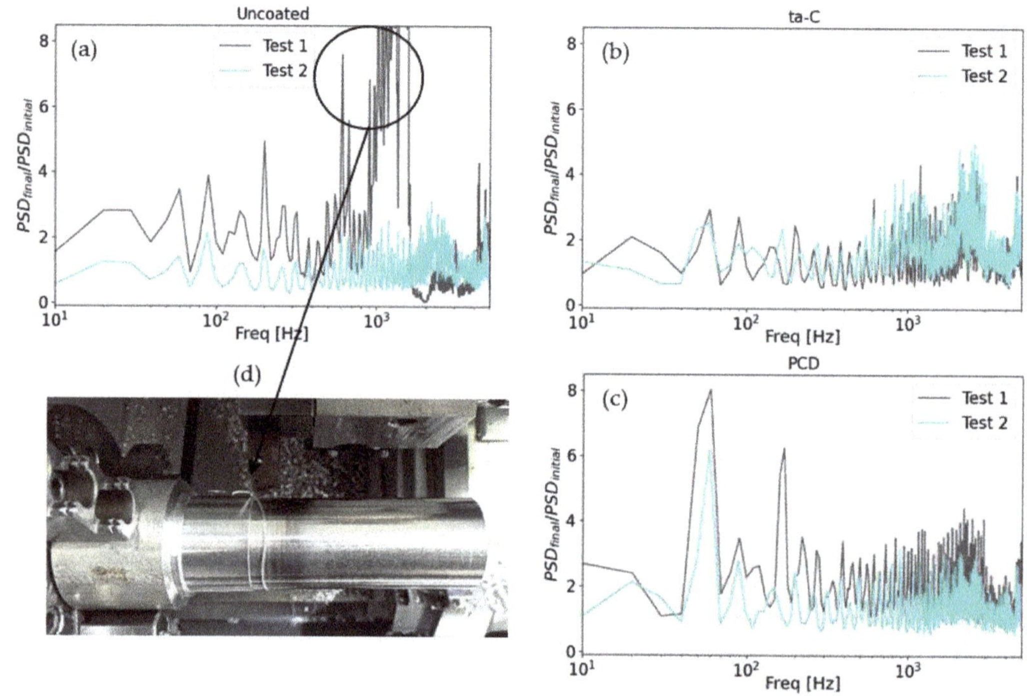

Figure 8. Ratio between the final and initial Power Spectral Density (PSD) of (**a**) un-coated, (**b**) ta-C coated, and (**c**) PCD-coated inserts. (**d**) Un-coated tool forming a large chip around the machined component.

From the set of images, it is possible to see that the un-coated inserts suffer the most extreme wear, where the vibrations increase about 20 times between the initial state and the final state in the frequencies between 1 and 2 kHz. This increase in vibrations can be attributed to the higher amount of stuck material at the cutting edge of un-coated insert, which favors the formation of large chips around the machined component. On the other side, for PCD and ta-C coated tools, the vibrations are more stable between the first and last operation. However, the PCD-coated tools generate a larger number of harmonics, increasing vibrations by more than three times compared to the new tool in all the measured spectrums. This behavior could be attributed to high roughness of PCD coating, which, although it does not delaminate, tends to increase the adhesion of aluminium material, which is more visible in the flank face.

4. Conclusions

In view of the results obtained, for the turning conditions used in this study, it can be concluded that the super hard ta-C coatings developed in this work using the filtered cathodic arc PVD technique offer optimal mechanical performance in terms of surface finish of machined aluminium parts. Although the low thickness of these coatings may limit tool durability compared to thicker PCD coatings, ta-C coatings may prove to be a potentially interesting alternative to PCD coatings for machining high-strength aluminium alloys, where surface finish may be as or more important than tool durability.

Vibration analysis on the tool wear supports the above hypothesis giving a clear superiority of coated over un-coated tools, and showing the need to work with coated tools. Between the ta-C and PCD coatings, the vibration shows small differences in the high-frequency range (above 1 kHz). However, for lower frequencies, the appearance of larger harmonics for the PCD coating might be generated by the higher tool surface roughness, which is also related to the higher roughness of the machined samples.

Author Contributions: Conceptualization, J.C., G.R. and J.M.G.C.; methodology, J.C., G.R. and J.M.G.C.; investigation, G.R., J.M.G.C., J.O.-P., R.B. and N.C.; resources, L.C.; writing—original draft preparation, J.C., G.R. and J.M.G.C.; writing—review and editing, J.C.; funding acquisition, J.C., M.V. and L.C. All authors have read and agreed to the published version of the manuscript.

Funding: This research was funded by the Ministry of Science, Innovation and Universities of Spain under the project REDUCA (RTC-2017-6479-4).

Institutional Review Board Statement: Not applicable.

Informed Consent Statement: Not applicable.

Data Availability Statement: Not applicable.

Conflicts of Interest: The authors declare no conflict of interest.

References

1. Tisza, M.; Czinege, I. Comparative study of the application of steels and aluminium in lightweight production of automotive parts. *Int. J. Lightweight Mater. Manuf.* **2018**, *1*, 229–238. [CrossRef]
2. Zainol, A.; Yazid, M.Z.A. Environmentally Friendly Approaches Assisted Machining of Aluminum Alloy 7075-T6 for Automotive Applications: A Review. *Int. J. Integr. Eng.* **2019**, *11*, 18–26. [CrossRef]
3. Afseth, A. Ultra-High-Strength Aluminium Alloys—Vehicle Production's Next Big Thing. *Lightweight Des. Worldw.* **2017**, *10*, 12–15. [CrossRef]
4. Coelho, R.T.; Yamada, S.; Aspinwall, D.K.; Wise, M.L.H. The application of polycrystalline diamond (PCD) tool materials when drilling and reaming aluminium based alloys including MMC. *Int. J. Mach. Tools Manuf.* **1995**, *35*, 761–774. [CrossRef]
5. Element Six. Diamond Handbook. Available online: https://www.e6.com/en/knowledge-base/brochures (accessed on 1 June 2022).
6. Erdemir, A.; Donnet, C. Tribology of diamond-like carbon films: Recent progress and future prospects. *J. Phys. D Appl. Phys.* **2006**, *39*, R311–R327. [CrossRef]
7. Hainsworth, S.V.; Uhure, N.J. Diamond like carbon coatings for tribology: Production techniques, characterisation methods and applications. *Int. Mater. Rev.* **2007**, *52*, 153–174. [CrossRef]
8. Oliver, W.C.; Pharr, G.M. An improved technique for determining hardness and elastic modulus using load and displacement sensing indentation experiments. *J. Mater. Res.* **1992**, *7*, 1564–1583. [CrossRef]
9. Orrit-Prat, J.; Bonet, R.; Rupérez, E.; Punset, M.; Ortiz-Hernández, M.; Guillem-Marti, J.; Lousa, A.; Cano, D.; Díaz, C.; García Fuentes, G.; et al. Bactericidal silver-doped DLC coatings obtained by pulsed filtered cathodic arc co-deposition. *Surf. Coat. Technol.* **2021**, *411*, 126977. [CrossRef]
10. Pujante, J.; Pelcastre, L.; Vilaseca, M.; Casellas, D.; Prakash, B. Investigations into wear and galling mechanism of aluminium alloy-tool steel tribopair at different temperatures. *Wear* **2013**, *308*, 193–198. [CrossRef]
11. Thomas, M.; Beauchamp, Y.; Youssef, A.Y.; Masounave, J. Effect of tool vibrations on surface roughness during lathe dry turning process. *Comput. Ind. Eng.* **1996**, *31*, 637–644. [CrossRef]
12. Abouelatta, O.B.; Mádl, J. Surface roughness prediction based on cutting parameters and tool vibrations in turning operations. *J. Mater. Process. Technol.* **2001**, *118*, 269–277. [CrossRef]

13. Quintana, G.; Garcia-Romeu, M.L.; Ciurana, J. Surface roughness monitoring application based on artificial neural networks fo ball-end milling operations. *J. Intell. Manuf.* **2011**, *22*, 607–617. [CrossRef]
14. Gonzalez Castro, J.M.; Ramirez Sandoval, G.; Vidales Coca, E.; Cuadrado Lafoz, N.; Bonada, F. Cutting Tool Wearing Identificatio Through Predictive Maintenance and Its Impact on Surface Quality. In *Artificial Intelligence Research and Development, Proceeding of the 23rd International Conference of the Catalan Association for Artificial Intelligence, Lleida, Spain, 20–22 October 2021*; Villaret, M Alsinet, T., Fernández, C., Valls, A., Eds.; IOS Press Ebooks: Amsterdam, The Netherlands, 2021; Volume 339, pp. 264–267.

lubricants MDPI

Article

Manufacturing of Aluminum Alloy Parts from Recycled Feedstock by PIG Die-Casting and Hot Stamping

Tatsuhiko Aizawa [1,*], Takeshi Kurihara [2] and Hiroki Sakayori [2]

[1] Surface Engineering Design Laboratory, Shibaura Institute of Technology, Tokyo 144-0045, Japan
[2] Sanko Light Kogyo, Co., Ltd., Kawasaki 211-0053, Japan
* Correspondence: taizawa@sic.shibaura-it.ac.jp; Tel.: +81-3-6722-8615

Abstract: PIG (Pin-Injection-Gate) die-casting and hot stamping was developed for fabrication of small-sized and thin-walled aluminum alloy parts from the recycled feedstock. The pure aluminum and aluminum alloy granules were utilized as a feedstock model of recycled materials. The measured mass of granules with the estimated weight from 3D-CAD (Computer Aided Design) of products was poured into the PIG-nozzles before injection. After quickly melting by induction heating inside the PIG-nozzle units, the aluminum melts were injected into a die cavity through the PIG-nozzle. No furnaces and no crucibles were needed to store the melt aluminum stock in different from the conventional die-casting system. No clamping mechanism with huge loading machine was also needed to significantly reduce the energy consumption in casting. Much less wastes were yielded in these processes; the ratio of product to waste, or, the materials efficiency was nearly 100%. Nitrogen supersaturation and TiAlN coating were used to protect the PIG-nozzle and the stamping die surfaces from severe adhesion from aluminum melt. The pure aluminum gears and thin-walled mobile phone case were fabricated by this process. X-ray tomography proved that both products had no cavities, pores and shrinkages in their inside. Using the hot stamping unit, the micro-pillared pure aluminum heatsink was fabricated to investigate the holding temperature effect on the aspect ratio of micro-pillar height to width.

Keywords: upward recycling; aluminum and aluminum alloys; pin injection gate; die casting; hot stamping; mechanical parts; heatsink; nitrogen supersaturation; TiAlN coating

1. Introduction

A green manufacturing has grown up as a keyword in the sustainable society with carbon neutrality [1]. In the solid-state recycling [2], the used aluminum alloy parts were mechanically refined and re-alloyed to the solid preform for metal forming to fabricate new products without remelting. In warm and hot extrusion processes [3], the aluminum alloy debris fragments by machining were used as a feedstock for forward and backward extrusion to shape the recycled preforms and to drive the aluminum material circulation. The conventional die-casting process [4] is effective to make near-net shaping of remelt aluminum alloy wastes into a new product; however, it is too much energy-consuming to be utilized for green recycling. The large furnaces or crucibles are necessary to remelt the aluminum alloy wastes and to hold those remelt materials. A pair of die-sets has to be closed to drive the casting and solidification processes in the die cavity by applying the huge clamping force. Many in-process wastes, such as spurs, runners, or bosses, are inevitably yielded in daily die-casting to significantly deteriorate the aluminum material efficiency in the recycling circulation.

The PIG (Pin-Injection-Gate) die-casting system was developed to save these demerits and weak points, which are intrinsic to the conventional die-casting [5–7]. This new type of aluminum die-casting processes is especially suitable for near-net shaping of small-sized parts and thin-walled members. In the similar manner to the plastic injection molding, a pin-injection-gate (PIG) design is employed to cast the aluminum melts into a die-cavity. No remelting furnaces and crucible are necessary; the used aluminum alloy debris fragments with the online measured weight are directly melt in the inside of PIG-nozzle. A slightly small clamping force is enough to drive the casting and filling processes in the die cavity. The amount of in-process wastes is minimized since the online-measured materials are automatically poured into the PIG-nozzle and remelt in it. After [8], the capacity of this PIG-die casting was proved to make upward recycling of the aluminum alloy debris articles. Certainly the near-net preforms can be PIG-die-cast to yield the small-sized products, but further working steps are needed to make fine shaping and near-net shaping.

In the present paper, this PIG die-casting system is further advanced to equip the hot stamping unit to make secondary working of the PIG-die-cast preforms. The total system including these PIG die-casting and hot-stamping units, is first explained as a suitable manufacturing system to circulation of used aluminum alloys. Two types of surface treatment are utilized to prevent the PIG-nozzle units as well as the dies from adhesion wear during the PIG-die casting and hot stamping; nitrogen supersaturation process and TiAlN coating. The PIG die casting behavior is described by free-injection and short-shot casting experiments. The miniature pure aluminum gear is die-cast by using a single PIG-nozzle unit. The aluminum alloy mobile phone case is also die-cast by using double PIG-nozzle units. The pure aluminum heatsink is fabricated by warm and hot stamping the PIG-die-cast aluminum preform to have the micro-pillared microtextures on its surface. This aluminum manufacturing process has a capacity to fabricate the small-sized and rib-structured aluminum alloy parts from the crashed fragment feedstock and to play a significant contribution to light-weight part design.

2. Materials and Methods

The PIG die-casting unit is combined with the hot stamping unit to build up the green aluminum processing for upward recycling from the aluminum machined chips and fragments to small sized parts and thin-walled products. Through alternative combination of two units, various light-weight metal forming procedures are available to fabricate the upward-recycled products. The plasma nitriding was mainly utilized for protection of PIG-nozzles and dies from galling by aluminum melts. In particular, the micro-textured die is prepared for warm and hot stamping by using the plasma printing method to prepare the mesh-patterned punch.

2.1. New Aluminum Manufacturing System by PIG Die-Casting and Hot Stamping

In the present aluminum manufacturing system, the PIG die-casting unit is combined with the hot stamping unit. The die-set is exchanged from one unit to another one by controlling the rotating bed. Various working procedures are possibly designed and accommodated to this system, as illustrated in Figure 1.

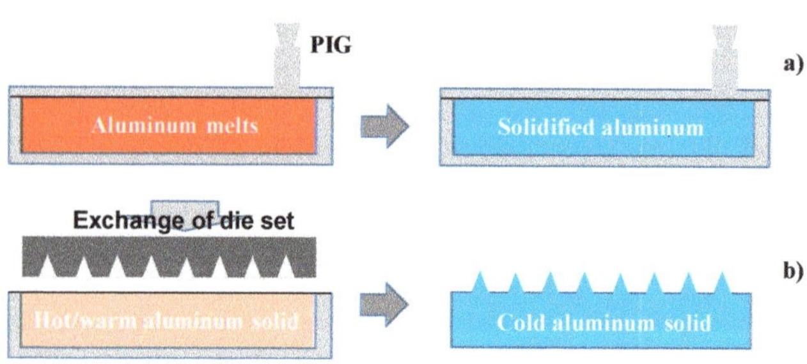

Figure 1. Two procedures to yield the aluminum products. (**a**) PIG die-casting procedure, and (**b**) Combined process of PIG die casting and hot stamping.

In the PIG die-casting in Figure 1a, the aluminum melt is cast into the die cavity and cooled down to yield the near-net shaped product. This is a standard procedure to make near-net shaping of small-sized and thin walled aluminum alloy parts. The PIG die-cast preform is further hot stamped to the micro-textured product under the specified holding temperature as illustrated in Figure 1b. Figure 2a depicts the overview of developed manufacturing system. Using the rotating bed in Figure 2b, the lower die-set including the primarily processed aluminum preform is transferred to the secondary step. In this secondary step, a new upper die-set is utilized for further processing. Let us explain the key techniques in this system.

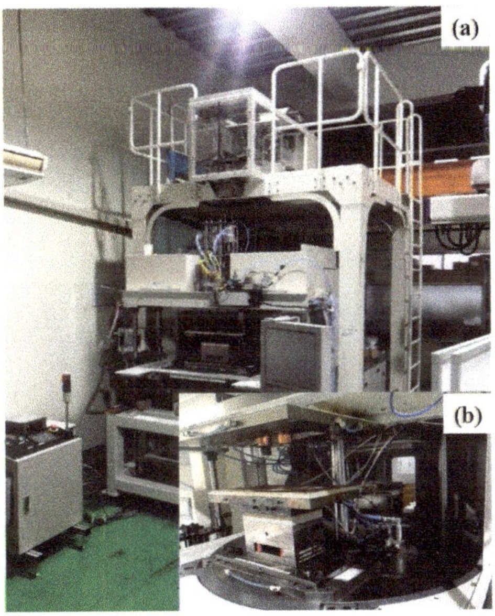

Figure 2. Manufacturing system including the PIG die-casting and hot stamping unit. (**a**) Overview of the total system, and (**b**) rotating bed to exchange the die set.

In the PIG die-casting unit, the aluminum melt is injected through the PIG-nozzle into the die cavity. As illustrated Figure 3a, the aluminum and aluminum alloy granules and their used fragments are utilized as a feedstock for casting. The weight of starting granular materials is on line measured with reference to the product volume data in the 3D-CAD (Computer Aided Design) model of product. The measured materials are poured and stacked into the PIG-nozzle unit. The stacked work is melt rapidly and uniformly by using the IH (Induction Heating)-coil with the designed high frequency of 500 kHz. After monitoring the temperature transients during heating, the work melt is pushed into a die cavity by using the pressurizing media. Figure 3b shows the actual PIG-nozzle unit.

In the warm and hot stamping unit, the upper die-set including the micro-textured punch is prepared for net-shaping. This die set is fixed into the upper bolster of stamping unit; its stoke is directly controlled to move down and up the upper die-set after programmable sequence for loading and unloading, respectively. In the following stamping experiment, the stroke speed is constant by 0.1 m/s.

As had been discussed on the hot extrusion studies of aluminum alloys [9], the aluminum remelt as well as hot preform were easy to adhere onto the die inner surfaces. Without surface treatments, the inner surface of PIG nozzle and dies could experience severe adhesive wear. To be free from galling and erosion by aluminum melts, the inner surface of PIG-nozzle was protected by the low temperature plasma nitriding [10] and TiAlN-base coating [11].

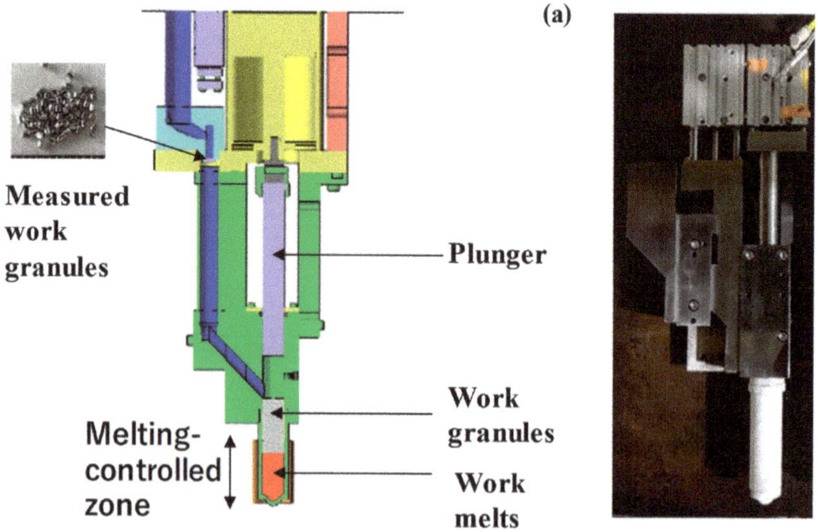

Figure 3. PIG-nozzle unit including the weight-measuring section, the plunger, the pressurizing tools, and the heating section. (**a**) Schematic view of PIG-nozzle unit, and (**b**) overview of PIG-nozzle unit.

2.2. Tribological Treatment for PIG-Nozzles and Dies

The low temperature plasma nitriding was first utilized to form a thick nitrogen supersaturated layer into the inner surfaces of PIG-nizzles and the die surfaces and to save them from the corrosion and erosion wear by chemical attack of aluminum melts. RF (Radio-Frequency)–DC (Direct Current) plasma nitriding system (YS-Electrical Industry, Co., Ltd.; Kofu, Japan) was utilized for this nitrogen supersaturation treatment of AISI420J2 PIG nozzles and dies at 673 K for 14.4 ks by 70 Pa. Figure 4 depicts this plasma nitriding system. As illustrated in Figure 4a, the inner surface of PIG-nozzle was directly nitrided by using the nitrogen and hydrogen mixture gas. The nitriding temperature was monitored by using the thermocouple, embedded into the DC-plate. The pressure was also controlled by using the gas flow meter. The deviation of temperature and pressure was reduced within ±0.1 K and ±0.05 Pa, respectively.

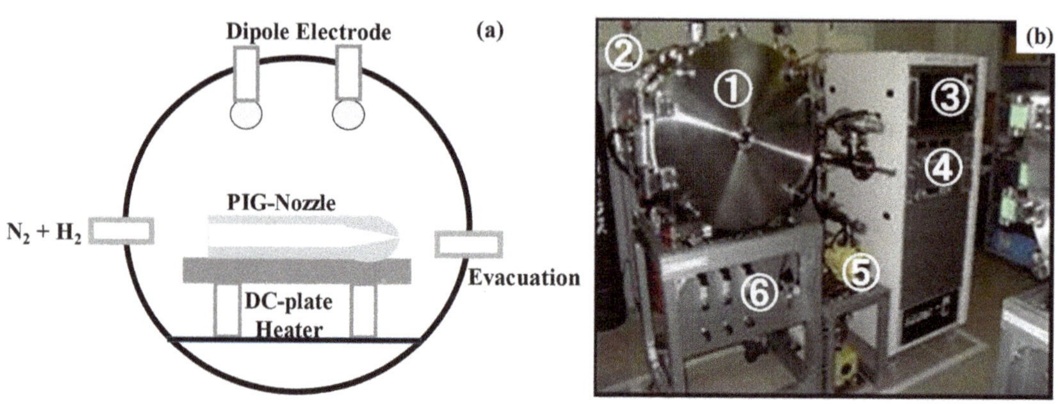

Figure 4. A low temperature plasma nitriding system. (**a**) Its schematic view, and (**b**) its overview with 1: vacuum chamber, 2: RF-generator, 3: control panel, 4: RF/DC controller, 5: evacuation unit, and 6: gas flow units.

The PIG nozzles were first nitrogen-supersaturated under the nitriding conditions. After nitriding, TiAlN film was deposited onto the PIG-nozzles by PVD process. Its thickness was 3 μm. The dies for casting and warm/hot stamping processes were only nitrogen supersaturated to harden their surfaces up to 1400 HV by the thick nitrided layer of 50 μm.

2.3. Warm and Hot Stamping Die for Microtexturing

An AISI316 punch for microtexturing the PIG-die-cast preform was prepared by plasma printing [12]. The screen printer was first used to draw the two dimensional pattern onto the punch surface as a mask. The low temperature plasma nitriding [10] was also employed to make nitrogen supersaturation onto the unmasked surface areas for selective hardening. The sand-blasting was utilized to mechanically remove the unmasked and un-nitrided AISI316 matrix of punch and to shape the designed microtextures onto the AISI316 punch.

In order to build up the AISI316 punch for hot-stamping the micro-pillared aluminum heatsink, the punch surface was shaped to have the mesh-patterned multi-heads with micro-cavities [13]. As depicted in Figure 5a, this mesh-pattern was designed by regular alignment of rectangular unit-cell in Figure 5b. The black-colored edges in Figure 5a,b, become a punch head while the white colored square corresponds to a micro-cavity, into which the aluminum work flows in the backward extrusion during the hot stamping. The screen film was first prepared to print the negative pattern to the designed microtexture in Figure 5a onto the punch surface. Its unit cell is shown in Figure 5c, where the white colored square in Figure 5b is printed in black and the black colored edge in Figure 5b are never printed. Since those printed zones work as a mask in the plasma nitriding, the white edge zone in Figure 5c is selectively nitrided in Figure 5d. After sand-blasting with the use of silica particles, the printed part in Figure 5d is mechanically removed to form the mesh-patterned punch heads, as depicted in Figure 5e.

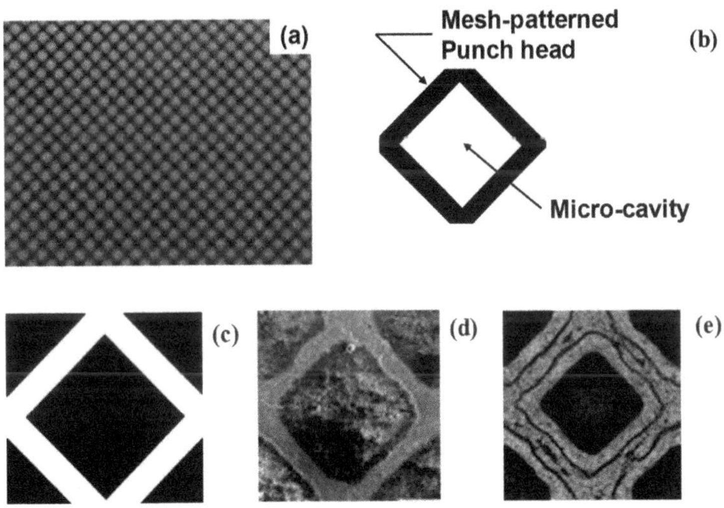

Figure 5. A plasma printing procedure with aid of the low temperature plasma nitriding. (**a**) CAD design on the mesh-patterned punch head, (**b**) a unit cell of mesh-pattern, (**c**) a unit cell on the screen-film for screen printing, (**d**) a nitrided unit cell of microtextures on the die surface at 673 K for 14.4 ks, and (**e**) a unit cell of mesh-patterned die after sand-blasting.

After the procedure in Figure 5, the mesh-textured AISI316 punch was shaped as shown in Figure 6a. Its surface consists of the mesh-textured multi-heads in Figure 6b; the maximum surface roughness on these punch heads was only 0.6 μm. This proves that the as-blasted AISI316 punch is directly used as a warm and hot stamping punch without further polishing. Three dimensional profile of this microtexture on the punch is shown in Figure 6c. The maximum depth of micro-cavities reaches 190 μm, which is deep enough for micro-extrusion of aluminum work during hot stamping.

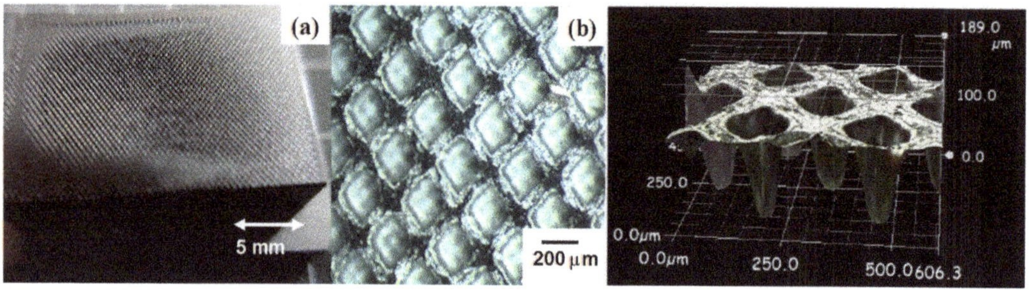

Figure 6. The mesh-textured AISI316 punch by the plasma printing. (**a**) Overview of textured punch, (**b**) SEM image on the punch surface, and (c) three dimensional profile of microtextures.

During this hot and warm stamping, the mesh-textured AISI316 punch has a risk of galling, where the hot aluminum preforms flow into the cavities in Figure 6. Through the plasma printing process in Figure 6, the AISI316 punch surfaces were also nitrogen supersaturated at 673 K for 14.4 ks by 70 Pa. After [10,14], these surfaces were expected to have sufficient corrosion and erosion toughness enough to be free from the adhesion of aluminum flow and chemical attack to the punch surfaces.

2.4. Work Materials

The aluminum granules with the purity of 99.7% and the size of 3 mm were prepared as a pure aluminum feedstock. AA5052 aluminum alloy particles were also utilized for mass-production of thin parts. The chemical composition of this alloy is listed in the following: e.g., 2.6 mass% Mg, 0.10 mass% Zn, 0.25 mass% Si, 0.4 mass% Fe, 0.10 mass% Cu, 0.10 mass% Mn, 0.25 mass% Cr, and Al in balance. The PIG-nozzle and casting dies were made from AISI420J2 while AISI316 was used for the warm/hot stamping dies.

2.5. Measurement and Characterization

A table-top X-ray tomography apparatus was employed to make non-destructive evaluation on the residual defects in the cast aluminum and aluminum alloy products. The laser microscopy and three dimensional profilometer were utilized to measure the concave and convex microtextures on the plasma-printed AISI316 die and the hot-stamped aluminum plates, respectively.

3. Results

Pure aluminum and A5052 granular particles were employed as a feedstock material for PIG die casting and hot stamping. Small-sized parts such as pure aluminum gears were die-cast to demonstrate that these are yielded with less amounts of waste and without residual pores and defects. The aluminum mobile phone case was also die-cast to prove that thin-walled parts are fabricated even in mass-production. The pure aluminum heatsink was fabricated by the two-step procedure.

3.1. Tribological Characterization of Nitrided AISI420J2

The hardness and nitrogen solute content were measured by the micro-Vickers hardness testing and SEM (Scanning Electron Microscopy)-EDX (Energy Dispersive X-ray spectroscopy), respectively. As depicted in Figure 7, the nitrogen content more than 3 mass% and higher hardness than 1000 HV were preserved in the nitrided layer with the thickness of 50 μm. In particular, most of octahedral vacancy sites in AISI420 crystalline cells were occupied by the nitrogen interstitial atoms in this high average nitrogen content of 3 mass%. After experimental studies in [10,14] and theoretical study in [15], the corrosion and erosion toughness is expected to be improved by this thick nitrided layer.

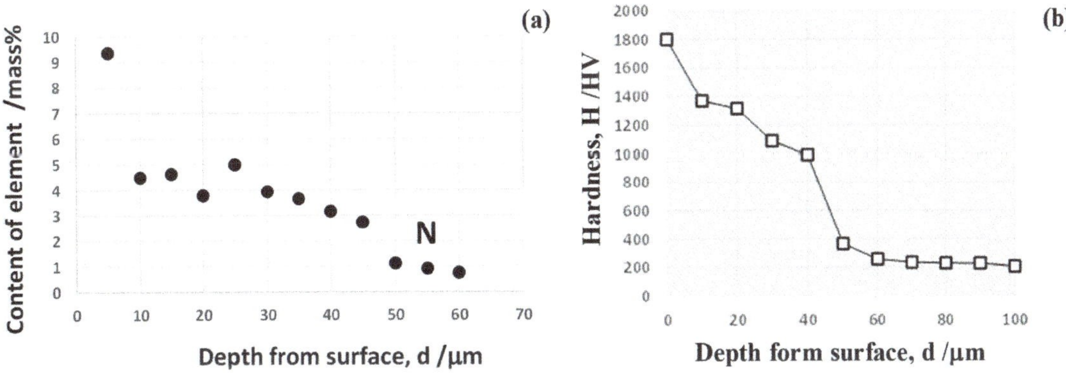

Figure 7. Nitrogen solute content and hardness depth profiles in the nitrided AISI20J2 materials for PIG-nozzles and dies. (**a**) Nitrogen solute depth profile measured by SEM-EDX, and (**b**) hardness depth profile measured by micro-Vickers hardness testing.

The hydrochloric acid with 20% concentration was used for dipping test at RT for 600 s to investigate the difference of corrosion and erosion toughness between the un-nitrided and nitrided AISI420J2. As depicted in Figure 8, the un-nitrided surfaces "A" were selectively etched while the nitrided surfaces "B" were never etched. After dipping test, SEM-EDX was also utilized to analyze the chromium and nitrogen content distribution from "A" to "B" across the boundary "X". Figure 9 reveals that no change was detected in the chromium profile across "X" but that the nitrogen content jumped up across "X" in correspondence to the microstructure in Figure 8b. This high corrosion and erosion toughness proves that both the nitrided PIG-nozzle and die surfaces have sufficient wear toughness against the chemically attacking gaseous and liquid components.

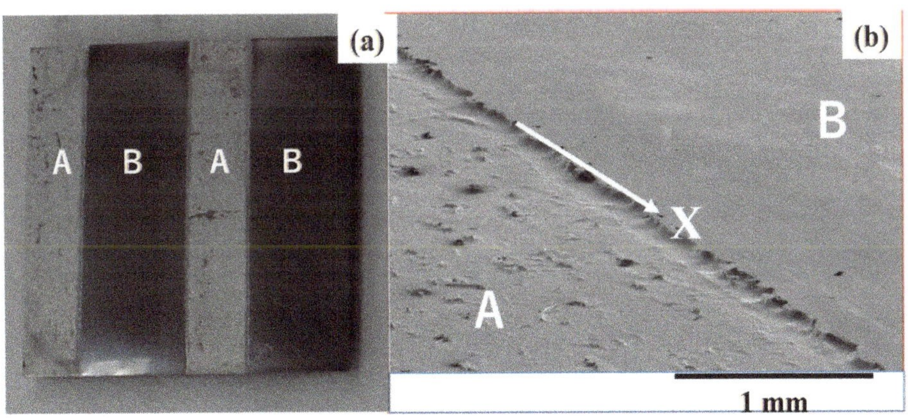

Figure 8. Comparison of the corrosion and erosion toughness between the un-nitrided and nitrided AISI420J2 surfaces. (**a**) Chemically etched specimen with un-nitrided zones (A) and nitrided zones (B), and (**b**) border between un-nitrided zone (A) and nitrided zone (B).

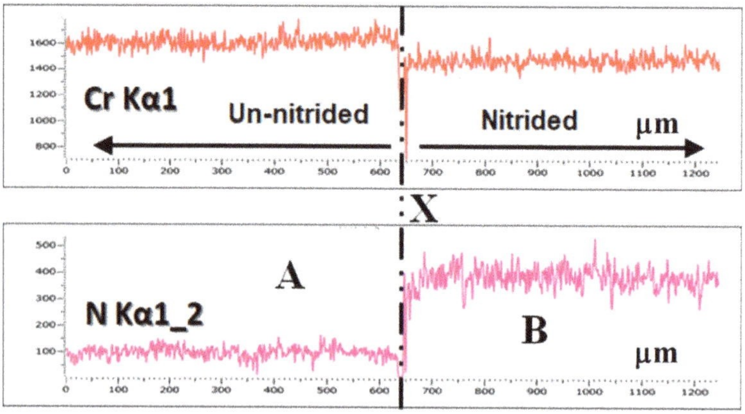

Figure 9. Distribution of chromium and nitrogen contents on the dipped AISI420J2 surfaces from "A" to "B" across "X".

3.2. *Free Injection Behavior through a Single PIG-Nozzle Unit*

In the similar manner to the normal die casting process, the PIG die casting process has a capacity to make fast-rate and stable injection of aluminum melts from the PIG-nozzle gates to the die cavity. The free injection experiment from the PIG-nozzle gate to the flat bed was performed to describe the dynamic flow behavior of pure aluminum melts. The high-speed video camera was utilized to monitor the streaming melt flow in this free injection experiment. A single PIG-nozzle unit was used in this experiment.

Figure 10 depicts a series of video frame-images from the onset of injection through the PIG-nozzle gate. An aluminum melt flow, streams from the outlet gate of PIG-nozzle straightforwardly against the flat bed. At the beginning, the melt started to stream in relatively low velocity due to the mechanical constraint by the PIG-nozzle gate. This flow rate was accelerated with time up to the terminal speed of 1450 mm/s. The average injection velocity was estimated to be 800 mm/s or 0.8 m/s. As discussed in [16], the slowest injection speed ranged from 0.1 to 0.8 m/s, and the highest injection speed reached 40 m/s. The present injection capacity is equivalent to the normal die casting class with the medium injection speed. Due to the gravity, a spherical ball was formed at the front end of aluminum melt stream in this free injection experiment. In practical die-casting, the aluminum jet stream shots onto the die walls and spreads into the die cavity.

Further CAE (Computer Aided Engineering) analysis is needed to optimize the aluminum and aluminum alloy melt flow in the die cavity. A hydrodynamic modeling provides a way to understand the effect of velocity profile of melts on the filling process as suggested in [17].

3.3. *Aluminum Flow into a Die Cavity*

The short-shot experiment [18] was used to investigate the aluminum melt flow in the die cavity. The double PIG-nozzle units were employed to pour the pure aluminum granules into two PIG-nozzles and to inject the aluminum melts from double PIG-gates into a die cavity. The die cavity was shaped to be rectangular. Figure 11a shows the aluminum flow pattern in the die cavity. The injected aluminum melts from the double gates flew from each end to the center of die cavity. As seen in Figure 11a,b, both ends of cavity near the inlets were first filled and both aluminum melts flew from both ends to the center of cavity with nearly the same speed. This flow pattern is similar to the potential flow mode of plastic melts in the plastic mold injection process as precisely stated in [19].

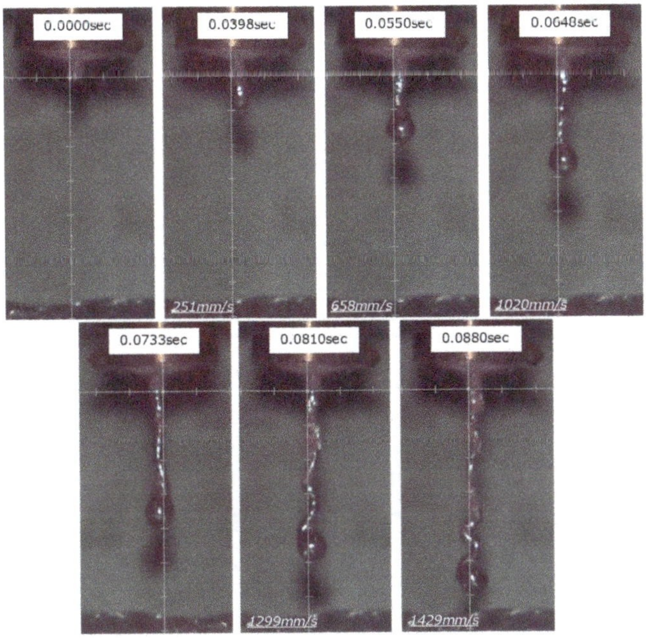

Figure 10. Dynamic behavior in the free injection of the aluminum melt from the PIG-nozzle outlet gate. Every snapshot was retrieved and selected from the vide frame images.

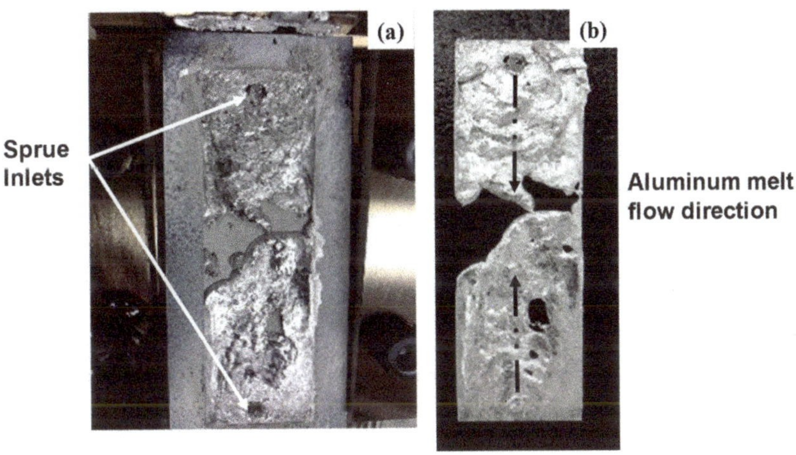

Figure 11. A short-shot experiment by using the double PIG-nozzle units for PIG die-casting. (**a**) Aluminum melt flow into the rectangular die cavity, (**b**) solidified aluminum melts.

As slightly stated in [16,17], the quality of die-cast aluminum alloys parts is dependent on the degassing of residual air from the die cavity and the lubrication on the die surfaces. In the present system, the nitrogen gas was blown before PIG die-casting to minimize the contamination of gaseous components into the cast aluminum. The inner surfaces of die cavity were nitrogen supersaturated at 673 K for 14.4 ks by 70 Pa. Due to its hardness and corrosion toughness stated in 3.1, the dies are free from adhesion of aluminum melts onto the dies even without the use of solid lubricants.

3.4. PIG Die-Casting of AA5052 Small Parts by Using a Single PIG-Nozzle Unit

This PIG die casting unit was first utilized to fabricate the small-sized aluminum parts using a single PIG-nozzle unit. A pure aluminum gear with eight teeth was selected as a targeting product. As reported in [20], most of automatic robotic arms require a light-weight, small-sized gears in the precise transmission system. Among them, the aluminum and aluminum alloy gears are demanded in the market together with the engineering plastic gears.

The die-set for PIG die casting was simply prepared by machining the lower die to have a gear-shaped cavity as well as the upper die-plate to have a sprue as an inlet of injection. Figure 12a depicts the pure aluminum gear which was fabricated by the PIG die casting. No burrs except for the sprue were formed with the gear product.

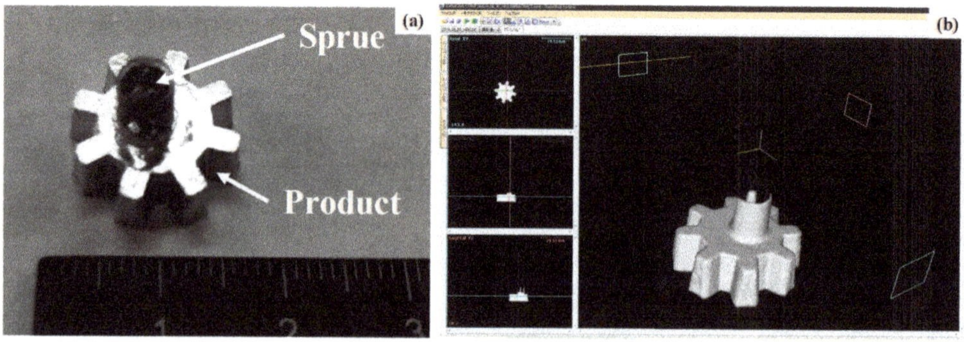

Figure 12. PIG die-cast, small-sized gear block by using a single PIG-nozzle unit. (**a**) The cast and solidified pure aluminum gear block, and (**b**) X-ray tomographic image of gear block.

The X-ray tomography was utilized to make nondestructive inspection on the inner defects and pores in the inside of the PIG-cast aluminum products. Since the X-ray penetration was much sensitive to the change of mass densities, the inner defects and pores were easily detected as a black zone against the white zones for dense materials. In addition, the sample was rotated under the X-ray beam irradiation from the fixed window; the three dimensional mass-density image was directly obtained for diagnosis of defects. Figure 12b depicts the three dimensionally constructed X-ray image in bird-eye view for the cast aluminum gear together with its three cross-sections. Each tooth is accurately shaped in trapezoidal to be working as a skewed gear. Three cross-sectional images prove that no pores and defects are left in the inside of cast gear. The near-net preforms of spur-gear and bevel-gear can be accurately formed by this PIG die-casting before finishing process.

3.5. PIG Die-Casting of Aluminum Mobile Phone Case Using a Double PIG-Nozzle Unit

Fine machining and die-casting have been utilized for mass production of aluminum mobile phone case with the shielding capacity of electromagnetic waves [21]. Due to a market demand for the recyclable low-weight parts, this aluminum mobile phone case was redesigned to reduce the weight as much as possible and to strengthen the stiffness for protection of electric parts. The thin-walled structuring design is often selected in manufacturing. The conventional metal forming processes were difficult to make thin-walled structures; the fine machining process was selected for mass-production of aluminum mobile-phone cases.

The mobile phone case model with the size of 200 mm x 120 mm and the minimum thickness of 0.5 mm, was employed to prepare the upper and lower die sets for PIG die-casting. Double PIG-nozzle units were utilized to inject the AA5052 aluminum alloy melts through two gates into the die cavity. Figure 13a depicts the cast and solidified AA5052 mobile phone case model. No burrs were also yielded except for the sprues and runners; the materials efficiency reaches 98% of poured AA5052 fragments into double PIG-nozzles.

X ray tomography was also utilized to make nondestructive evaluation on the integrity of four sections in the case model. The clay was utilized as a support of thin-walled aluminum case. No pores were also detected in every four section in Figure 13b. As depicted in Figure 13b$_2$, the flat support panel with the thickness of 0.5 mm was uniformly shaped by this forming. Figure 13b$_4$ proves that curved thin-walled panel was also net-shaped.

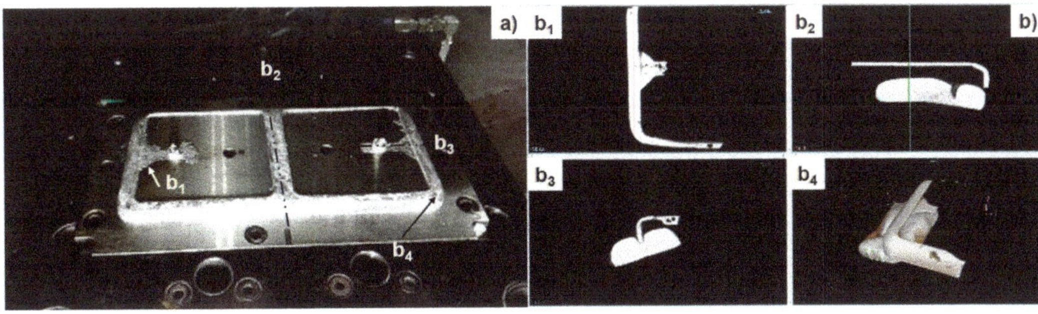

Figure 13. PIG die-cast, thin-walled AA5052 mobile phone case by using the double PIG-nozzle units. (**a**) The cast and solidified AA5052 mobile phone case, and (**b**) X-ray tomographic images of case.

3.6. Warm and Hot Stamping of PIG Die-Cast Preform to Micro-Pillared Heatsink

The PIG die-cast aluminum and aluminum alloy preforms are subjected to forging, blanking and shaping by using the hot stamping unit. This metal forming unit works well to make fine shaping and finishing of preform surfaces. As stated in [22], the temperature control is essential in the hot stamping to improve the dimensional accuracy of products. In the present system, the thermocouple was embedded into the lower cassette die to monitor the thermal transients through every step. In the following experiment, the warm and hot stamping experiments were performed at the specified holding temperature (T_H). A PIG die-cast pure aluminum plate was used as a specimen to make microtexturing onto its surface.

Various developments have been reported in the literature on the pure aluminum heatsink with the tailored microtextures to each heat-transferring application [23]. In this study, a micro-pillared pure aluminum heatsink is fabricated by hot stamping with the use of mesh-textured AISI316 punch. Figure 14a depicts the micro-pillared pure aluminum heatsink, warm stamped at 323 K. The pure aluminum work was backward extruded into the micro-cavity array on the AISI316 punch. As depicted in Figure 14b,c, the micro-pillars with the root size of 200 μm × 200 μm and the height of 76 μm are shaped in regular alignment on the heatsink. The depth of micro-cavity in the AISI316 punch is 190 μm in Figure 6c; the extruded height of work is still less than the micro-cavity depth.

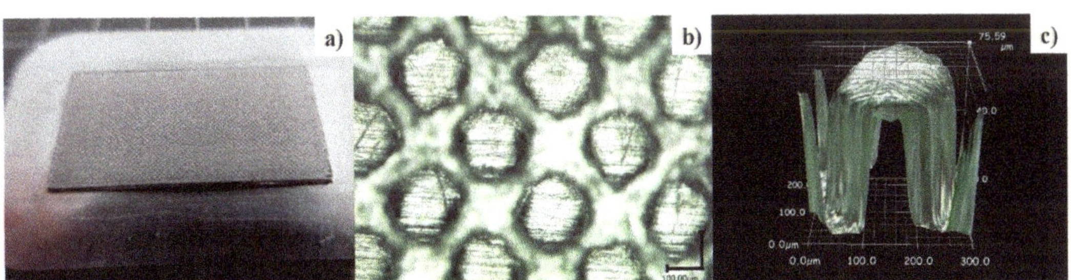

Figure 14. Warm stamping of pure aluminum solidified preform for microtexturing its surface. (**a**) Warm stamped pure aluminum heatsink at 323 K, (**b**) alignment of micro-pillars formed by warm stamping, and (**c**) three dimensional profile of a micro-pillar shaped onto the aluminum preform.

The holding temperature was varied to be T_H = 373 K, 473 K and 573 K, respectively, in the hot stamping with reference to the monitored lower die temperature. As depicted in Figure 15, the extruded micro-pillar heights (H) at 373 K and 473 K, were nearly the as 76 μm at 323K; e.g., H = 86 μm at 373 K, and H = 73 μm at 473 K. On the other hand, the plastic flow in the backward extrusion was enhanced at 573 K. The extruded micro-pillar height reached 110 μm. This reveals that hot stamping under T_H > 573 K enables to make fulfilling of pure aluminum work into the micro-cavity array on the AISI316 punch and to yield the micro-pillared heatsink with relatively high aspect ratio.

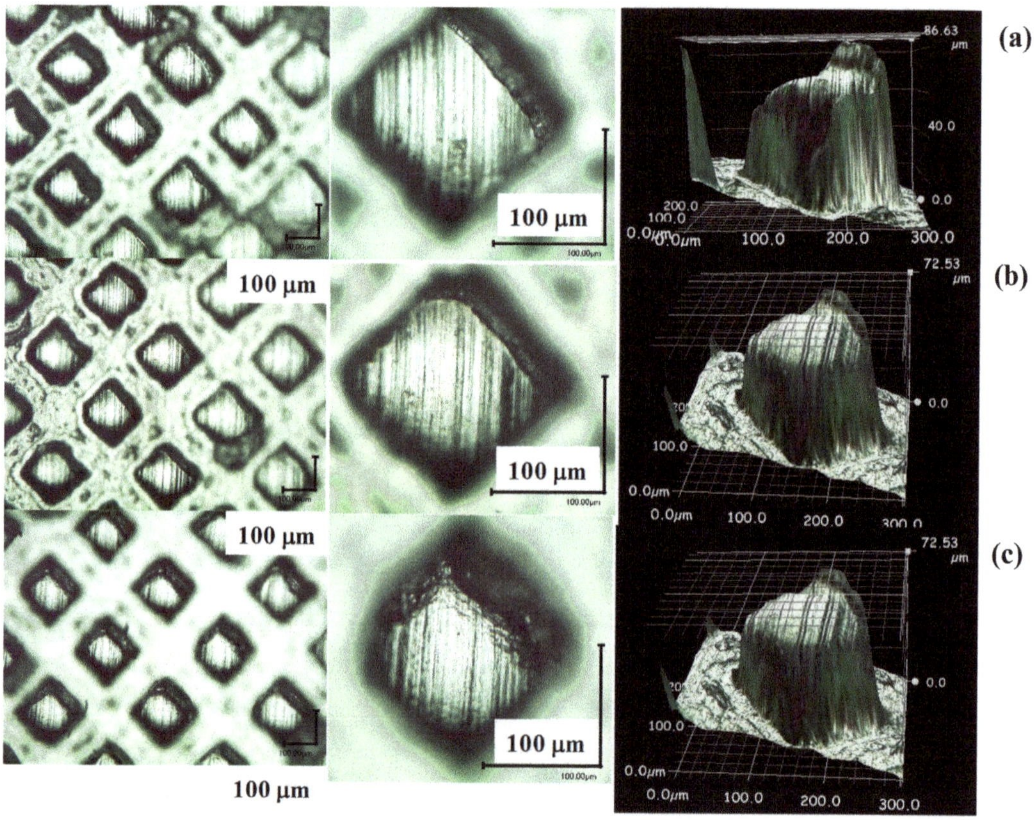

Figure 15. Effect of the holding temperature during the hot stamping on the microtexture formation. (**a**) Micro-pillar array, top surface of extruded micro-pillar, and its three dimensional profile at 373 K, (**b**) at 473 K, and (**c**) at 573 K.

4. Discussion

In the sustainable manufacturing, the weight of parts and tools is reduced by exchanging the steels and other structural materials with the aluminum and aluminium alloys. In parallel with this light-weight structuring, the stiffness of aluminium alloy parts must be improved by the tailored thin-walled structure design. Among several structuring designs, the rib-structuring method was highlighted in recent years [24,25]. The rib works as a reinforcing member to increase the bending and torsional rigidities as additive structures to the aluminum alloy plate [24]. As stated in [25], the rib, which is structured against the bending and torsional directions, works well to improve the rigidities. Let us evaluate on the rigidities of PID-die cast gear teeth and curved case through the X-ray tomography. As seen in Figure 12b, the gear tooth surfaces were formed in rigid against the rotational direction. In Figure 13b, the curved surface of A5052 case was shaped to have uniform thickness of 0.5 mm and its corner was formed to be a smooth panel. These prove that various rib structures are directly shaped into the small-sized and thin-walled products by using the PIG die-casting.

As reported in [26], the effect of holding temperature on the flow stress of aluminum works plays a role to determine the hot stamping schedule. In the hot stamping process for microtexturing onto the cast aluminum plate, the pure aluminum work is difficult to fully fill the micro-cavity array on the AISI316 punch when $T_H < 573$ K. As shown in Figure 11, the excluded micro-pillar has top width of 100 μm and height of 111 μm, and, the distancing space between two adjacent micro pillars becomes 80 μm at $T_H = 573$ K. This reveals that a micro-pillared pure aluminum heatsink is yielded to have high-aspect-ratio micro-fins with the similar profiles of micro-cavities on the textured punch by using the hot stamping at $T_H \sim 600$ to 650 K.

In the normal die-casting operation, the aluminum alloy preform was shaped by the die casting, and the completely solidified cold preform was pressed to punch out the sprues, the runners or the bosses, and to get out the final preform before finishing and polishing. In the present system, the PIG die-casting unit is linked with the hot stamping unit by rotating the lower die-set on the flat bed. The precise shaping and finishing is performed to fabricate the net-shaped products with less electric energy consuming. The high holding temperature of cast preforms is utilized in the hot stamping; its low flow stress and high deformability results in net-shaping of products. In case of shaping the micro-pillared heatsink, the micro-pillar height increases monotonously with holding temperature in Figure 15. The hot stamping pf PIG-die cast preforms is useful to reduce the energy consumption and to make nee-shaping of aluminum heatsink.

The degassing and solid lubrication played a keyword in the conventional die-casting. In the present PIG-die casting, the nitrogen gas blowing was performed before casting to minimize the contamination of residual air in the cast products. As proved in the non-destructive diagnosis by the X-ray tomography, no residuals and defects were detected to demonstrate the effectiveness of the present degassing procedure. An evacuation process before PIG-die casting might be necessary to further improve the quality of miniature aluminum alloy products in mass production.

The nitrogen supersaturation process was mainly utilized for protection of the PIG-nozzle and die surfaces from the adhesion of aluminum remelts in the PIG-die casting and hot stamping. TiAlN coating onto the nitrided PIG-nozzle also wrought well to protect the PIG-nozzle inner surfaces from severe adhesion of aluminum melts even after successive PIG-die casting processes. This high performance first comes from less chemical compatibility between the aluminum and the bound/unbound nitrogen in the die. As listed and suggested in [27], the supersaturated nitrogen solutes in the nitrided layer as well as the bound nitrogen in TiAlN have excellent chemical resistance to aluminum melts. In addition, after [28–30], the present PVD coated TiAlN has high oxidation toughness never to be eroded by pure aluminum and aluminum alloy melts during the PIG die casting. In this manufacturing of aluminum and aluminum alloy parts and members, the nitrogen super-saturation process and TiAlN coating are recommended as an effective surface treatment for galling-free die-casting and hot stamping.

5. Conclusions

A new aluminum manufacturing was proposed to link the PIG die casting unit with warm and hot stamping process for net-shaping of aluminum and aluminum alloy parts and devices. The PIG die-cast preform was stamped at the specified holding temperature to make full use of its low flow stress and to make net-shaping of the product. The small-sized parts and thin-walled members were directly fabricated by the PIG die-casting. The PIG die-cast case thickness reached 0.5 mm together with rib-structuring for reinforcement of aluminum alloy parts and members. Their dimensional accuracy is more improved by warm and hot stamping as a finishing step. In the present study, this new manufacturing process was performed only in two steps. The PIG die-casting process plays the primary step to yield the preform, and, the warm/hot stamping plays the secondary step to fabricate the near net shaped product. Various manufacturing process with several steps can be designed and performed by exchanging the upper die-sets in each step. To be noticed, the nitrogen supersaturation and TiAlN coating protect the PIG-nozzle, casting and stamping dies from severe chemical attacking of aluminum melts without galling in successive usages.

A micro-pillared pure aluminum heat sink was fabricated by this manufacturing system. Higher aspect ratio of micro-pillar height to root width is attained by hot stamping. Due to the tailored alignment in CAD of micro-pillar textures, the distancing space among the aligned micro-pillars works as a coolant channel to increase the heat penetration factor without increase of the pressure drop in cooling. The nitrogen supersaturated punch was useful for anti-galling hot stamping of aluminum preforms.

Toward the sustainable manufacturing in the carbon neutral society, the expecting forming system of aluminum and aluminum alloys has to stand on the upward recycling procedure of used materials. The present manufacturing process has a capacity to utilize the used aluminum alloy fragments as a feedstock for casting into preform and net-shaping into products without significant loss of work materials and electric power.

Author Contributions: Conceptualization, T.A.; methodology, T.A.; software, T.A. and T.K.; validation, T.A., T.K. and H.S.; formal analysis, T.A.; investigation, T.A., T.K. and H.S.; resources, H.S.; data curation, T.A. and T.K.; writing—original draft preparation, T.A.; writing—review and editing, T.K.; supervision, T.A. and H.S.; project administration, H.S. All authors have read and agreed to the published version of the manuscript.

Funding: This research was funded by METI-program on the Supporting Industry Project, Kanto-Division, METI with the contract number of #18062375 (31 March 2021).

Data Availability Statement: Not applicable.

Acknowledgments: Authors would like to express their gratitude to S-I. Kurozumi (Nano Film Coat, llc.) and the late T. Yamaguchi (Sanko Light Kogyo, Co., Ltd.) for their help in experiment.

Conflicts of Interest: The authors declare no conflict of interest.

References

1. Edwards, N. With green aluminum flowing, manufactures struggle to make product ESG top priority. *Forbes* 22 July 2020.
2. Aizawa, T.; Luangvaranunt, T.; Kondoh, K. Solid state recycling from green wastes to aluminum alloys with high material efficiency. *J. JIM* **2001**, *65*, 581–588.
3. Jeong, M.-S.; Lee, S.-Y.; Lee, I.-K.; Lee, S.-K.; Kim, D.; Cho, Y.-J.; Ko, D.-C. Green alternative aluminum extrusion process through process convergence. *Int. J. Prec. Eng. Manuf.* **2014**, *15*, 1173–1177. [CrossRef]
4. Hartlieb, M.; Anderson, K. Die casting. 2A. In *Aluminum Science and Technology*; ASM International: Almere, The Netherlands, 2018.
5. Aizawa, T.; Yamaguchi, T.; Sakayori, H. JP-Patent 2019 with P19HS-001, 31 December 2019.
6. Aizawa, T.; Yamaguchi, T.; Sakayori, H. Die Casting System and Manufacturing Procedure of Die Cast Products. JP-Patent 2021 with P21SL-001, 31 December 2021.
7. Aizawa, T.; Yamaguchi, T.; Sakayori, H. Metal Forming System and Production Procedure. JP-Patent 2021 with P21SL-002, 31 December 2021.
8. Aizawa, T.; Kurihara, T.; Yamaguchi, T.; Sakayori, H. Near-net shaping of aluminum alloy parts by pin-injection-gate die-casting. In Proceedings of the 13th AWMFT, Shanghai, China, 6 December 2021; pp. 253–262.
9. Funazuka, T.; Dohda, K.; Takatsuji, N.; Hu, C.; Ngernbamrung, S. Effect of die coating on surface crack depth of hot extruded 7075 aluminum alloy. *Friction* **2022**, *143*, 21–32. [CrossRef]
10. Aizawa, T. *Low Temperature Plasma Nitriding of Austenitic Stainless Steels. Chapter 3 in Stainless Steels and Alloys*; IntechOpen: London, UK, 2019; pp. 31–50.
11. Aizawa, T.; Dohda, K.; Shiratori, T. Nano-structured tribo-coating for cold and hot stamping. In Proceedings of the 52nd ICFG Plenay Meeting, San Sebastian, Spain, 16 September 2019; pp. 116–123.
12. Shiratori, T.; Aizawa, T.; Saito, Y.; Dohda, K. Fabrication of micro-punch array by plasma printing for micro-embossing into copper substrates. *Materials* **2019**, *12*, 2640. [CrossRef] [PubMed]

13. Shiratori, T.; Aizawa, T.; Saito, Y.; Wasa, K. Plasma printing of an AISI316 micro meshing punch array for micro embossing onto copper plates. *Metals* **2019**, *9*, 396. [CrossRef]
14. Borgioli, F.; Galvanetto, E.; Bacco, T. Low temperature nitriding of AISI300 and 200 series austenitic stainless steels. *Vacuum* **2016**, *12*, 51–60. [CrossRef]
15. Domain, C.; Becquart, C.S.; Foct, J. Ab initio study of foreign interstitial atom (C, N) interactions with intrinsic point defects in α-Fe. *Phys. Rev. B* **2004**, *69*, 144122. [CrossRef]
16. ECO, Aluminum Die Casting. Available online: https://www.ecodiecasting.com/aluminum-die-casting/ (accessed on 3 June 2022).
17. Xie, Z.; Jiao, J.; Yang, K. Theoretical and experimental study on the fluid-structure—Acoustic coupling dynamics of a new water lubricated bearing. *Tribol. Int.* **2023**, *177*, 107982. [CrossRef]
18. Cleary, P.W.; Ha, J.; Prkash, M.; Nguyen, T. Short shots and industrial case studies: Understanding fluid flow and solidification in high pressure die casting. *Appl. Math. Model.* **2010**, *34*, 2018–2033. [CrossRef]
19. Seow, L.W.; Lm, Y.C. Optimizing flow in plastic injection molding. *J. Mater. Process. Technol.* **1997**, *72*, 333–341. [CrossRef]
20. JST, Aluminum Alloy Gear. Available online: http://www.precise-gear.com/aluminum-alloy-gear.html (accessed on 3 June 2022).
21. Gangopadhyay, S.; Acharya, R.; Chattopadhyay, A.K.; Sargade, V.G. Effect of cutting speed and surface chemistry of cutting tools on the formation of bul or bue and surface quality of the generated surface in dry turning of AA6005 aluminum alloy. *Mach. Sci. Technol.* **2010**, *14*, 208–223. [CrossRef]
22. Maeno, T.; Mori, K.-I.; Yachi, R. Hot stamping of high-strength aluminum alloy aircraft parts using quick heating. *CIRP Ann.* **2017**, *66*, 269–272. [CrossRef]
23. Kumaraguruparan, G.; Sornakumar, T. Development of testing of aluminum micro channel heat sink. *J. Therm. Sci.* **2010**, *19*, 245–252. [CrossRef]
24. Gong, M.F.; Li, Z.; Dong, J. Mechanics analysis of reinforcing rib structure in aluminum heating-plate automatic casting system. *Adv. Mater. Res.* **2012**, *538–541*, 1755–1761. [CrossRef]
25. Ohashi, T.; Tabetabei, H.M.; Nishimura, T. Rib-structure on A5083 aluminum alloy sheet generated friction stir forming. *Preced. Eng.* **2017**, *207*, 1153–1158. [CrossRef]
26. Ma, Z.; Ji, H.; Huang, X.; Xiao, W.; Tang, X. Research on high temperature stamping forming performance and process parameters optimization of 7075 aluminum alloy. *Materials* **2021**, *14*, 5435. [CrossRef]
27. Chemical Resistance Reference Chart. Available online: https://www.calpaclab.com/content/chemical-charts/Aluminum.pdf (accessed on 6 November 2022).
28. Kumar, S.; Maity, S.R.; Patnaik, L. Friction and tribological behavior of bare nitrided, TiAlN and AlCrN coated MDC-K hot work tool steel. *Ceram. Int.* **2020**, *46*, 17280–17294. [CrossRef]
29. Tollmann, W.; Grisales, D.; Stangier, D.; Butzke, T. Tribomechanical behavior of TiAlN and CrAlN coatings deposited onto AISI H11 with different pre-treatments. *Coatings* **2019**, *9*, 519. [CrossRef]
30. Dohda, K.; Aizawa, T.; Funazuka, T. Process tribology of PVD nitride coated dies in hot metal forming. *Friction* 2022; *in press.*

lubricants

MDPI

Article

Tribological Performance of Esters, Friction Modifier and Antiwear Additives for Electric Vehicle Applications

Gerard Cañellas [1,*], Ariadna Emeric [1], Mar Combarros [1], Angel Navarro [1], Lluis Beltran [1], Montserrat Vilaseca [2,*] and Jordi Vives [3]

1 Research & Development Department, Industrial Química Lasem, Avinguda De la Indústria 7, Polígon Industrial Pla del Camí, 08297 Castellgalí, Spain
2 Eurecat, Centre Tecnològic de Catalunya, Unit of Metallic and Ceramic Materials, Plaça de la Ciència 2, 08243 Manresa, Spain
3 Department of Mining, Industrial and ICT Engineering, Escola Politècnica Superior d'Enginyeria de Manresa, Universitat Politècnica de Catalunya, Avinguda Bases de Manresa 61-73, 08242 Manresa, Spain
* Correspondence: g.canellas@iql-nog.com (G.C.); montserrat.vilaseca@eurecat.org (M.V.)

Abstract: The replacement of conventional lubricants with esters is an alternative to provide a low environmental impact and at the same time excellent lubricity features, the high solubility of additives, good viscosity index, low volatility, and high thermal stability. Friction modifiers and antiwear/extreme pressure additives are extensively used to save energy and increase operational life in machine components. In this study, the lubricity of a Group IV base oil containing ester and various benchmark friction modifiers and/or antiwear/extreme pressure additives is measured to evaluate the influence of the ester on the tribological performance of the mixture components. The tribological performance is discussed based on the tabulation of the traction coefficient using a Mini-Traction-Machine and on the measurement of the specific wear rate from the wear scar of the experimental studies using an optical profilometer. In general, results show synergies between the ester and the additive formulations, reducing the wear rate to 75% and decreasing the traction coefficient a 20 to 50%, depending on the evaluated additive.

Keywords: friction modifier; antiwear; additives; wear rate; traction coefficient; boundary lubrication; lightweight

1. Introduction

Along with the advancement of the modern automotive industry, and the exponential transformation of the internal combustion engine vehicles (ICEVs) fleet into hybrid and electric vehicles, fluids have to offer higher levels of performance and lower the life cycle cost (LCC) [1]. Current automotive lubricants are characterised to be low-viscous [2] and they are usually composed of a base oil containing a large amount of additives. One of the main objectives of using lubricants in automotive is to improve fuel efficiency by reducing frictional losses and increase the operational life by minimising wear, in increasingly harsh operating conditions, improving the life cycle assessment (LCA) of some mechanical parts [3,4]. Some studies show that for a model of a fully electric car from 2017, the frictional losses generated in the engine and transmission represent an intake of 2% of the total energy delivered from the batteries, including the recovered energy [3]. Taking action to decrease friction could be reflected in savings of up to 9% of global energy use [5]. Regarding the base oil, the conventional choices are synthetic-based, mainly polyalphaolephines (PAOs), which belong to Group IV base stocks according to the American Petroleum Institute (API) classification, and mineral oil-based lubricants due to their good performance on final application [6]. One of the biggest challenges is the development of a new generation of eco-friendly base stocks that could replace the current ones which suppose threats to the environment with the increased environmental awareness as a primary driving force.

Synthetic esters (SEs) are gaining prominence as environmentally acceptable lubricants due to their no toxicity, excellent biodegradation and the possibility to achieve tailor-made properties at the same or higher level of performance as conventional lubricants [7].

The driving force in this study is the increasing demand of SEs, which belong to Group V base stocks, to form semi-synthetic lubricants that can meet and fulfil the upcoming requests from the higher performance electric vehicles operation specifications, such as the higher torque in low-temperature operative ranges compared with conventional vehicles [8]. Specifically, on bearing systems where the temperature range is comprised between 30 and 160 °C on some types of motor rotors [9]. It is well known that substantial improvements and optimisation of the engine oils and transmission fluids can be achieved when these compounds are incorporated into PAO formulations. Some biodegradable engine lubricants composed of 1:1 mixtures of esters with PAOs and conventional additives have been tested demonstrating better tribological performance and environmental compatibility [10]. Compared with low molecular weight PAOs, esters offer less volatility and higher solubility due to higher polarity. Additionally, PAOs tend to shrink rubber seals. On the other side, esters are less hydrolytically stable and may show lower oxidative stability. In comparison with mineral oil-based lubricants, SEs offer higher polarity and seal compatibility [11]. In addition, they offer better lubricity and antiwear properties, high thermal stability and conductivity, higher specific heat, environmental compatibility, and high-temperature operation range, among others. In general, SEs have analogous performance as mineral oils [12]. These properties offer promising applications in the automotive sector. For instance, by increasing the electric vehicles (EV) autonomy, and reducing the weight and waste by unifying the vehicle fluids in innovative drivetrain configurations [13]. On the contrary, the cost of esters limits its use, but considering that SEs may require less maintenance, the total energy and machine downtime savings, the cost difference can be misleading. An optimum point between environmental compatibility, performance and cost often has to be searched [14].

This paper is focused on the behaviour of an ester in presence of friction modifier additives, which can drastically reduce friction and keep it stable [15], and on antiwear additives due to the use of slim lubricants, in order to assure assembly integrity under the applied conditions [16]. Since lubricant additives are designed to optimise work with conventional base oils and improve tribological performance, a great number of experimental studies are available to investigate the compatibility between additives and base oils. However, to the best of our knowledge, studies aimed at understanding the effect of the esters in friction modifiers and antiwear additives have not been widely reported and the research completed cannot be generalised owing to the broad range of additives. For instance, it was found that the lubrication performance of various fatty acids and amines formulated in a polyol ester oil, is highly influenced by structural differences, such as chain length or unsaturations [17] but the influence of the ester performance on the mixture has not been evaluated. Furthermore, the tribological compatibility of a trimethylolpropane (TMP) ester as an additive in a PAO formulation containing glycerol mono-oleate (GMO), molybdenum dithiocarbamate (MoDTC) or zinc dialkyldithiophosphate (ZDDP) was also studied but the main scope was on the influence on DLC coated steel where the ester is more effective by itself [18]. In addition, it was found that under reciprocating conditions, controlled amounts of the ester can improve the lubricant performance, but they can also decrease the activity of the MoDTC additive [19]. The friction performance of molybdenum dithiophosphate (MoDTP) was also studied in an ester-containing lubricant by means of a reciprocating contact. Additionally, high concentrations of ester reduce the additive activity [20].

The aim of this paper is to evaluate the effect of a Group V ester on the tribological performance of some conventional benchmark antiwear and/or friction modifier additives, in front of steel, when it is incorporated into a Group IV base stock PAO 6 in comparison with a pure PAO 6 formulation in sliding/rolling conditions. This study is performed by means of a Mini-Traction-Machine, reproducing the possible mechanical contact mechanisms found in an EV with the test conditions, and profilometry analysis. Some studies, using this technique on steel surfaces, have been previously reported. For instance, in this study [21] where the most effective conditions of fourteen friction modifiers of various families were identified. In addition, this technique was used in a study where a synergistic tribochemical impact of some aminic friction modifiers in combination with ZDDP in PAO 4 on tribological performance was determined through severe sliding/rolling conditions [22].

2. Materials and Methods

2.1. Test Base Fluids and Additives

The base stocks used throughout the study consist of a Group IV polyalphaolefine 6 (PAO 6), and a blend (BO) Group V ester (from Industrial Química Lasem, Castellgalí, Spain) with PAO 6 at the same weight proportion. The kinematic viscosity properties of the base stocks are summarised in Table 1. The base oil was specifically designed to solubilise a 1% (%w/w) of all the additives under study. In order to just study the ester effect and not interfere with the evaluation, the final kinematic viscosity of both, PAO 6 and BO formulations, are not adjusted. The final formulations had the same kinematic viscosity as the raw base stocks.

Table 1. Oil used.

Code	Description	Viscosity at 40 °C	Viscosity at 100 °C	Viscosity Index
PAO 6	PAO 6	30.3 cSt	5.9 cSt	143
BO	Base Oil: Group V ester + Group IV PAO 6 (1:1)	22.9 cSt	4.8 cSt	136

Five commercially available AW and/or FM benchmarking additives were evaluated, see detailed description in Table 2. The molecular structure of each additive used is depicted in Figure 1. All these additives are intended for the automotive sector among other applications.

Table 2. Additives tested.

Code	Description	Function
ZDDP	Zinc Bis(2-Ethylhexyl)Phosphorodithiodate	AW
MoDTP	2-Ethylhexyl-MoDTP	AW + FM
OFM	Bis-(2-hydroxypropyl) Tallowamide	FM + low AW
PFM1	Polyisobutylsuccinate-polyetheramine polymer	FM
PFM2	Polymeric product from reaction of C8-C18 unsat FA, DEA and Propyleneoxide	FM

2.2. Test Materials

AISI 52100 was used as reference material for steel (MTM specimens are provided by PCS Instruments Ltd, London, UK; SRV specimens are provided by Optimol Instruments Prüftechnik GmbH, Munich, Germany). This steel grade is used in several mechanical components of engines and transmissions. AISI 52100 steel balls and discs were used to simulate steel–steel contacts. The chemical composition and main properties are summarised in Tables 3 and 4 correspondingly. The surface roughness and hardness of SRV specimens are provided by the supplier. Conversely, the surface roughness value of MTM specimens is experimentally measured with the optical 3D measurement instrument AliconaInfiniteFocusSL (Bruker Corp., Billerica, MA, USA). Additionally, materials hardness is experimentally measured, in triplicate, with a microhardness tester FM-700 (Future-Tech Corp., Kawasaki, Japan).

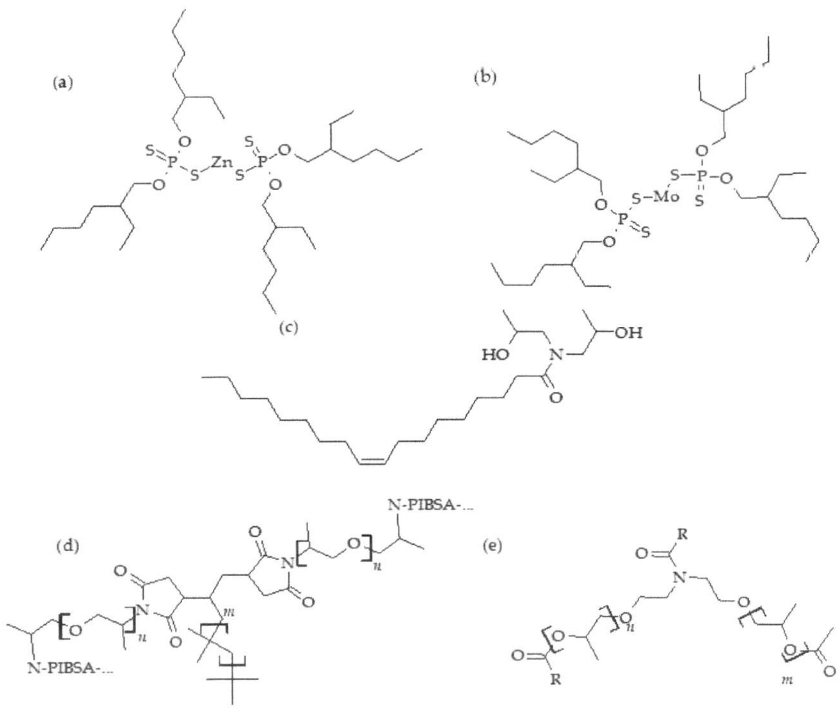

Figure 1. Additives' molecular structures. (**a**) ZDDP; (**b**) MoDTP; (**c**) OFM; (**d**) PFM1; (**e**) PFM2.

Table 3. AISI 52100 steel chemical composition. Reprinted/adapted with permission from Ref. [23]. 1996, ASM International ®.

Element	Composition, % (*w/w*)
Carbon, C	0.98–1.1
Chromium, Cr	1.3–1.6
Iron, Fe	96.5–97.32
Manganese, Mn	0.25–0.45
Phosphorous, P	≤0.025
Silicon, Si	0.15–0.30
Sulphur, S	≤0.025

Table 4. AISI 52100 steel specimens properties.

Property	MTM Specimen		SRV Specimen	
	Disc	Ball	Disc	Ball
Hardness, Vickers 1 kg	755 ± 10	876 ± 8	790 ± 50	805 ± 25
Roughness, Ra (mm)	<0.020		0.046 ± 0.003	0.020 ± 0.001
Young's Modulus (GPa) [23]	210		210	
Poisson's Ratio [23]	0.30		0.30	

2.3. *Test Method and Conditions*

2.3.1. Tribological Behaviour

The characterisation of the tribological performance of the additives was performed by monitoring the traction coefficient through a set sliding/rolling contact generated between a steel disc and a steel ball using a Mini-Traction-Machine (MTM2; PCS Instruments,

London, UK), represented in Figure 2. This apparatus is controlled with software which allows setting the load, the fluid temperature and the ball and disc rotational speeds.

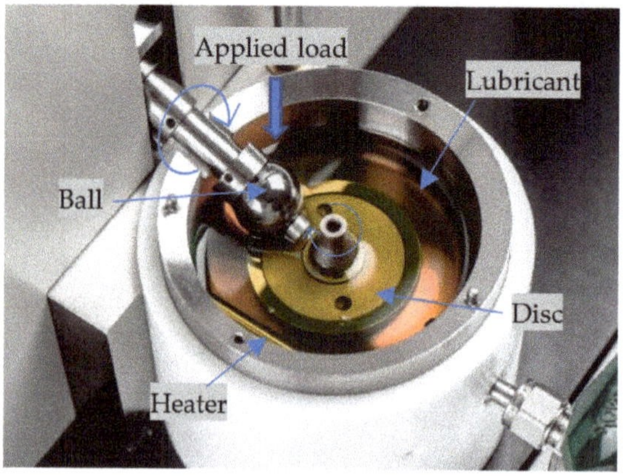

Figure 2. MTM2 assembly representation.

Test conditions are shown in Table 5. The temperature was set at 80 °C, a value comprised on the bearing's operative temperature and a reference value on lubricity and bearing studies [24–26]. Fresh ball and disc specimens were used for each test. Before each test, they are immersed in heptane in an ultrasonic bath for six minutes. The test was run for 240 min. Each additive was tested in triplicate in order to ensure good repeatability.

Table 5. MTM test parameters [27].

Contact Conditions	
Load	31 N
Sliding rotation ratio, SRR (%)	50%
Mean speed	50 mm·s^{-1}
Temperature	80 °C

All tribotests were carried out at an applied load to ensure the same theoretical contact diameter of 250 μm. The initial contact pressure for the 19 mm diameter balls employed, based on Hertzian law calculous, is 0.95 GPa.

Traction coefficient is obtained from the relation of measured friction force and applied load.

2.3.2. Evaluation of ZDDP Activation Conditions

The activation study of the extreme pressure and antiwear additive was performed following the DIN Standard 51834-2. It was performed by subjecting it under reciprocating sliding conditions with a punctual contact between a steel disc and a steel ball using an SRV® 5 tribometer (Optimol Instruments Prütechnik GmbH, Munich, Germany), represented in Figure 3.

Test conditions are in Table 6. Fresh 10 mm diameter ball and disc specimens were used for each test. Before each test, they are immersed in heptane in an ultrasonic bath for six minutes. The test was run for 120 min. Each load was tested in triplicate in order to ensure good repeatability.

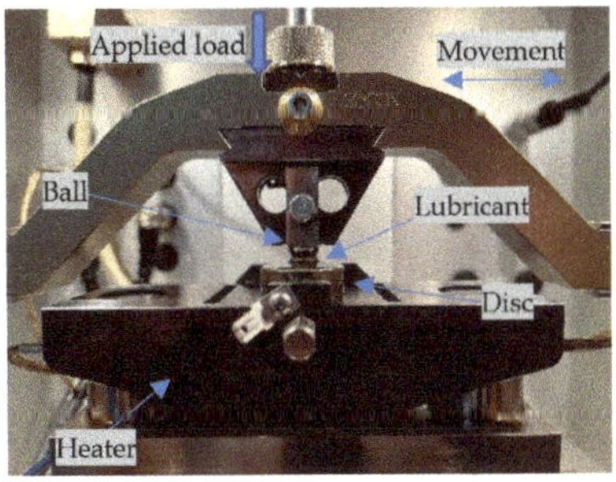

Figure 3. SRV assembly representation.

Table 6. SRV test parameters.

Contact Conditions	
Load	55 and 275 N
Pressure *	1.70 and 2.94 GPa
Stroke	2 mm
Frequency	50 Hz
Temperature	80 °C

* Based on Hertzian law.

2.3.3. Wear Behaviour

The characterisation of wear behaviour of the final formulations was performed. After the tests, specimens are cleaned by immersing them in an ultrasonic bath with heptane and analysed with a S neox 090 3D optical profilometer instrument (Sensofar Metrology, Terrassa, Spain). The scar profiles from disc specimens were obtained from 2 × 2 mm topographic images of the worn surfaces with a 100× magnification. Topographic images were processed, via SensoView v1.9 software (Sensofar Metrology, Terrassa, Spain), to obtain a mean profile of the surface where the cross-section area of the modified surface can be measured. Four topographic images are recorded for each disc, as represented in Figure 4.

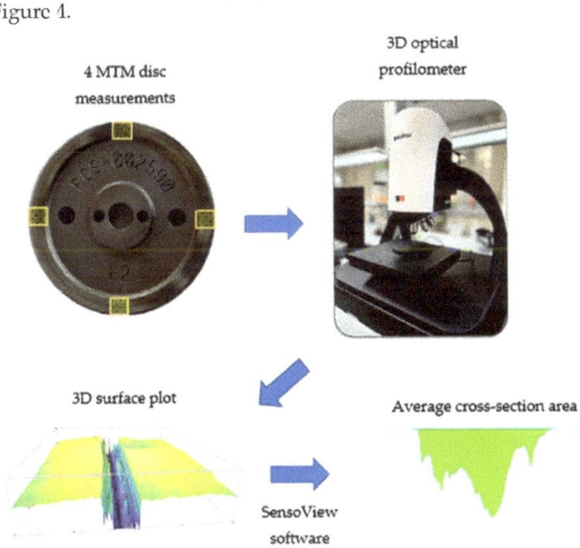

Figure 4. Wear measurement sequence.

The mean cross-section area is multiplied by the corresponding track perimeter to obtain the scar volume. Finally, by dividing the worn volume by the total sliding distance and the applied load, the specific wear rate is obtained.

3. Results

The friction behaviour of the different additives is summarised in Figure 5, where a representative traction curve for each formulation is represented, due to no significant differences between replicates.

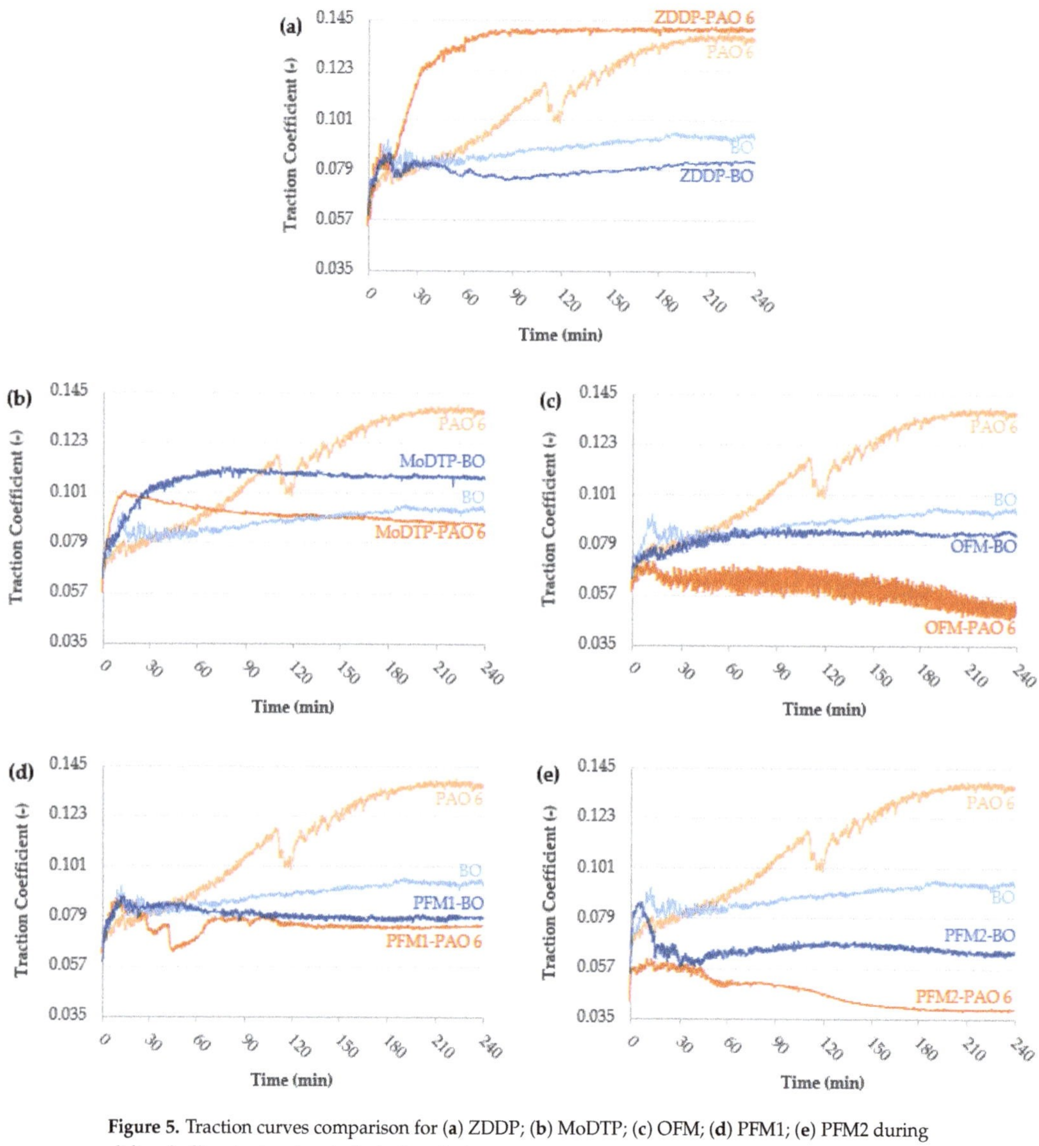

Figure 5. Traction curves comparison for (**a**) ZDDP; (**b**) MoDTP; (**c**) OFM; (**d**) PFM1; (**e**) PFM2 during sliding/rolling test under steel–steel contact.

After the sliding/rolling test, the surface profilometries of the cleaned discs and balls, used with the final formulations, were completed. The overall average wear rates are represented in Figure 6.

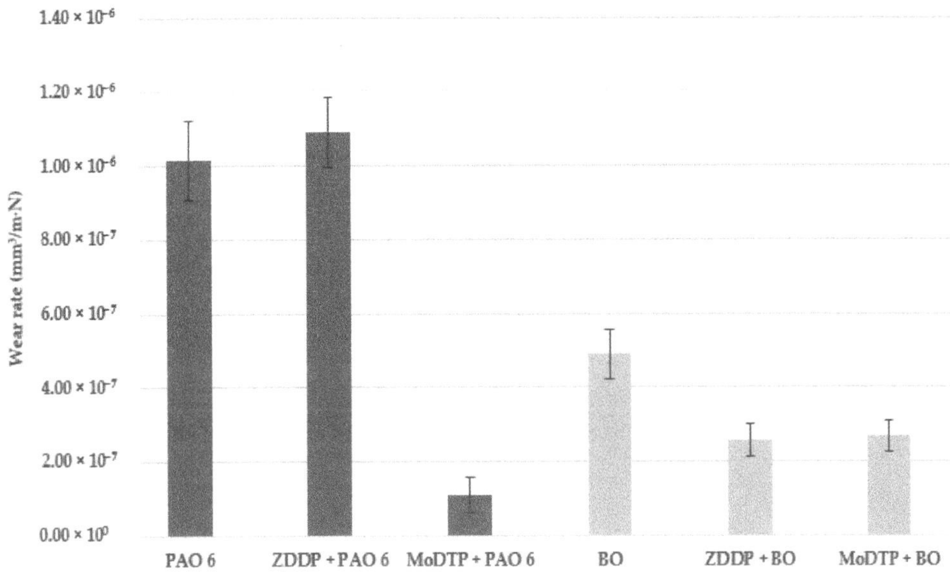

Figure 6. Steel discs specific wear rates. No measurable wear scars are obtained with OFM, PFM1 and PFM2 additives.

The traction curve of PAO 6 shows a constant increase in the traction coefficient until 0.135 with a sharp oscillation around 120 min. When the ester is added, the traction curve lows the slope, stabilising the traction coefficient around 0.093, since the minute 60 (Figure 5). The evaluation of the disc surface used in PAO 6 test, shows a wear rate value of 1.01×10^{-6} $mm^3 \cdot m^{-1} \cdot N^{-1}$. When ester is added, the wear rate decreases to 4.89×10^{-7} $mm^3 \cdot m^{-1} \cdot N^{-1}$ (Figure 6).

When ZDDP additive is added to PAO 6, a noticeable increase in the traction coefficient is observed during the first 60 min to finally stabilise around 0.141, the highest value of this study. When an ester is added to the formulation, the traction coefficient is reduced, and it is stabilised around 0.083 (Figure 5a). The behaviour of the traction curve is similar to the observed in the BO traction curve, but the result is even lower. The evaluation of the disc surface after the test under ZDDP in PAO 6 gives a wear rate value of 1.09×10^{-6} $mm^3 \cdot m^{-1} \cdot N^{-1}$, showing no significant change, compared with the PAO 6 wear rate. Conversely, when the ester is added, the obtained wear rate value is 2.55×10^{-7} $mm^3 \cdot m^{-1} \cdot N^{-1}$ (Figure 6) improving the results obtained in all the previous formulations, PAO 6; BO; and ZDDP + PAO 6. In order to check the activity of the ZDDP additive, the wear behaviour of the same ZDDP in BO formulation was tested using the DIN 51834-2 as standard at 55 and 275 N by triplicate, observing the antiwear performance of this additive improves when conditions are harsher, see Figure 7.

In the case of the MoDTP additive in PAO 6, the traction curve shows a sharp increase during the first 20 min and then it decreases slowly until a final value of 0.087. A clear traction reduction and a change in the behaviour of the traction curve compared with PAO 6 is observed. When ester is added, the traction coefficient increases to 0.110 (Figure 5b), a higher value compared with the previous formulation and the BO. The behaviour of the traction curve is like the observed in the ZDDP in PAO 6 mixture. The wear rate values of the MoDTP in PAO 6 and MoDTP in BO are 1.09×10^{-7} and 2.68×10^{-7} $mm^3 \cdot m^{-1} \cdot N^{-1}$

correspondingly (Figure 6). A great protection against wear is observed and it is similar in both cases, improving the performance of both base stocks.

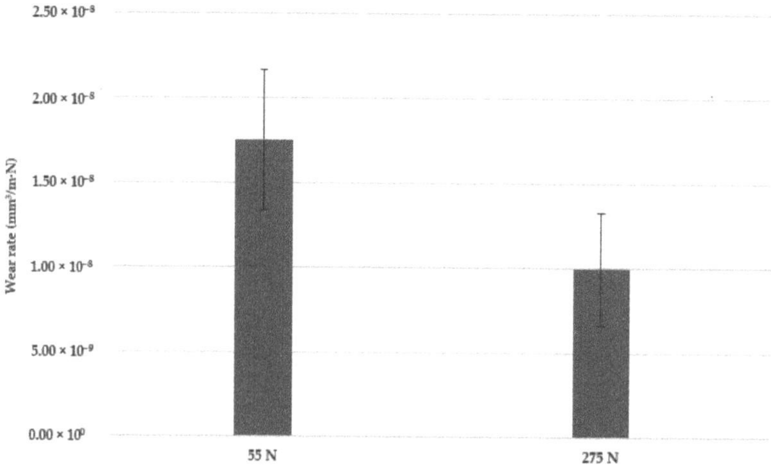

Figure 7. Antiwear performance of ZDDP additive in BO.

When OFM additive is added to PAO 6, a clear decrease in the traction coefficient is observed until a final value of 0.048. During the test, the traction coefficient decreases following a slight slope and shows a broad oscillation, an undesired result. When an ester is added, the traction coefficient increases to 0.083 (Figure 5c). Even with this increase, the OFM in the BO mixture gives better results compared with both reference base stocks and also, the traction curve is more stable than the one obtained with the OFM additive in PAO 6. The traction curve is similar to the result with ZDDP in PAO 6 but with a lower increase in the traction at the beginning of the curve. Negligible wear rates are obtained.

The traction coefficient behaviour and the final traction value of the PFM1 additive mixtures with PAO 6 and BO are similar in both cases. The final traction coefficient values are 0.075 for the additive in PAO 6 and 0.079 in BO (Figure 5d). Only a slight difference is observed around the 45 min of the PFM1 in the PAO 6 test, where a sudden destabilisation of the curve takes place. Negligible wear rates are obtained.

The last study case is the PFM2 additive. It presents a solubility problem under PAO 6 which is solved when the ester is added. Regarding the traction curve in PAO 6, the lowest traction coefficient of this study is obtained with a value of 0.039. The traction curve has a similar behaviour to the MoDTP in PAO 6, but with a more pronounced slope. When an ester is added the traction coefficient increases. A sudden decrease in the traction curve takes place during the first 30 min to finally get stabilised to 0.063 (Figure 5e). It is the lowest value among the different additives in BO mixtures. Negligible wear rates are obtained.

4. Discussion

When an ester is added to the pure base stock, a decrease of 31% in the traction coefficient is observed. In addition, the wear behaviour is drastically improved, reducing the wear rate ratio a 52%. The ester interacts with the surface and creates a tribofilm which is able to reduce the traction coefficient and also protects the surface against wear.

Regarding the traction coefficient and wear rate of the ZDDP additive, when it is blended with PAO 6, the highest values of this study are obtained. ZDDP (Figure 1a) is described as one of the most effective and broadly used antiwear additives under steel–steel contact. It also offers other advantages such as antioxidants, corrosion inhibitors and extreme pressure properties [28]. It interacts with the surface through a chemical process through the formation of a sacrificial bilayer [29]. This bilayer has a solid-like morphology which is able to inhibit the entrainment of liquid lubricant between the rubbing surfaces [30]. This effect keeps the boundary friction value of the tests up to 4% above the reference PAO 6 friction value. Conversely, when the ester is blended with the previous formulation, the obtained traction coefficient is lower, decreasing the formulation traction coefficient by 41%. This result clearly shows that the ester has a detergency effect on the additive tribofilm. Regarding the wear behaviour, the addition of ZDDP additive in PAO 6 has a worsening effect, increasing the wear rate value to around an 8% (Figure 6). Even though, when the ester is added, the wear rate is reduced a 77% and 48% compared to ZDDP + PAO 6 and pure BO samples, correspondingly (Figure 6). The antiwear behaviour of the additive ZDDP in PAO 6 has not been seen under MTM test conditions which are not harsher enough (Figure 6), whereas under SRV harsher conditions, ZDDP shows a decrease in the wear rate values (Figure 7). The ester has synergistic effects with this additive and improves considerably the capacity of protection against wear of the formulation under both MTM and SRV test conditions.

MoDTP has been investigated as well (Figure 1b). It is classified as a highly effective friction modifier from the family of organomolybdenum compounds. MoDTP is also able to reduce wear and act as an antioxidant additive by itself and also presents synergism with ZDDP [31]. Friction and wear reduction on steel surfaces are achieved through an oxidation and reduction process, resulting in an amorphous phosphate glass tribofilm, containing MoS_2 composites among others [32] which usually requires harsh conditions of pressure and temperature to be generated [33].

Even observing a decrease of 36% of the traction coefficient with respect to the PAO 6 reference, probably the additive is not working under its optimum conditions. When an ester is added, a worsening effect takes place, and the traction coefficient increases by 26%. The friction performance is even worse than the BO reference. The friction modifier properties of this additive are inhibited in presence of an ester. The antiwear behaviour of formulations with and without ester is maintained. Yet the addition of an ester has a worsening effect and an increase of 45% in the wear rate value is observed. The ester and the additive have competitive surface interaction mechanisms and these results could also be explained due to ester–additive interaction.

The OFM additive is classified as a friction modifier, belonging to the nitrogen-derivated organic friction modifiers subclass. To understand its action mechanism, it is necessary to deeply understand the molecular structure (Figure 1c). A polar head which includes the amide and hydroxyl functional groups can be differentiated from a lipophilic hydrocarbon chain, originally from the tallow oil fatty acids. The polar group interacts with polar metal surfaces while non-polar chains are extended to the base oil. Molecules interact with each other creating a protective layer on the metal surface. Furthermore, a multilayer system is formed via a conglomeration process [34]. These layers exhibit low shear-off strength between them, boosting friction reduction.

As can be expected from an organic friction modifier, after surface conditioning, the OFM additive keeps the traction coefficient constant, and the traction coefficient is decreased a 64% compared with PAO 6. When an ester is added, the traction coefficient shows a substantial increase of 73% but it continues reducing the traction coefficient of both base stocks. A kind of surface adsorption competition exists between both, but at this concentration are compatible.

PFM1 is classified as a friction modifier from the block copolymer family. Two main types of blocks can be identified in its structure: on one side, polar blocks, constituted by polyetheramine monomers and on the other side, non-polar blocks, which in this additive are the polyisobutyl chains from polyisobutylsuccinate. This configuration is responsible for the friction reduction mechanism [35]. Polar blocks remain anchored with the polar surface, while lipophylic blocks are extended to the oil and give lubricity to the system (Figure 1d) [36,37].

When the PFM1 additive is added to the PAO 6 base stock, the traction coefficient is decreased by 44% compared to the reference oil. The addition of the ester has a neutral effect on the additive performance. The lubricity effects provided by the structure of the polymer prevail rather than the possible surface competition effects between ester and additive.

Finally, PFM2 is classified as a friction modifier and structurally can be identified as a complex amide/ester compound, a polymer of low molecular weight. The interaction with the surface is similar to the PFM1 additive behaviour. The polar part includes the polypropyleneglycol, the amide and the ester functional groups. The non-polar blocks are formed by the hydrocarbon chains from fatty acids (Figure 1e).

The evaluation of the PFM2 additive in PAO 6 shows better friction modifier properties, decreasing the traction coefficient to 71% compared to pure PAO 6. The addition of the ester has a clear worsening effect on the traction coefficient which increases by 62%. Still, the traction coefficient decreased by around 32% compared with the BO reference. In this case, the surface competition has a greater impact than the lubricity provided by the polymer structure. Probably, the polymer is not large enough to avoid the contact between surfaces, oppositely to the PFM1 additive.

5. Conclusions

The combination of the PAO, as base stock, and the additives under study show great efficiency. The obtained traction curves of the PAO–ester formulations, show better performance than the base oil traction curve, so ester addition has a great effect on the evaluated parameters. The friction modifying properties of the additives have also improved the friction behaviour of the used base oil, showing synergism or neutral effect of the ester. One exception is observed, MoDTP additive formulation shows a larger traction coefficient, but it keeps its antiwear properties. The surface competition between ester and those additives which physically interacts with the surface has been proven. Further understanding of ester–additive interactions is required. In addition, it has been demonstrated that under the test conditions, without high temperature and high pressure, the ZDDP additive is unactive. Even though, when the base oil contains a part of ester, ZDDP additive seems to be more effective, while ester is able to decrease the traction coefficient. Synergism between both ester and additive takes place. Either way, the addition of an ester into the studied formulations has had, in general, a positive effect on the tribological behaviour of the final formulations, decreasing the PAO 6 traction coefficient a 20 to 50% depending on the additive. The partial substitution of mineral oils and synthetic base oils for SEs is possible. Further investigations are being performed to study the influence of the additives in tribology for different metal contacts, especially those where aluminium is present.

Author Contributions: Conceptualisation, G.C., A.E., M.C., A.N., L.B., M.V. and J.V.; methodology, G.C.; investigation, G.C., A.E., M.C., A.N., L.B., M.V. and J.V.; resources, A.N. and L.B.; writing—original draft preparation, G.C.; writing—review and editing, G.C. and M.V.; funding acquisition, L.B. and M.V. All authors have read and agreed to the published version of the manuscript.

Funding: This work received the support of the Pla de Doctorats Industrials from Secretaria d'Universitats i Recerca del Departament d'Empresa i Coneixement of the Generalitat de Catalunya.

Conflicts of Interest: The authors declare no conflict of interest.

References

1. Chen, Y.; Jha, S.; Raut, A.; Zhang, W.; Liang, H. Performance Characteristics of Lubricants in Electric and Hybrid Vehicles: A Review of Current and Future Needs. *Front. Mech. Eng.* **2020**, *6*, 571464. [CrossRef]
2. Tormos, B.; Ramírez, L.; Johansson, J.; Björling, M.; Larsson, R. Fuel consumption and friction benefits of low viscosity engine oils for heavy duty applications. *Tribol. Int.* **2017**, *110*, 23–34. [CrossRef]
3. Holmberg, K.; Erdemir, A. The impact of tribology on energy use and CO_2 emission globally and in combustion engine and electric cars. *Tribol. Int.* **2019**, *135*, 389–396. [CrossRef]
4. Czerwinski, F. Current Trends in Automotive Lightweighting Strategies and Materials. *Materials* **2021**, *14*, 6631. [CrossRef] [PubMed]

Holmberg, K.; Andersson, P.; Erdemir, A. Global energy consumption due to friction in passenger cars. *Tribol. Int.* **2012**, *47*, 221–234. [CrossRef]
Nagendramma, P.; Kaul, S. Development of ecofriendly/biodegradable lubricants: An overview. *Renew. Sustain. Energy Rev.* **2012**, *16*, 764–774. [CrossRef]
Pettersson, A. High-performance base fluids for environmentally adapted lubricants. *Tribol. Int.* **2007**, *40*, 638–645. [CrossRef]
Greaves, M. Oil Soluble Synthetic Polyalkylene Glycols A New Type of Group V Base Oil. *Lube-Tech* **2013**, *104*, 1–23.
Zhou, A.; Du, C.; Peng, Z.; Peng, Q.; Qin, D. Rotor Temperature Safety Prediction Method of PMSM for Electric Vehicle on Real-Time Energy Equivalence. *Math. Probl. Eng.* **2020**, *2020*, 3213052. [CrossRef]
10. Siegel, R.; Skidd, C. Case studies utilizing mobile on-site recycling of industrial oils for immediate reapplication. *Lubr. Eng.* **1995**, *51*, 767–770.
11. Brown, P. Synthetic basestocks (Groups IV and V) in lubricant applications. *Lubr. Eng.* **2003**, *59*, 20–22. Available online: https://www.proquest.com/scholarly-journals/synthetic-basestocks-groups-iv-v-lubricant/docview/226966160/se-2?accountid=15300 (accessed on 3 December 2022).
12. Shah, R.; Gashi, B.; Rosenkranz, A. Latest developments in designing advanced lubricants and greases for electric vehicles—An overview. *Lubr. Sci.* **2022**, *34*, 515–526. [CrossRef]
13. Tung, S.C.; Woydt, M.; Shah, R. Global Insights on Future Trends of Hybrid/EV Driveline Lubrication and Thermal Management. *Front. Mech. Eng.* **2020**, *6*, 571786. [CrossRef]
14. Pettersson, A. Tribological characterization of environmentally adapted ester based fluids. *Tribol. Int.* **2003**, *36*, 815–820. [CrossRef]
15. Spikes, H. Friction Modifier Additives. *Tribol. Lett.* **2015**, *60*, 5. [CrossRef]
16. Bunemann, T.F.; Kenbeck, D. Lubricants and additives Chapter 7. Organic friction modifiers. In *Lubricant Additives: Chemistry and Applications*, 3rd ed.; CRC Press, Taylor & Francis Group: Boca Raton, FL, USA, 2017; pp. 195–210.
17. Cyriac, F.; Tee, X.Y.; Poornachary, S.K.; Chow, P.S. Influence of structural factors on the tribological performance of organic friction modifiers. *Friction* **2021**, *9*, 380–400. [CrossRef]
18. Zahid, R.; Mufti, R.A.; Gulzar, M.; Hassan, M.B.H.; Alabdulkarem, A.; Varman, M.; Kalam, A.; Zulkifli, N.W.B.M.; Yunus, R. Tribological compatibility analysis of conventional lubricant additives with palm trimethylolpropane ester (TMP) and tetrahedral amorphous diamond-like carbon coating (ta-C). *Proc. Inst. Mech. Eng. Part J J. Eng. Tribol.* **2018**, *232*, 999–1013. [CrossRef]
19. Trindade, E.D.; Durango, A.Z.; Sinatora, A. Friction and wear performance of MoDTC-containing and ester-containing lubricants over steel surfaces under reciprocating conditions. *Lubr. Sci.* **2015**, *27*, 217–229. [CrossRef]
20. Wang, W.; Li, C.; Yang, J.; Shen, Y.; Xu, J. Friction performance of MoDTP and ester-containing lubricants between CKS piston ring and cast iron cylinder liner. *Lubr. Sci.* **2018**, *30*, 33–43. [CrossRef]
21. Guegan, J.; Southby, M.; Spikes, H. Friction Modifier Additives, Synergies and Antagonisms. *Tribol. Lett.* **2019**, *67*, 83. [CrossRef]
22. Soltanahmadi, S.; Esfahani, E.A.; Nedelcu, I.; Morina, A.; van Eijk, M.C.P.; Neville, A. Surface Reaction Films from Amine-Based Organic Friction Modifiers and Their Influence on Surface Fatigue and Friction. *Tribol. Lett.* **2019**, *67*, 80. [CrossRef]
23. Davis, R.; Davis, J.R. *Cast Irons*; AMS International: Materials Park, OH, USA, 1996.
24. Fujita, H.; Spikes, H.A. Study of Zinc Dialkyldithiophosphate Antiwear Film Formation and Removal Processes, Part II: Kinetic Model. *Tribol. Trans.* **2005**, *48*, 567–575. [CrossRef]
25. Salmeron, G.C.; Leckner, J.; Schwack, F.; Westbroek, R.; Glavatskih, S. Greases for electric vehicle motors: Thickener effect and energy saving potential. *Tribol. Int.* **2022**, *167*, 107400. [CrossRef]
26. Gonda, A.; Capan, R.; Bechev, D.; Sauer, B. The Influence of Lubricant Conductivity on Bearing Currents in the Case of Rolling Bearing Greases. *Lubricants* **2019**, *7*, 108. [CrossRef]
27. Fujita, H.; Glovnea, R.P.; Spikes, H.A. Study of Zinc Dialkydithiophosphate Antiwear Film Formation and Removal Processes, Part I: Experimental. *Tribol. Trans.* **2005**, *48*, 558–566. [CrossRef]
28. Spikes, H. The History and Mechanisms of ZDDP. *Tribol. Lett.* **2004**, *17*, 469–489. [CrossRef]
29. Ratoi, M.; Niste, V.B.; Alghawel, H.; Suen, Y.F.; Nelson, K. The impact of organic friction modifiers on engine oil tribofilms. *RSC Adv.* **2014**, *4*, 4278–4285. [CrossRef]
30. Miklozic, K.T.; Forbus, T.R.; Spikes, H.A. Performance of Friction Modifiers on ZDDP-Generated Surfaces. *Tribol. Trans.* **2007**, *50*, 328–335. [CrossRef]
31. Bobzin, K.; Brögelmann, T.; Kalscheuer, C.; Thiex, M. Formation of tribochemical reaction layers on a metal modified amorphous carbon coating a-C:H:Zr (ZrCg). *Tribol. Int.* **2019**, *135*, 152–160. [CrossRef]
32. Grossiord, C.; Martin, J.; Le Mogne, T.; Palermo, T. In situ MoS2 formation and selective transfer from MoDPT films. *Surf. Coat. Technol.* **1998**, *108–109*, 352–359. [CrossRef]
33. De Barros'Bouchet, M.; Martin, J.; Le-Mogne, T.; Vacher, B. Boundary lubrication mechanisms of carbon coatings by MoDTC and ZDDP additives. *Tribol. Int.* **2005**, *38*, 257–264. [CrossRef]
34. Tang, Z.; Li, S. A review of recent developments of friction modifiers for liquid lubricants (2007–present). *Curr. Opin. Solid State Mater. Sci.* **2014**, *18*, 119–139. [CrossRef]
35. Ye, J.; Burris, D.L.; Xie, T. A Review of Transfer Films and Their Role in Ultra-Low-Wear Sliding of Polymers. *Lubricants* **2016**, *4*, 4. [CrossRef]

36. Martini, A.; Ramasamy, U.S.; Len, M. Review of Viscosity Modifier Lubricant Additives. *Tribol. Lett.* **2018**, *66*, 58. [CrossRef]
37. Müller, M.; Topolovec-Miklozic, K.; Dardin, A.; Spikes, H.A. The Design of Boundary Film-Forming PMA Viscosity Modifier *Tribol. Trans.* **2006**, *49*, 225–232. [CrossRef]

lubricants

MDPI

Article

Wear Mechanisms in Press Hardening: An Analysis through Comparison of Tribological Tests and Industrial Tools

Jaume Pujante [1,2,*], Eduard Garcia-Llamas [1], Giselle Ramírez [1], Nuria Cuadrado [1,2], Agim Ademaj [3], Montserrat Vilaseca [1] and Daniel Casellas [1,4]

1 Eurecat, Centre Tecnològic de Catalunya, Unit of Metallic and Ceramic Materials, Plaça de la Ciència 2, 08243 Manresa, Spain; eduard.garcia@eurecat.org (E.G.-L.); giselle.ramirez@eurecat.org (G.R.); nuria.cuadrado@eurecat.org (N.C.); montserrat.vilaseca@eurecat.org (M.V.); daniel.casellas@eurecat.org (D.C.)

2 Department of Science and Material Engineering, Universitat Politècnica de Catalunya, Av. Eduard Maristany 10-14, 08019 Barcelona, Spain

3 METAKUS Automotive GmbH, Fehrenberger Straße 1a, 34225 Baunatal, Germany; agim.ademaj@metakus.com

4 Division of Solid Mechanics, Luleå University of Technology, 971 87 Luleå, Sweden

* Correspondence: jaume.pujante@eurecat.org

Abstract: Press hardened components have become widespread in the automotive industry in structural and crash-resistant applications, thanks to the combination of the complex shapes and high mechanical properties obtained. However, the press hardening of coated boron steel results in severe adhesive-based wear, with tool maintenance being required in as few as 3000 cycles. The current industrial implementation of press hardening is defined to work around this phenomenon. While this aspect has been studied by different authors, most of the literature deals with laboratory-scale tribosimulators, leaving an open question into how this knowledge transfers to macroscopic effects on the industrial process. In this work, wear in press hardening is studied by comparing the results obtained in laboratory conditions with a pilot-scale line, and finally, with wear mechanisms observed on industrial tools. The aim of this study is to consolidate the current knowledge about the micro-mechanisms involved, and to understand to what extent the existing tests reproduce the actual mechanisms observed in the press floor. The results show how material transfer mainly happens as an accumulation of dust compacted into initial defects on the tool surface. Moreover, this mechanism is effectively reproduced in laboratory tribosimulators and pilot environments, showing a similar morphology to wear on industrial tools. The work sheds light on the underlying causes of wear, and its potential mitigation strategies.

Keywords: AlSi-coated boron steel; press hardening; tribology; wear

1. Introduction

Press hardening, also known as hot stamping, is a non-isothermal forming process for sheet metals, where forming and quenching take place in the same step [1]. In direct press hardening, blanks are austenitized at a temperature between 900 and 950 °C for 4–10 min in a furnace. The hot blank is then transferred to a set of cooled dies where it is formed in a single stroke. During this step, the cooled dies quench the formed component at a cooling rate higher than 50 °C/s, ensuring a full martensitic microstructure. The total cycle time, including transfer, forming, and quenching, typically takes 15 to 25 s [2].

The use of press hardening allows for the production of lightweight components with very high mechanical properties and complex shapes while avoiding problems such as spring back on the component or damage including fracture on the tools, associated with the cold forming of Ultra-High Strength Steels (UHSSs). For this reason, press hardened components have mainly found uses in structural and crash reinforcement components for the automotive industry [3,4].

The most commonly used press hardening material is aluminized boron steel sheet. This sheet steel is coated with a layer of Al-10%Si with a thickness of between 20 and 36 μm, depending on the specifications [5,6]. During heat treatment, the coating develops a complex sub-layer structure of Al-Si-Fe intermetallic phases plus a metallic Fe-based diffusion layer in the coating–substrate interface. These intermetallics have been reported to be hard and brittle [6–8].

The use of AlSi coated steel sheet offers significant advantages in the direct press hardening process: it prevents decarburization and scaling of the steel sheet, provides barrier corrosion protection [9], and improves component paintability [6].

On the other hand, the presence of the coating dominates tool–workpiece contact during forming, and greatly affects the wear mechanisms observed. The main active wear mechanism appears to be material transfer from the sheet metal coating to the tool. This material accumulates in irregular lumps on the tool surface, which affect component quality [10,11]. The accumulation of the material transfer mechanism is fast and severe, and results in press hardening tools requiring maintenance in the form of re-polishing in as few as 3000 production cycles [12].

Wear in press hardening is a well-researched topic, with relevant contributions coming from different research groups. These contributions often include studies of wear mechanisms using different tribosimulators or laboratory test configurations. Without the intent of compiling a comprehensive list, and following an approximate order of decreasing abstraction, some examples are provided here: Hardell [13] and then Pelcastre [14] used a pin-on-disk reciprocating setup to analyze wear in press hardening, Bruschi and Ghiotti et al. [12] used discontinuous pin-on-disk sliding tests, and Venema et al. [15] presented strip drawing-based simulators, as did later works of Pelcastre et al. [16]. Recent works by Venema [17] and these same authors, Pujante et al. [18], presented semi-industrial tests on omega configurations. All of these experiments obtained remarkably coincident results, identifying a combination of abrasive and adhesive wear mechanisms and proposing the generation of particles from the coating as playing an important role.

However, the analysis of actual press hardening industrial tools is scarce, and this is ultimately needed to verify that the assumed mechanisms and trends are correct. Examples include works such as [19–21], which mainly describe wear features without being yet able to describe the acting micromechanisms.

Filling this gap between laboratory tests and industrial reality is highly interesting. First, it would allow for the development of fundamental knowledge on the actual wear micromechanisms acting in the industrial environment, and understanding to what extent the phenomena observed in laboratory tests translate into the industrial process. Second, it would bring to understanding the relative importances of the different observed mechanisms, and therefore allow for mitigation strategies or solutions to be developed. Finally, it would frame precisely to what extent laboratory tests can be used for the development of these solutions.

In this work, the mechanisms of wear on press hardening tools processing AlSi coated sheet steel are discussed by comparing the wear on actual industrial tools, measured through surface replication, with results from laboratory tribological experiments with two different levels of abstraction: a strip drawing test and an omega tool configuration. An analysis of the results allow for the proposal of a solid extrapolation on how the mechanisms of material transfer act on an industrial process, and to what extent they are reproduced in simplified laboratory setups.

2. Materials and Methods

2.1. Materials

All tests were performed on commercial AlSi-coated 22MnB5 steel sheet. Details on the material format and characteristics are offered for each individual test.

2.2. Evaluation of Coating Mechanical Properties

The mechanical properties of the sub-layers in the AlSi coating were studied at room temperature via nanoindentation, in order to determine the mechanical properties of the individual sub-layers.

Nanoindentation measurements were performed using an XP Nanoindenter (MTS Corp, Oak Ridge Tennessee (EEUU)) equipped with a Berkovich indenter. Samples were analyzed in cross-section and mechanically polished to a mirror surface finish with a 0.05 µm colloidal silica suspension. The indentations were performed at a maximum penetration depth of 200 nm. Hardness values were computed using the Oliver and Pharr methodology [??]

2.3. Strip Drawing Tests

Tool wear was compared to inserts worn in a strip drawing-based tribometer system for press hardening (Figure 1). In this test, sheet metal samples are austenitized in a furnace. The strip is transferred to the strip drawing tool, where it is clamped between two flat tool steel specimens under controlled pressure and pulled by a sled at constant velocity. The frictional force is measured during the test. A detailed description of the test setup can be found elsewhere [23]. While the main aim of this test is to be used to obtain Coefficient of Friction values, it is still a valid reproduction of the tool steel-sheet metal contact in press hardening and its tribological interaction.

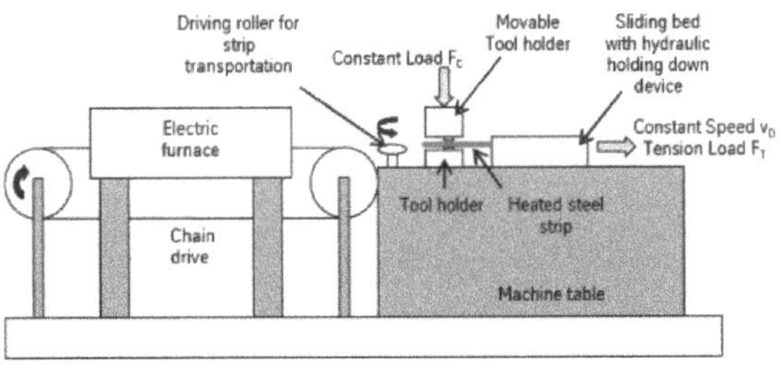

Figure 1. Strip drawing simulator for friction studies in press hardening.

The tool inserts were manufactured from tool steel 1.2367, hardened to 48 ± 1 HRC. The surface of the tool specimens was polished to a roughness Ra of 0.26 µm. The strips of AlSi-coated boron steel were subject to 390 s austenitization at a 900 °C furnace temperature. In these heat treatment conditions, the sub-layer structure in the coating was completely formed [8].

Strips were slid at a velocity of 60 mm/s, while the tool steel inserts were clamped at 6 MPa of pressure. The temperature of the strip at the moment of the beginning of sliding was approximately 800 °C; an average Coefficient of Friction of 0.428 was measured during the sliding test.

The test was repeated five times using the same tool inserts, in order to generate a state of initial wear on the tool steel.

2.4. Pilot Environment: Hot Stamping of Omega Components

A laboratory-scale press hardening line was set up for these trials, aiming to reproduce the main conditions of an industrial environment. Set up of this experimental configuration was discussed in references [18,24]; samples produced in these two references are discussed in the current work.

Heating was performed in a 3 m long continuous roller-hearth convection-radiation furnace, working in an open (oxygen-containing) atmosphere. Austenitized sheets are manually transferred to a 150 t hydraulic press equipped with a set of cooled dies, where hot stamping is performed.

The component geometry was based on an omega-shape (Figure 2), a common concept in press hardening literature. The final geometry was defined based on a survey of cross-section measurements obtained from production B-pillars. The dimensions thus defined were considered to combine a relevant example of the working conditions of industrial tooling with ease of manufacture and inspection. Tool laterals were designed as two symmetrical replaceable inserts. One side was manufactured in commercial hot work tool steel QRO90 heat treated to 50–52 HRC. The other side was left untreated at 36 HRC; a comparison between the two sides may aid in the identification of tribological behavior.

Using this setup, 130 mm × 250 mm, 1.7 mm thick USIBOR 1500 AlSi blanks were stamped with a cycle time of 35 s. Austenitization was performed at 930 °C with a total 360 s furnace time. Tools were measured to reach 80–90 °C at their surface after 25 blanks, assumed as close to steady state. This range of temperatures is representative for industrial press hardening tools.

The campaign discussed in this study spanned 800 cycles, which were divided into several batches. After 270 cycles, replicas of the tools were obtained and analyzed, along with the tool used at the end of production. After the end of the study, one of the lateral inserts was subject to destructive analysis.

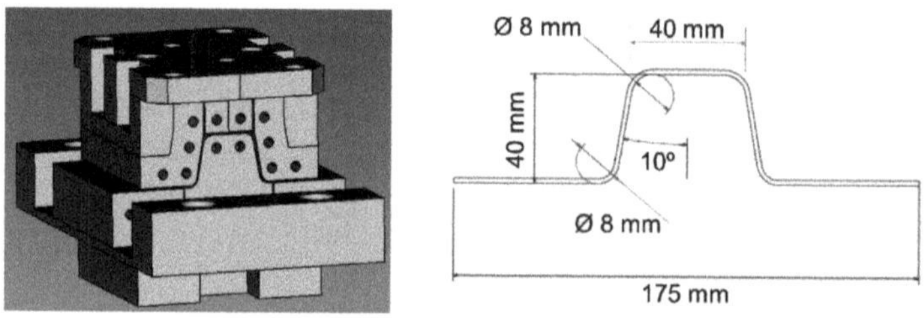

(**a**) (**b**)

Figure 2. Pilot hot stamping environment. (**a**) Schematic of the tool assembly, including replaceable inserts; (**b**) Component geometry.

2.5. Study of Industrial Tools

Using surface replication, it has been possible to detect and to measure wear on industrial press hardening tools while removing the complications associated with either taking the industrial tool to the lab, or taking dedicated analysis equipment to the industrial environment.

Industrial tools were analyzed at various points of their service life, using production stops to acquire replicas. Details of the tools included in this study are presented in Table 1. Due to production and confidentiality issues, no exact cycle numbers can be given. Instead, order-of-magnitude approximations will be offered for each case. All tools were analyzed in the context of the research project RFCS TestTool.

Table 1. Industrial tools analyzed via surface replication.

Component Produced	Tool Material	Hardness	Sheet Metal Material
B-Pillar	DIN 1.2367	48 HRC	AlSi-coated Boron Steel
B-Pillar	DIN 1.2344	48–50 HRC	AlSi-coated Boron Steel
Frame Dash Panel	DIN 1.2367	48 HRC	AlSi-coated Boron Steel

Surface Replication

Surface replication has been proposed as a methodology for the analysis of wear in industrial production tools [19,21]. In this technique, the surface topographies of the tools are transferred onto a polymeric replica that can be studied in the laboratory without interrupting production. The Accuracy of replication compounds is good enough for the evaluation of the wear processes, with some formulations showing precisions of better than 0.1 µm, and accuracies of better than 10% [21,25,26].

Replicas were analyzed in the laboratory via optical profilometry techniques, using a Sensofar Plµ 2300 confocal microscope.

3. Results

3.1. Properties of the Al-Si Coating

3.1.1. Micromechanical Analysis of the Coating

AlSi-coated boron steel samples 1.7 mm in thickness were subject to different heat treatments in a convection-radiation furnace. In all cases, the heat treatment consisted in introducing the sample inside the furnace for a set time, followed by extraction and cooling between a pair of steel blocks for an approximate 15 s.

After heat treatment at 900 °C, the coating on the samples, initially metallic Al-Si, developed a microstructure consisting of five sub-layers (Figure 3a). Grigorieva et al. [8] identified the various sub-layers in the system via transmission electron microscopy (TEM). According to [8], the outermost sub-layer (sub-layer 1 in Figure 3a) corresponds to Fe_2Al_5. Sub-layer 2 corresponds to the Al-Si-Fe ternary phase τ_1, sub-layer 3 corresponds again to Fe_2Al_5, and sub-layer 4 to phase τ_1. Finally, a diffusion layer (sub-layer 5) can be observed between coating and substrate, consisting mainly of an iron matrix with $FeAl_3$ inclusions.

In this work, two sets of samples were prepared. The first, under "nominal" conditions, used a furnace temperature of 900 °C and a total furnace time of 390 s. The second was an extremely long furnace time of 1800 s at 900 °C. The aim of this second test was to verify whether changes in the furnace time modify the properties of the sub-layers, or only their relative proportion on the sample.

Nanoindentation studies were performed on cross-sectional samples, as described in Section 2.2. Figure 3a shows a reference image of the indented layers. The hardness results are summarized in Figure 3b.

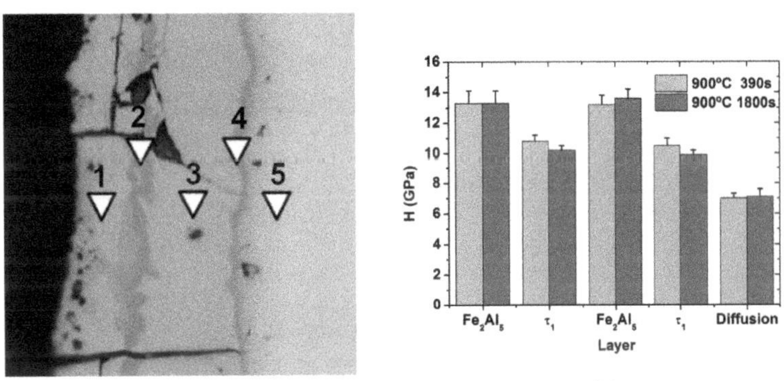

(a) (b)

Figure 3. Mechanical characterization of the sub-layers of the heat treated AlSi coating. (**a**) Sub-layer structure; triangles indicate the different indented sub-layers identified as follows: (1) topmost Fe_2Al_5, (2) τ_1, (3) Fe_2Al_5, (4) τ_1, (5) Diffusion layer. (**b**) Comparison of hardness of the five layers after different heat treatments.

All layers were much harder than the base heat treated sheet steel, presenting hardnesses of between 7 and 14 GPa. Beginning from the outside of the coating, sub-layers one and three, corresponding to the binary phase Fe_2Al_5, were slightly harder (13 GPa) than layers 2 and 4 (10 GPa), corresponding to the Al-Si-Fe ternary phase τ_1. The outermost part of the diffusion layer (layer 5) showed a lower hardness than the rest of the layers (7 GPa)—approximately in the range of the martensitic 22MnB5 substrate.

This observation agrees with the existing literature, where the coating is reported to be hard and brittle [27]; Pelcastre estimated its overall hardness to be between 511 and 837 HV [14].

It is worth observing that the layer properties did not appear to be modified by the heat treatment, but rather, the hardness was the same for each phase in both of the studied samples. This correlates well to observations from other authors about the phase evolution in AlSi coatings for hot stamping: even though the proportions of the different phases in the coating evolve with time, the chemical composition of each phase remains approximately constant during the process, as discussed in [8], as they are stoichiometric compounds.

Finally, a study was performed to determine the exact properties of the diffusion layer. According to [8], this layer consists mainly of steel with a gradient of aluminum alloyed content. Therefore, it is possible that the properties of the layer are not constant or homogeneous, as they do not correspond to a particular stoichiometric composition.

A detailed analysis of the diffusion layer showed a decrease in hardness inside this layer (Figure 4); an effect also reported by Fan and De Cooman [27]. For the sample subject to the recommended heat treatment, 390 s at 900 °C, an indent located 5 µm under the bottommost τ_1 showed a considerable drop of more than 4 GPa; a 10 µm diffusion layer had been measured for this sample. In the case of a long austenitization sample, 1800 s at 900 °C, a steady decrease in hardness was observed along the 25 µm wide diffusion layer. In this case, a minimum of 3 GPa was found approximately 18 µm under the bottommost τ_1.

This change in mechanical properties could be explained by the diffusion of Al into the Fe-based substrate. In general terms, aluminum is known to adversely affect steel quenchability. The authors Kiani-Rashid and Edmonds [28] observed in a series of dilatometry tests on cast iron that 6.16 % of Al dissolved in a Fe matrix inhibited martensite formation in Fe alloyed with C and Si. In any case, the growth of this diffusion layer is a well known and reported phenomenon. Most relevantly, it is identified as one aspect to control regarding the weldability of the finished products.

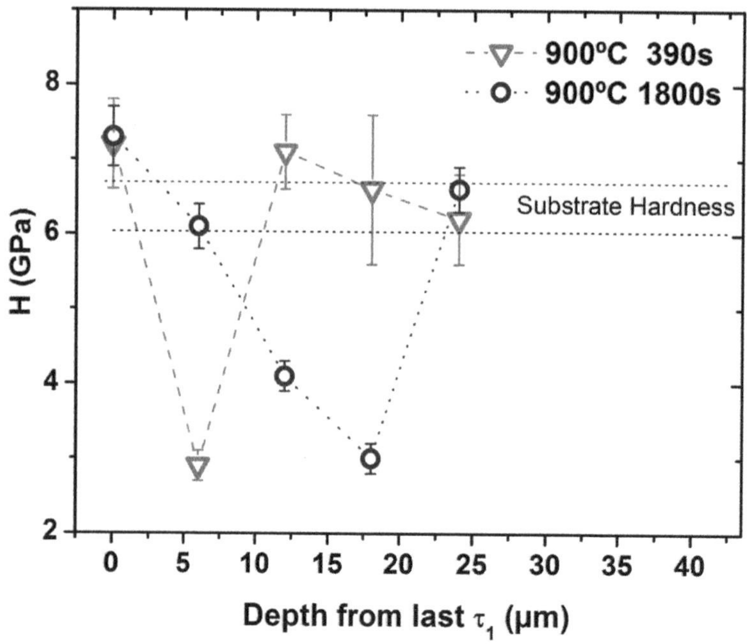

Figure 4. Microhardness profile obtained across the diffusion zone.

3.1.2. Failure of the Coating during Forming

The literature on press hardening reports that adhesion on tools consists of material from the AlSi coating [13,17,19]. The mechanisms through which this material is removed can be related to the properties discussed in the previous section. The coating is formed by intermetallic phases, shows a high degree of hardness, and is reported to be brittle and prone to flaking [27]. This suggests that particles may break off from the coating during forming, generating loose particles that can be then compacted onto the tool.

One cause for this can be friction: the forces and pressures involved in tool-sheet metal sliding can cause particles from the coating to break. This can be clearly seen in the sheet metal used in the strip drawing simulator displayed in Figure 5. When non-slid sections of the sheet metal are observed, the coating is whole and continuous, showing only typical cracks and defects commonly reported for this material (Figure 5a).

However, if a sample slid against the tool steel inserts is analyzed, the coating shows a much worse status, including severe cracking and broken particles (Figure 5b). these mechanisms have been observed and reported by other authors; notably, Ref. [17] observed an almost complete coating removal in slid sheet metal on a tribosimulator drawing radius.

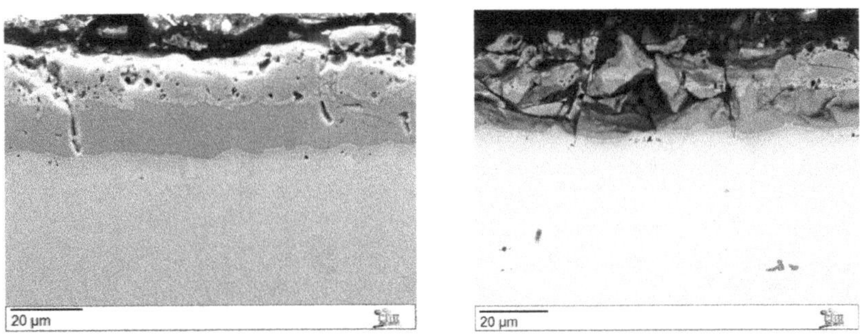

(a) (b)

Figure 5. Effect of sliding on coating integrity, generated during testing in a strip drawing tribosimulator; (**a**) coating outside the slid zone; (**b**) coating situated on the slid zone, showing severe damage and flaking.

A second mechanism considered in this work is the fracture of the AlSi intermetallic coating when severe plastic deformation takes place on the steel substrate. Qualitatively speaking, the coating should not be able to accommodate significant length changes. In order to prove this, sheet metal samples have been analyzed from the pilot omega components, analyzing the component radius (Figure 6). It could be observed that cracks form readily on the outside of the radius; these cracks correspond to separations or "gaps" in the coating that cannot elongate to cover the increased length (Figure 6a,b). Similar behaviors have been described by Gui et al. [29] in a study performing tensile tests on this same material. Moreover, in that work, very large deformations caused coating particles to flake off from the surface.

Deformation on the inside of the radius (Figure 6c,d) results in coating material breaking from the surface. It is believed that, subject to sufficient compression, the brittle intermetallic coating flakes to accomodate deformation. These cracks look bright when they are observed on the surface, presumably due to the flaking off of the topmost layers, including surface oxides.

As a general conclusion, these two mechanisms (friction and plastic deformation of the substrate) constitute active mechanisms that can generate coating debris during forming, regardless of further interactions with the tool.

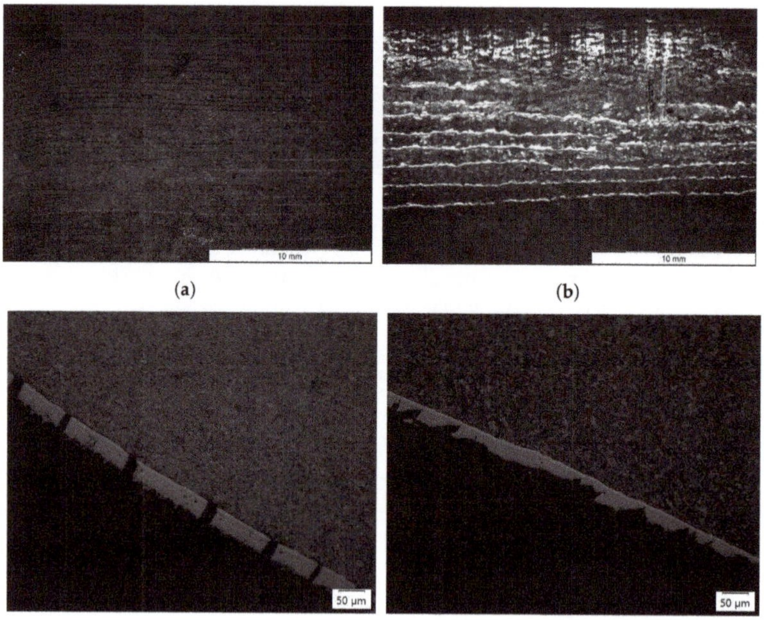

(**a**) (**b**)

(**c**) (**d**)

Figure 6. Appearance of cracks on the radius of a formed pilot plant component: (**a**,**c**) 10× stereomicroscopy images of the surface of an external (**a**) and internal (**c**) radii; (**c**,**d**) OM of the cross-sections corresponding to external (**b**) and internal (**d**) radii.

4. Wear Mechanisms on Press Hardening Environments

4.1. Wear on Strip Drawing Inserts

Strip drawing inserts already show the initiation of wear, despite the low amount of material processed in the test. An overview of the inserts, seen in Figure 7, shows a patch of irregular adhesion on the center of the sample, surrounded by abrasive wear tracks.

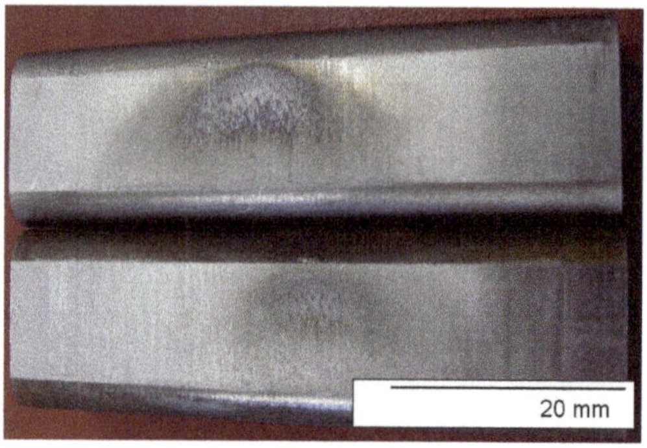

Figure 7. Inserts of the strip drawing facility after testing.

A detailed SEM analysis of the inserts sheds light on the active micromechanisms. First of all, abrasive wear and surface plastic deformation can be observed on the originally polished tool steel (Figure 8a). The appearance of abrasive wear is reasonable, considering that the mechanical analysis of the coating presented in Section 3.1 showed that the hardness of the coating is well above the hardness of the tool steel. In these conditions, sliding under pressure generates material removal and ploughing on the tool steel.

The newly generated grooves appear to act as nucleation points for material transfer, in the form of the compaction of loose dust flaked off from the coating though friction, as was discussed in Section 3.1.2. Figure 8a shows the frontal part of the adhesion patch. Dust can be seen accumulating in this point; moreover, this dust is still at least partially loose on the accumulation side.

Based on this analysis, a mechanism can be proposed where loose debris accumulates on topographical features on the tool surface. As the amount of accumulated dust grows, particles are compacted and partially sintered into a compact layer.

Ultimately, this layer is made of compacted brittle dust, anchored to the tool steel mainly through mechanical means. As a result, the adhesion layer can break off. Figure 8b shows the appearance of the compacted layer at its trailing edge. Cracks can be seen on this zone, showing that the compacted layer is indeed breaking and shedding particles as a result of repeated sliding with the sheet metal.

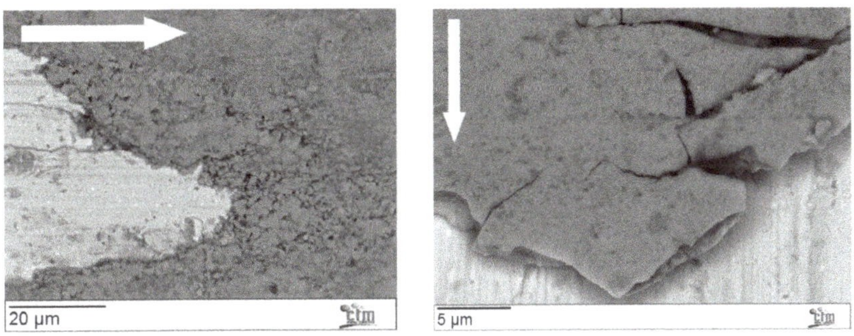

(a) (b)

Figure 8. Micromechanisms of material transfer on strip drawing inserts; arrow shows sliding direction. (**a**) Accumulation of dust at the front side of the surface features, leading to compaction; (**b**) Fracture of compacts at the rear side. Abrasive wear grooves parallel to the sliding direction can be observed on the bright tool steel surface.

4.2. *Wear on Pilot Environment Inserts*

After approximately 800 production cycles, tool inserts on the pilot environment showed a complex surface, with different areas where different wear mechanisms dominate. An overview is offered in Figure 9.

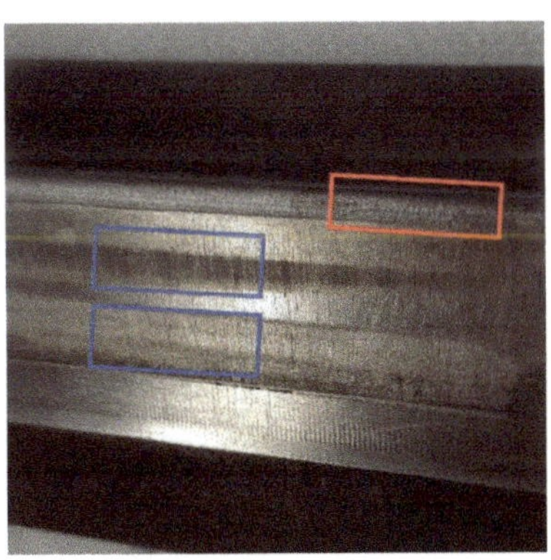

Figure 9. Overview of a pilot environment insert, showing a variety of zones with different wear mechanism combinations: gross material transfer in the drawing radius (red square), fringes of combined wear with sparse material transfer on the tool wall (blue squares), and overall abrasive wear on the rest of the surface.

4.2.1. Gross Material Transfer

Gross material transfer was observed mainly on the drawing radius of the tool (Figure 10). This phenomenon took the shape of thick transfer layers with an homogeneous, almost polished surface. These layers could be recognized thanks to observed rough patches: when analyzed using profilometry, as described in [24], these patches could be confirmed to be holes in the layer. SEM analysis (Figure 10b,c) confirms that the rough surface observed is still transferred material: the patch corresponds to a large flake falling off from the transfer layer, but the fracture takes place inside of the layer itself, and not by dislodging from the original tool surface.

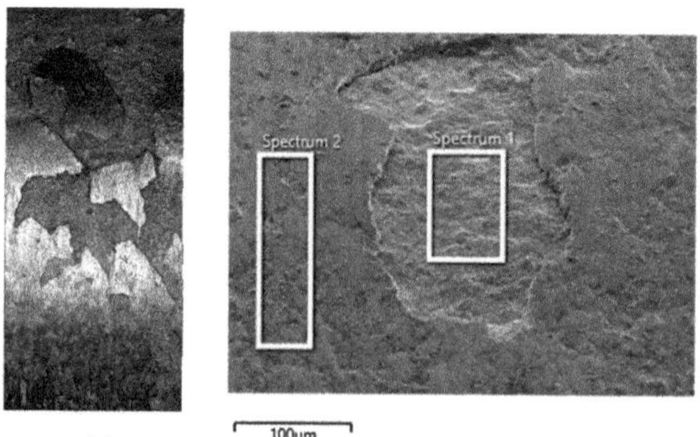

(a) (b)

Sample	Al	Si	Fe	O
1	42.78	3.77	42.21	11.24
2	42.22	3.07	41.67	13.04

(c)

Figure 10. Material transfer on the drawing radius of the pilot omega tool. (**a**) Overview of the aspect of the radius (photography); (**b**) SEM detail of a rough patch in a polished zone; and (**c**) EDX analysis corresponding to zones 1 and 2.

An overview of the surface through SEM imaging confirms that the tool radius is covered on a compact transfer layer; backscatter electron imaging, offering contrast based on chemical mass, reveals a chemical composition that is clearly differentiated from the tool steel (Figure 11a). The coverage is mostly homogeneous, and only some specific instances can be observed of the bare tool steel showing amid the adhesion. It is worth noticing that this wear state was formed after only 500 production cycles, showing that the material transfer mechanism is extremely active.

Figure 11b shows the front border of this transfer layer, where a combination of wear micromechanisms appear, including plastic deformation, abrasive wear, and the accumulation of debris. This is very similar to the observations in the strip drawing inserts shown in Figure 8 and discussed in Section 4.1, showing that the proposed wear mechanisms of dust accumulation on the surface topography features also fit with the observations in this test configuration.

When observed in cross-section (Figure 11c), this transfer layer is clearly seen to correspond to an accumulation of particles. The morphologies of these particles can be compared to the flakes generated during coating sliding (Figure 5) or severe plastic deformation (Figure 6). Indeed, it appears that the most likely mechanism is the formation of powder and its subsequent accumulation on the tool through geometrical and mechanical means, as opposed to a mechanism of chemical/metallurgical adhesion of the coating on the tool, leading to subsequent coating failure. This is further supported by the fact that no clear welding or transition phase was observed on the adhesion–tool interface: instead, the tool surface appeared to have suffered from a combination of plastic deformation and wear, and the features thus formed acted as anchor points for the transfer material to accumulate.

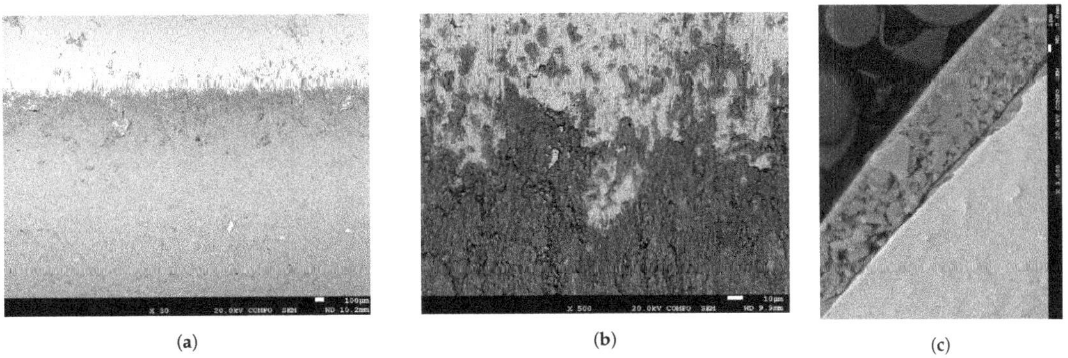

Figure 11. Surface SEM analysis of the transfer layer on the pilot plant tool radius; sliding direction is top to bottom. (a) Overview showing homogeneous coverage of the tool (lighter shade) via compact material transfer (darker shade); (b) Upper border of the transfer layer: accumulation of debris, together with mechanical damage on the substrate can be observed; (c) Cross-section of the radius, showing a layer of accumulated material.

4.2.2. Abrasive Wear and Plastic Damage

Pilot environment inserts suffer moderate surface damage in the form of abrasion and plastic deformation.

Abrasive wear can appear through two-body micromechanisms, as described by Archard et al. [30], or three-body interaction, and they involve a loss of material on the tool surface. In the case of the studied pilot plant inserts, only mild abrasion is observed. Most interestingly, abrasion marks appear in a clear and obvious way when soft (36 HRC; image in Figure 12a) tool steels are employed in the work referenced in [18]. This clear effect of hardness on the appearance of the mechanism is perfectly consistent with the Archard model, establishing a linear relationship between abrasive wear and hardness.

Even on harder tools, surface damage appears even at very low cycle numbers. As an illustrative example, Figure 13b shows details from the slid surface on a 52 HRC pilot plant tool insert after 500 cycles. Clear evidence can be observed of abrasion on the surface in the form of channels parallel to the top-to-bottom sliding direction. Moreover, a series of ridges and small cracks can be seen, consistent with local plastic deformation generated in an intense contact. Finally, an adhesion of coating particles has begun to form on these ridges, indicating an initiating mechanism for gross material transfer. It must be noticed that these abrasive mechanisms were able to modify the surface roughness of the tool [18,24]. However, they did not result in macroscopic changes to the tool geometry.

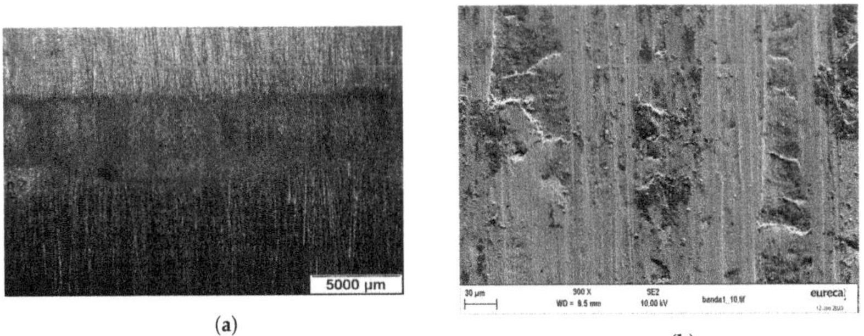

Figure 12. Examples of mechanical damage observed in a pilot plant insert: (a) wear marks parallel to the sliding direction (top to bottom) on a die wall; (b) SEM detail of wear tracks (top-to-bottom sliding lines) and plastic deformation (bumps and ridges in the surface) in this zone.

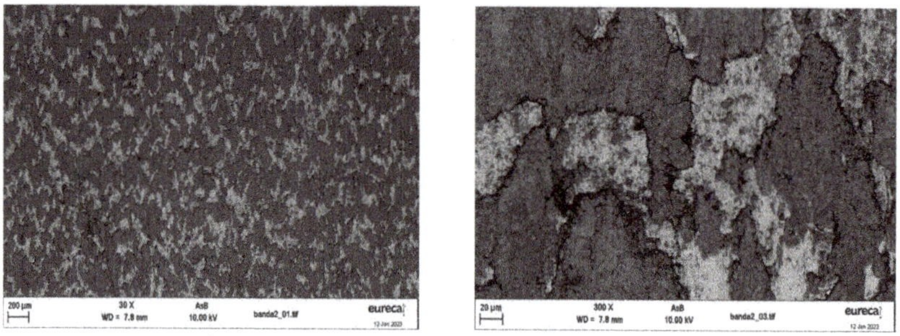

(**a**) (**b**)

Figure 13. Sparse material transfer appearing on fringes on the vertical tool wall; sliding direction is top to bottom: (**a**) SEM overview showing material transfer (dark) on the tool surface (bright); (**b**) Detail of material transfer features.

4.2.3. Combined Wear and Sparse Material Transfer

The flat vertical wall on the omega shape showed horizontal fringes with a matte finish (see Figure 13a). These fringes are reminiscent of material transfer in the strip drawing inserts, consisting of sparse accumulations of material transfer, as opposed to a thick, continuous compact layer.

An SEM analysis of this fringe (Figure 13b) shows a complex combination of wear mechanisms. The tool steel surface is covered on drag marks and plastic deformation, with an appearance that is similar to that described in the previous section. Material transfer takes place on these surface features. In this case, this accumulated material is not as regular as in the radius. First, no significant thickness can be measured via profilometry. Instead, only changes in roughness can be observed. This was discussed in reference [24]. Moreover, the area coverage is not as homogeneous, and the surface appears matted in recognizable individual features, as seen in Figure 13a. Each of these features (Figure 13b) show an appearance that is consistent with an accumulation of dust. Moreover, cracking and fracture are observed on these sparse compacts; again, similar to that observed in the strip drawing inserts (Figure 8).

4.3. Wear on Industrial Tools

4.3.1. Gross Material Transfer on Industrial Tools

As reported in the literature, the tools used in the press hardening of AlSi coated boron steel mainly show adhesive wear-related phenomena, namely, the transfer of material from the workpiece to the tool [21]. The adhered material formed macroscopic features on the tool surface, which could be recognized through naked eye inspection (Figure 14).

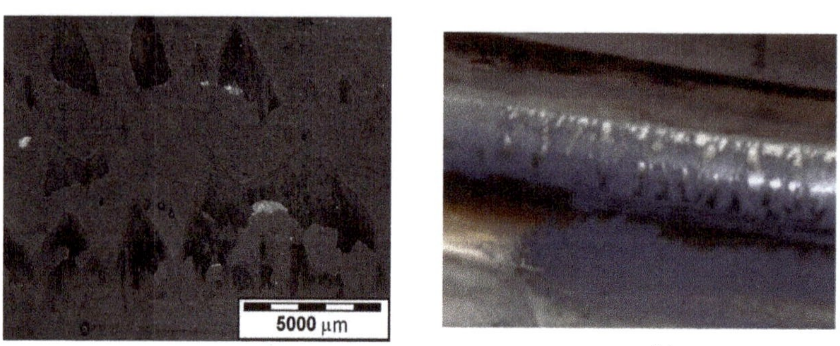

(**a**) (**b**)

Figure 14. Gross material transfer on industrial tools. (**a**) Stereomicroscopy image obtained from a surface replica of a tool radius; (**b**) Picture obtained from a radius of a production tool.

Fragments of adhered material broken off from the tool surface were found entrapped in the replicas, as can be seen in Figure 14a. These particles were hard and brittle, with texture, aspect, and properties similar to compacted powder. Semi-quantitative chemical analysis was performed on these particles via SEM/EDX. All of the particles analyzed showed similar chemical compositions, including primarily aluminum, silicon, and iron (Table 2). This chemical composition roughly corresponds to the coating after heat treatment [6,8]. Based on these results, it is possible to conclude that adhesion in press hardening tools is related to the build-up of material from the coating.

Table 2. EDX analysis performed on samples of adhered dust extracted from a press hardening tool; elements in mass %.

Sample	Al	Si	Mn	Fe
1	37.3	11.9	0.6	41.8
2	45.6	6.1	0.6	44.1
3	36	7	0.7	49.1

4.3.2. Abrasive Wear and Mechanical Damage

Abrasive wear is reported to be a relevant factor in the durability of press hardening tools.

Material transfer imposes a tight maintenance schedule, requiring the removal of transfer layers in as little as 3000 production cycles [12]. Abrasive wear, on the other hand, implies a loss of tool geometry and eventually leads to a macroscopic loss in tool shape, leading to insufficient tool–workpiece contact and a consequent loss in cooling and tolerances.

Abrasive wear can be observed on a microscopic scale on tools with high roughnesses or machining patterns. In this case, the removal of the asperities takes place as the production cycles accumulate, beginning on the areas of the most intense contact. This was reported in [19] for tools working on uncoated material. In this case, the removal of the asperities takes place as the production cycles accumulate, beginning with the areas of most intense contact (Figure 15).

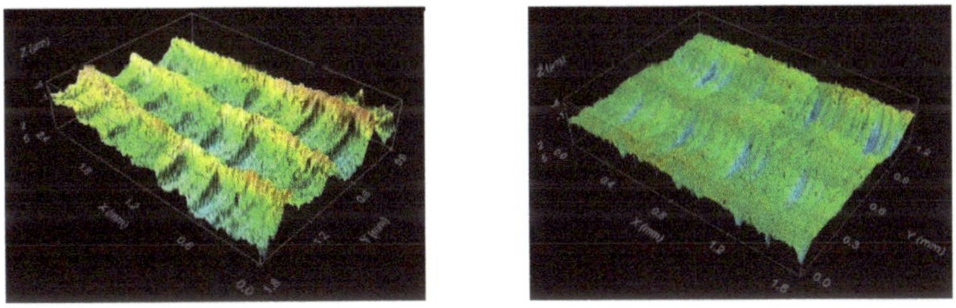

(a) (b)

Figure 15. Abrasive wear on an industrial press hardening tool. The original surface finish (**a**) has been completely polished away after a production campaign (**b**).

Informal reports from the industry point at cycles of maintenance of approximately 50,000–80,000 cycles, where the tool surface needs to be regenerated via overlay welding on worn spots, and subsequent machining of the surface is needed. A set of production tools could be analyzed using surface replication at the beginning of its production and after approximately 150,000 production cycles. A comparison of the radius profile, presented in Figure 16, shows a considerable loss of macroscopic shape.

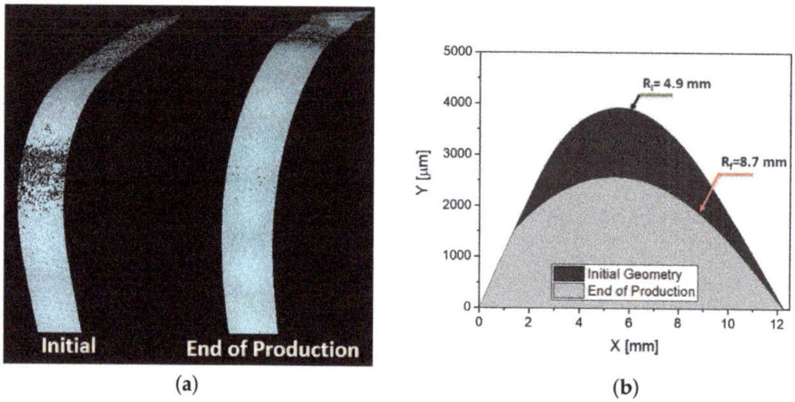

Figure 16. Profilometry analysis on a drawing radius from a production tool at the begining and end of production (approx. 150,000 cycles); (**a**) Overview and (**b**) comparison of profiles; please notice that X- and Y axes are not to scale.

4.3.3. Combined Wear and Sparse Material Transfer

Industrial tools show instances of sparse material transfer, with similar morphologies to that observed in the pilot environment. Again, these features appear as matte patches on the tool surface, and they typically appear at some point of the vertical walls in the deep drawing sections.

Destructive analysis of the tools was not possible. Therefore, studies could only be performed indirectly, through the extraction of surface replicas (Figure 17). The overall morphology of this feature is basically equivalent to that in the pilot Environment inserts. When analyzed through confocal microscopy, it can be seen that the rough area of the fringe is formed by discontinuous adhesion lumps, surrounded by bare tool steel with evidence of abrasive wear. An analysis of the surface outside this zone shows an initial smooth surface finish.

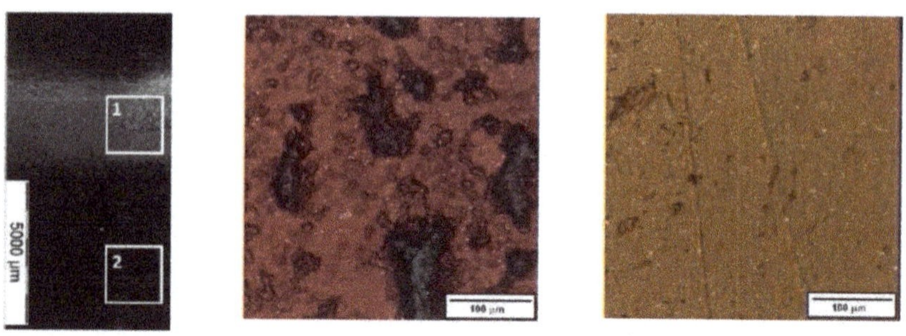

Figure 17. Replica analysis of a combined wear fringe on a vertical wall of an industrial tool.

4.3.4. Ploughing

In the mechanism of ploughing, protrusions on a hard surface interact mechanically with a soft or brittle counterpart during sliding, resulting in the removal of material from the weaker surface. Material thus removed can become loose wear debris, or remain adhered on the asperity (Figure 18). Ploughing is a well known tribological interaction mechanism. Material transfer through ploughing has been reported in the scientific literature for press hardening tribosimulators [31–33].

Ploughing can take place in industrial press hardening tools whenever there are protrusions on the tool steel surface: the locally increased pressure and deformation of the coating results in locally severe tool–component interaction.

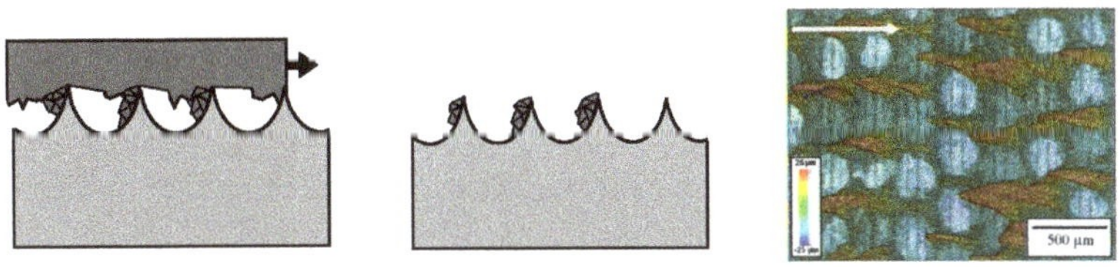

(a) (b) (c)

Figure 18. Mechanism of material transfer through ploughing. (**a**,**b**) Schematic of the micromechanism at work; (**c**) Example of material transfer through plowing observed on an industrial tool with rough surface finish.

Figure 18c shows a relevant example, where material transfer due to ploughing was observed on an industrial press hardening tool with high roughness. In this case, the tool surface showed a machining pattern with a roughness Ra of over 2 µm. Replicas obtained at low cycle numbers from tool walls showed an accumulation of transferred material on the tip of each of the surface finish features. As of 2023, the current trend on industrial press hardening tools is to generate a relatively smooth surface finish precisely to avoid this phenomenon. However, this image is still an illustrative case on how surface features generate material transfer through mechanical means.

5. Discussion

5.1. Material Transfer as the Compaction of Loose Wear Debris

In the mechanism of debris compaction, loose wear debris becomes trapped inside the tool–component contact, and is pressed against irregularities in the tool surface. Due to the high temperatures and the locally high pressures involved, these particles become compacted into a glaze-like layer (Figure 19). This mechanism has been proposed by several authors. Pelcastre et al. [10,33,34] have discussed the important role of dust compaction on the formation of macroscopic material transfer features on press hardening tools via tribological tests in different configurations. Boher [11] also observed in a tribosimulator that macroscopic material transfer takes the form of compacted powder with a morphology and chemical composition that is consistent with the coating particles.

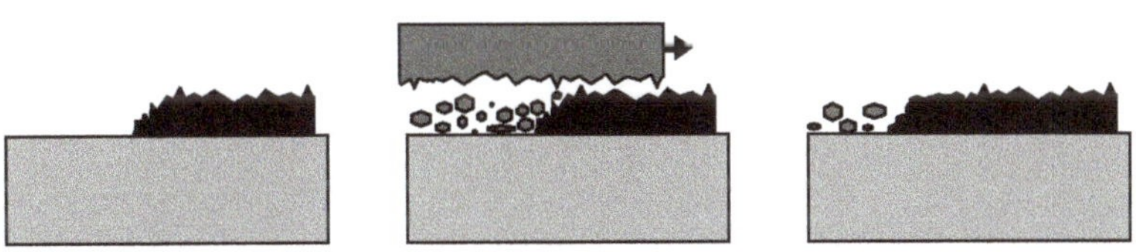

(a) (b) (c)

Figure 19. Mechanism of compaction of wear debris. Given a tool with surface features (**a**), loose debris becomes trapped between the tool and the component during forming, and pressed against these features (**b**), under conditions of pressure and temperature. This leads to further material being consolidated onto the existing feature (**c**).

The results discussed in Section 4 show strong evidence of this mechanism. Powder compaction on the front side of surface features has been directly observed in the strip drawing simulator (Figure 8a) and in the pilot environment (Figures 11 and 13). The resulting macroscopic features have been directly observed for the pilot environment (Figures 11 and 13), including verification through cross-sectional analysis and EDX chemical composition measurement. The observed morphology fits well with the proposed mechanisms in the published literature. Notably, Venema et al. [35] observed very similar results in a strip drawing configuration, including a cross-sectional analysis of transfer layers, with the results being essentially identical to those obtained here. Similar observations were made [17] in a U-shaped stamping tribosimulator resembling the omega tool configuration used in this work. Material transfer as the mechanical compaction of wear debris is therefore well documented in abstract laboratory tests, and it has also been confirmed to appear in the hot stamping of omega components in this work.

It is reasonable to extrapolate that the same mechanism is taking place on industrial tools. On the one hand, the morphology of the gross material transfer layers is the same as for the industrial tools and the pilot environment; this can be seen by comparing Figures 11 and 14. This is further confirmed through the chemical analysis of particles extracted from the transfer layer in an industrial tool (Table 2, compared to the chemical composition of the transfer layer in the pilot environment (Figure 10c).

The main implication is that in press hardening, the majority of material transfer occurs through geometric and mechanical causes. When severe sliding or plastic deformation occur, dust particles will fall from the coating (see Section 3.1). This mechanism is most intense over the drawing radius, where the two factors are combined. These loose particles will then accumulate onto any geometrical feature, leading to the formation of compacts. While the role of chemical affinity and metallurgical adhesion cannot be completely disregarded, this implies that surface modification leading to reduced affinity will not be able to completely prevent material transfer. This is consistent with observations on [36], where the material transfer during sliding was observed to be governed more by mechanical interaction than by chemical affinity (in that case, in the aluminium-tool steel tribopair).

This observation also offers a possible strategy for reducing the severity of material transfer. By tailoring the properties of the coating at the moment of formation, it may be possible to reduce flaking, and therefore, the amount of dust available to be compacted. In this regard, Kondratiuk et al. [23] reported a decrease in galling at high sliding temperatures. Venema [15] also report differences in the Coefficient of Friction at different sliding temperatures, as do Ghiotti et al. [37] and also Pelcastre et al. [16]. While these works do not offer a common explanation on the specific phenomena, it is still relevant that different setups and test conditions detect differences in friction behavior depending on the sheet temperature.

5.2. Compact vs. Sparse Material Transfer

The difference between compact, gross, and sparse material transfer is highly arbitrary, and is only proposed for the sake of discussion. However, it appears to be relevant that in some cases, material transfer appears in the shape of thick, compact layers, and in others, in sparse lumps with no macroscopic geometrical impact.

Sparse material transfer has been observed on industrial tools since the beginning of the press hardening of coated steel, and it has been reported in investigations concerning industrial tools [19,20]. This feature typically appears as a dull fringe on a flat tool surface, typically at middle-height on deep drawing walls. To this date, detailed observations of these fringes has only been performed through surface replication, as seen in Figure 17. The appearance of these fringes on pilot plant inserts resembling the deep drawing section of a B-Pillar allows them to be studied via SEM, as seen in Figure 13.

Sparse material transfer appears to take place in flat surfaces. This has been observed to be true for industrial tools, but also for pilot tools (Figure 13), and even for tribosimulator inserts (Figure 7). Works in the published literature using flat tribosimulators [16,23,35] also show wear mechanisms that are more similar to this sparse wear. On the contrary, test configurations based on drawing through a radius, such as the work by Boher et al. [11] resulted in thick, compact layers of dust accumulation.The main cause leading to the formation of thick, compact adhesion as opposed to this sparse material transfer appears to be purely geometrical/topological: gross material transfer appears in the radius, while sparse transfer appears on flat surfaces.

While intuitive, the underlying causes for this are not easy to pinpoint. On the one hand, it can be tied to the higher stress, pressure, and intensity of sliding generated by a drawing radius; indeed, increased material transfer has been classically reported, even in cold forming [38]. In the particular case of press hardening, this has been analytically studied by Deng et al. [39], using an Archard-based phenomenological model that accurately predicted increased material transfer (galling) in the radius of a pilot-scale tool.

On the other hand, the specific system of press hardening introduces an additional factor: the availability of dust to be compacted onto transfer layers. According to Section 3.1, coating particles are generated in higher numbers under intense sliding and plastic deformation, both taking place on a drawing radius: a greater availability of particles to compact will lead to more material transfer.

Finally, as discussed above, the contribution of temperature cannot be discarded. Works in the literature [15,35,40] consistently report on a change in dominant wear mechanisms at different temperatures, with increased "sparse" adhesion at lower temperatures. While this has not been directly observed in this work, it can be noticed that the zones of the omega component where sparse adhesion is observed contact later than the radius, implying a certain loss of temperature in the sheet. The exact reason for this possible change of behavior is unclear, and can be due to changes in the relative hardness levels of tool and coating, leading to different dominances of abrasion, or even to temperature-dependent properties of the coating [35], leading to different dust generation trends.

5.3. Initiation of Material Transfer

It is extremely difficult to design a single experiment that can isolate contributions from mechanical/topological and chemical/adhesive transfer mechanisms in an industry-resembling environment. Therefore, the initiation mechanism for this material transfer cannot be unequivocally demonstrated, but an educated guess can be deduced from the available information.

A discussion on ploughing (Figure 18) reveals how surface features will result in a dramatic acceleration of material transfer mechanisms. While a tool with a smooth surface finish does not offer anchoring points, surface damage mechanisms will quickly generate such features. This has been observed in the strip drawing simulator (Figure 8), where the bare tool steel showed sliding marks. It has also been observed in the pilot environment (Figure 13b), where bare tool steel shows a catalog of sliding damage as well as plastic deformation crimps and cracks.

Works in the literature report similar findings. Pelcastre et al. [16] investigated material transfer in strip drawing tests, performing a cross-sectional analysis of the worn components. Abrasive wear and plastic deformation of the tool surface played a significant role, with transfer layers forming on abrasion grooves, and even generating "intermixed transfer layers" where severe tool plastic deformation resulted in the interlocking of tool steel and material transfer. Similar mixed layer morphology was observed by Venema et al. [35], particulary when sliding at low temperature.

These same authors observed different severities of sparse material transfer when comparing hard (52 HRC) and soft (36 HRC) tool steel QRO90 in a pilot environment, in reference [18]. While micromechanisms could not be explored as the destructive analysis of the tools was not possible at that point, the observations are consistent with a lower degree of hardness leading to more surface damage, and therefore, more initiation points. Notably, this change in surface hardness led to an increased area coverage of sparse adhesion, but not to an increased thickness of material transfer in the vertical tool wall, while gross material transfer in the radius was basically equivalent. The main explanation for this is that a lower hardness leads to an increased amount of material transfer sites, leading to more coverage. On the other hand, the total amount of gross material transfer in a single location once anchor points are formed is governed by the generation of dust, which is most intense in the drawing radius.

This can be contrasted with other tests run in this same configuration by the same authors. In [41], the same pilot test configuration is used to run an 800-cycle production campaign comparing QRO90 and additive manufactured 1.2709 steel, both at an approximate hardness of 50 HRC. The results in that case show very similar sparse wear fringes, despite the significant differences in the chemical compositions of the two steels. Again, material transfer in the radius is similar in both cases, with thickness occurring in the 400 µm range. Finally, Ref. [24] uses the same configuration to compare QRO90 bare steel to hot work tool steel with a wear-resistant coating. While gross material transfer is mostly equivalent, sparse material transfer is much less severe on the coated tool. In this case, however, the coating itself is a confounding factor. On the one hand, the much harder surface resists abrasive wear and surface plastic deformation, leading to less initiation sites being formed. On the other, however, the coating is explicitly designed as an anti-adhesion coating, including a carbonaceous top-coat. As no destructive analysis was performed in that occasion, it is impossible to clearly understand which of the two factors contributes the most.

Surface damage leading to material transfer is consistent with discussions in the literature about similar tribosystems. As an example, in [36], it was observed that AlCrN-coated tool steel showed higher chemical-metallurgical adhesion with pure aluminum than bare tool steel. However, in sliding tests, material transfer was more severe for the latter. An observation of the acting micromechanisms showed that oxide particles had generated abrasion grooves on the bare tool steel, where material transfer began via mechanical means. On AlCrN-coated samples, this abrasive wear had not taken place, and material transfer could only happen through smearing of the soft aluminium, leading to thin blobs of smeared material transfer instead of the accumulation of dust.

6. Conclusions

In this work, wear mechanisms appearing in press hardening tools working with AlSi-coated boron steel have been studied by comparing the inspection of industrial tools, with results being obtained through laboratory tribology tests. Observations of these analyses have been compared across the different studied systems, and also with the open literature. The main conclusions which can be drawn are the following:

- The main wear mechanism appearing in the press hardening of AlSi-coated boron steel is material transfer. This material transfer is not only related to chemical interaction, but is also generated via mechanical means. The contribution of the latter is decisive in the growth of macroscopic wear features.
- The coating in AlSi-coated boron steel is hard and brittle, showing hardness levels (up to 14 GPa) that are well above a bare tool steel. Moreover, the heat treatment time does not affect the properties of each individual layer, but only their relative proportions.
- Particles can flake off the coating under different conditions, notably during sliding with an external surface (namely, the tool), or in locations of severe plastic deformation. The natural consequence of this is that drawing radii and slope changes are locations where dust generation will be particularly active.
- The compaction of wear debris from the coating is the main micromechanism resulting in the formation of macroscopic features on industrial tools. This phenomenon can readily occur on nucleation spots on the tool surface, even for those generated by other wear mechanisms. As a consequence, this material transfer is mostly based on mechanical and topological reasons. While it has been largely assumed that chemical affinity plays a role, it has not been possible to demonstrate this in the current work.
- The hard sheet metal coating and particles flaking from it are very aggressive towards the tool steel. Abrasion marks are generated extremely fast when direct sliding under pressure takes place, and local plastic deformation can also be observed. These defects will serve as anchor points for material transfer in the form of the accumulation of dust particles.

A follow-up of these conclusions is that material transfer is inherent in the system, and it will take place as long as the current configuration (alloyed AlSi coating, sliding at high temperature, and the absence of lubrication) is kept. However, at least according to the observations of this work, tweaking some of the aspects of the process will affect the severity of wear.

- Tool roughness needs to be kept to a minimum. This does not only involve a good initial surface finish, but also the means to protect this surface from the aggressive abrasive wear generated by the coating particles.
- Maximizing tool surface hardness should have a beneficial impact. On the one hand, it will slow down macroscopic abrasive wear and a loss of geometry. On the other hand, it will also hamper local abrasion and plastic deformation from creating anchor points for material transfer.
- While chemical adhesion has not been demonstrated in this work, a tool surface with low affinity for the coating will make it easier to dislodge adhesion during maintenance, or even through the cycles.

Ultimately, some aspects are left unanswered and could be the focus of future research. One such topic is the effect of system temperature on the dominance of the different wear mechanisms; while some discussion already exists in the literature, translating these observations into macroscopic behavior would be of great industrial relevance. Finally, this work could not prove the effects of chemical/metallurgical adhesion in material transfer in press hardening. Further research into this topic is of great interest, as it has major implications for the design of solutions to wear in press hardening.

Author Contributions: Conceptualization, J.P.; methodology, J.P., E.G.-L., N.C. and A.A.; validation, J.P.; formal analysis, J.P. and E.G.-L.; investigation, J.P., E.G.-L., N.C., G.R. and A.A.; resources, J.P.; data curation, J.P.; writing—original draft preparation, J.P.; writing—review and editing, E.G.-L., G.R., N.C., A.A., M.V. and D.C.; visualization, J.P., E.G.-L. and G.R.; supervision, J.P. and M.V.; project administration, J.P., M.V. and D.C.; funding acquisition, J.P., M.V. and D.C. All authors have read and agreed to the published version of the manuscript.

Funding: The results were partially obtained within the framework of Retos-Colaboración 2017 Project RTC-2017-6448-4 INNOESTAMP (Innovación en recubrimientos PVD para aplicaciones de estampación en caliente), granted by the Spanish Ministry of Science Innovation and University (Ministerio de Ciencia, Innovación y Universidades). Strip drawing simulator tests and the analysis of industrial tools was mainly performed in the framework of the project TestTool, which received funding from the European Commission Research Fund for Coal and Steel under grant agreement no RFCS-CT-2011-00023.

Data Availability Statement: Not applicable.

Conflicts of Interest: The authors declare no conflict of interest.

References

1. Merklein, M.; Lechler, J. Investigation of the thermo-mechanical properties of hot stamping steels. *J. Mater. Process. Technol.* **2006**, *177*, 452–455.
2. Nagathan, A.; Penter, L. Chapter 7: Hot Stamping. In *Proceedings of the Sheet Metal Forming—Processes and Applications* ; Altan, T., Tekkaya, A., Eds.; ASM International: Detroit, MI, USA, 2012; pp. 153–163.
3. Georgiadis, G.; Tekkaya, A.; Weigert, P.; Weiher, J.; Kurz, H. Investigations on the Manufacturability of Thin Press Hardened Steel Components. *Procedia CIRP* **2014**, *18*, 74–79.
4. Neugebauer, R.; Altan, T.; Geiger, M.; Kleiner, M.; Sterzing, A. Sheet metal forming at elevated temperatures. *CIRP Ann.-Manuf. Technol.* **2006**, *55*, 793–816. [CrossRef]
5. Karbasian, H.; Tekkaya, A. A review on hot stamping. *J. Mater. Process. Technol.* **2010**, *210*, 2103–2118. . atprotec.2010.07.019. [CrossRef]
6. Suehiro, M. Properties of Aluminum-coated Steels for Hot-forming. *Nippon. Steel Tech. Rep.* **2003**, *88*, 16–21.
7. Fan, D. A review of the physical metallurgy related to the hot press forming of advanced high strength steel. *Steel Res. Int.* **2009**, *80*, 241–248.
8. Grigorieva, R.; Drillet, P.; Mataigne, J.M.; Redjaïmia, A. Phase Transformations in the Al-Si Coating during the Austenization Step. *Solid State Phenom.* **2011**, *174–174*, 748–790.
9. Allély, C.; Dosdat, L.; Clauzeau, O.; Ogle, K.; Volovitch, P. Anticorrosion mechanisms of aluminized steel for hot stamping. *Surf. Coatings Technol.* **2014**, *238*, 188–196. . [CrossRef]
10. Pelcastre, L.; Hardell, J.; Prakash, B. Galling mechanisms during interaction of tool steel and Al–Si coated ultra-high strength steel at elevated temperature. *Tribol. Int.* **2013**, *67*, 263–271. [CrossRef]

11. Boher, C.; Roux, S.L.; Penazzi, L.; Dessain, C. Experimental investigation of the tribological behavior and wear mechanisms of tool steel grades in hot stamping of a high-strength boron steel. *Wear* **2012**, *294–295*, 286–295. [CrossRef]
12. Ghiotti, A.; Sgarabotto, F.; Bruschi, S. A novel approach to wear testing in hot stamping of high strength boron steel sheets. *Wear* **2013**, *302*, 1319–1326.
13. Hardell, J. High Temperature Tribology of High Strength Boron Steel and Tool Steels. Ph.D. Thesis, Luleå Tekniska Universitet, Luleå, Sweden, 2009.
14. Pelcastre, L. High Temperature Galling: Influencing Parameters and Mechanisms. Ph.D. Thesis, Luleå University of Technology, Luleå, Sweden , 2013.
15. Venema, J.; Atzema, E.; Hazrati, J.; Matthews, D.; van den Boogaard, T. Modelling of Friction in Hot Stamping. *Procedia Manuf.* **2020**, *47*, 596–601.
16. Pelcastre, L.; Kurnia, E.; Hardell, J.; Decrozant-Triquenaux, J.; Prakash, B. High temperature tribological studies on hardfaced tool steels for press hardening of Al-Si coated boron steel. *Wear* **2021**, *476*, 203728. [CrossRef]
17. Venema, J.; Stache, R.; Kotzian, M. Hot Strip Draw Wear Test and Influence of Tool Material on Adhesive and Abrasive Wear. In Proceedings of the 8th International Conference on Hot Sheet Metal Forming of High-Performance Steel CHS2 2022, Barcelona, Spain, 30 May–2 June 2022 ; Oldenburg, M., Hardell, J., Casellas, D., Eds.; Verlag Wissenschaftliche Scripten: Auerbach, Germany, 2022; pp. 171–178, ISBN 978-3-95735-150-0.
18. Pujante, J.; Garcia-Llamas, E.; Casellas, D. Study of Wear in Press Hardening Using a Pilot Facility. In Proceedings of the 7th International Conference on Hot Sheet Metal Forming of High-Performance Steel CHS2 2019, Luleå, Sweden, 2–5 June 2019; Oldenburg, M., Hardell, J., Casellas, D., Eds.; Verlag Wissenschaftliche Scripten: Auerbach, Germany, 2019; pp. 151–158, ISBN 978-3-95735-104-3.
19. Pujante, J.; Vilaseca, M.; Eriksson, K.; Clobes, J.; Alsmann, M.; Casellas, D. Wear mechanism identification on hot stamping tools. In Proceedings of the 3rd International Conference on Hot Sheet Metal Forming of High-Performance Steel CHS2 2011, Kassel, Germany, 13–16 June 2011; Oldenburg, M., Prakash, B., Steinhoff, K., Eds.; Verlag Wissenschaftliche Scripten: Auerbach, Germany, 2011; pp. 377–384, ISBN 978-3-942267-17-5.
20. Pelcastre, L.; Hardell, J.; Herrera, N.; Prakash, B. Investigations into the damage mechanisms of form fixture hardening tools. *Eng. Fail. Anal.* **2012**, *25*, 219–226. [CrossRef]
21. Vilaseca, M.; Pujante, J.; Ramírez, G.; Casellas, D. Investigation into adhesive wear of PVD coated and uncoated hot stamping production tools. *Wear* **2013**, *308*, 148–154. [CrossRef]
22. Oliver, W.; Pharr, G. An improved technique for determining hardness and elastic modulus using load and displacement sensing indentation experiments. *J. Mater. Res.* **1992**, *7*, 1564–1583. [CrossRef]
23. Kondratiuk, J.; Kuhn, P. Tribological investigation on friction and wear behaviour of coatings for hot sheet metal forming. *Wear* **2011**, *270*, 839–849. [CrossRef]
24. Azkona, I.; Pujante, J.; Garcia-Llamas, E.; Caro, J.; Orrit-Prat, J.; Bonet, R. A Novel PVD Coating for Wear Reduction in Press Hardening Tools. In Proceedings of the 8th International Conference on Hot Sheet Metal Forming of High-Performance Steel CHS2 2022, Barcelona, Spain, 30 May–2 June 2022; Oldenburg, M., Hardell, J., Casellas, D., Eds.; Verlag Wissenschaftliche Scripten: Auerbach, Germany, 2022; pp. 513–520, ISBN 978-3-95735-150-0.
25. Nilsson, L.; Ohlsson, R. Accuracy of replica materials when measuring engineering surfaces. *Int. J. Mach. Tools Manuf.* **2001**, *41*, 2139–2145. [CrossRef]
26. Jonsson, P. Sheet Metal Trimming Dies-Characterisation Methods of Geometry and Surface Topography and Influence on Wear. Diploma Thesis, Chalmers University of Technology, Gothenburg, Sweden, 2010.
27. Fan, D.; de Cooman, B. State-of-the-Knowledge on Coating Systems for Hot Stamped Parts. *Steel Res. Int.* **2012**, *83*, 412–433. [CrossRef]
28. Kiani-Rashid, A.; Edmonds, D. Phase transformation study of aluminium-containing ductile cast irons by dilatometry. *Mater. Sci. Eng. A* **2008**, *481–482*, 752–756. [CrossRef]
29. Gui, Z.X.; Wang, K.; Zhang, Y.S.; Zhu, B. Cracking and interfacial debonding of the Al–Si coating in hot stamping of pre-coated boron steel. *Appl. Surf. Sci.* **2014**, *316*, 595–603. [CrossRef]
30. Archard, J. Contact and Rubbing of Flat Surfaces. *J. Appl. Phys.* **1953**, *24*, 981–988. [CrossRef]
31. Heinrichs, J.; Jacobson, S. The influence from shape and size of tool surface defects on the occurrence of galling in cold forming of aluminium. *Wear* **2011**, *271*, 2517–2524.
32. Menezes, P.; Kishore, S.; Kalias, V. Studies on friction and transfer layer: Role of surface texture. *Tribol. Lett.* **2006**, *24*, 265–273. [CrossRef]
33. Pelcastre, L.; Hardell, J.; Courbon, C.; Prakash, B. Tribological behaviour of Al-Si-coated ultra-high-strength steel during interaction with tool steel at elevated temperatures: Influence of tool steel surface topography parameters on galling. *Proc. Inst. Mech. Eng. Part B J. Eng. Manuf.* **2015**, *229*, 1373–1384. [CrossRef]
34. Pelcastre, L.; Hardell, J.; Prakash, B. Investigations into the occurrence of galling during hot forming of Al–Si-coated high-strength steel. *Proc. Inst. Mech. Eng. Part J J. Eng. Tribol.* **2011**, *225*, 487–498. [CrossRef]
35. Venema, J.; Hazrati, J.; Matthews, D.; Stegeman, R.; van den Boogaard, A. The effects of temperature on friction and wear mechanisms during direct press hardening of Al-Si coated ultra-high strength steel. *Wear* **2018**, *406–407*, 149–155. [CrossRef]

6. Pujante, J.; Vilaseca, M.; Casellas, D.; Riera, M.D. The Role of Adhesive Forces and Mechanical Interaction on Material Transfer in Hot Forming of Aluminium. *Tribol. Lett.* **2015**, *59*, 10. [CrossRef]
7. Ghiotti, A.; Bruschi, S.; Medea, F.; Hamasaiid, A. Tribological behavior of high thermal conductivity steels for hot stamping tools. *Tribol. Int.* **2016**, *97*, 112–122. [CrossRef]
8. Schedin, E. Galling mechanisms in sheet forming operations. *Wear* **1994**, *179*, 123–128. [CrossRef]
9. Deng, L.; Pelcastre, L.; Hardell, J.; Prakash, B.; Oldenburg, M. Numerical investigation of galling in a press hardening experiment with AlSi-coated workpieces. *Eng. Fail. Anal.* **2019**, *99*, 85–96. [CrossRef]
0. Medea, F.; Venturato, G.; Ghiotti, A.; Bruschi, S. Tribological performances of new steel grades for hot stamping tools. *J. Phys. Conf. Ser.* **2017**, *896*, 012047. [CrossRef]
1. Pujante, J.; González, B.; Garcia-Llamas, E. Pilot Demonstration of Hot Sheet Metal Forming Using 3D Printed Dies. *Materials* **2021**, *14*, 5695. [CrossRef] [PubMed]

lubricants

MDPI

Article

Mechanical Properties and Tribological Study of Bottom Pouring Stir-Cast A356 Alloy Reinforced with Graphite Solid Lubricant Extracted from Corn Stover

Vavilada Satya Swamy Venkatesh * and Pandu Ranga Vundavilli

School of Mechanical Sciences, Indian Institute of Technology Bhubaneswar, Bhubaneswar 752050, India; pandu@iitbbs.ac.in

* Correspondence: pdfme24003@iitbbs.ac.in

Abstract: The present work epitomises extracting the graphite (Gr) solid lubricant from the corn stover. The extracted Gr was incorporated as reinforcement in the A356 alloy (Al-7Si), and the effect of the Gr particles on the mechanical and tribological properties was investigated. In spite of this, the input process parameters for the dry sliding wear test at room temperature against the EN31 steel disc were optimised through ANOVA analysis. The fabricated A359—X wt% (X = 0, 2.5, 5, 7.5) composite through bottom pouring stir casting techniques was analysed microstructurally by using XRD and FESEM analysis. The micro Brinell hardness and tensile strength were investigated per ASTME10 and ASTME8M standards. A wear test was performed for the composite pins against the EN31 steel disc according to ASTM G99 specifications. The XRD analysis results depict the presence of carbon (C), aluminium (Al), and silicon (Si) in all the wt% of the Gr reinforcement. However, along with the elements, the Al_2Mg peak was confirmed for the A356—7.5 wt% Gr composite and the corresponding cluster element was confirmed in FESEM analysis. The maximum micro Brinell hardness of 92 BHN and U.T.S of 123 MPa and % elongation of 7.11 was attained at 5 wt% Gr reinforcement due to uniform Gr dispersion in the A356 alloy. Based on the ANOVA analysis, the optimal process parameters were obtained at 20 N applied load, 1 m/s sliding velocity, and 1000 m sliding distance for the optimal wear rate of 0.0052386 g/km and 0.364 COF.

Keywords: A356 alloy; Gr reinforcement; corn stover; bottom pouring stir casting; tribological analysis

1. Introduction

Wear is commonly observed in all the components moving relative to one another such as automobile connecting rods, pistons, bearings, and cylinder bores. Hence, the consideration of wear is crucial while designing with these elements to ensure superior and accurate performance in all tribological applications [1]. Metal matrix composites (MMCs) have been shown to have improved wear properties, as well as higher specific strength and stiffness than monolithic alloys. As a result, it is important in tribology to investigate MMC wear characteristics from several different perspectives [2].

Aluminium alloys are widely used in the automobile sector given their inherent properties like being lightweight and having high specific strength, better corrosion resistance, and malleable characteristics [3]. The hypereutectic Al-Si alloy with a silicon content of 17% is known to exhibit remarkable wear resistance and load-bearing properties [4]. The mechanical and tribological properties of the Al-Si alloy are modified by reinforcing the primary and secondary reinforcements such as B_4C, Al_2O_3, MgO, SiC, Zirconium Silicate ($ZrSiO_4$), and Gr [5–7] in the matrix. Among these reinforcements, graphite reinforcement was abundantly used to synthesize the aluminium composite to fabricate the brake discs and bearing surface due to the possessiveness of the excellent self-lubricating properties, enhanced durability, and corrosion resistance [5,8,9]. Jaswinder [10] incorporated 6 wt% graphite in the A359-10%SiC composite, and results reveal that the added graphite lowered the wear and frictional properties of the A356-Gr-SiC composite. Chou Shang-Nan [11] analysed the squeeze-casted Al-Si-Al_2O_3 hybrid composite and concluded that a maximum bending strength and fracture toughness of 443 MPa and 11.35 MPa $m^{1/2}$ were observed at 5 wt% Al_2O_3 reinforcement. Haji Zamanai [12] studied the stir-cast Al-Si-Al_2O_3-10%ZrO_2 composite. The experimentation confirms that the incorporation of alumina oxide and zirconium oxide improves the compressive strength, yield strength, and tensile strength of the hybrid composite.

Several researchers studied the mechanical and wear properties of the composites reinforced with self-lubricated solid lubricants such as graphite, magnesium, molybdenum disulphide, boron nitride, and other particles. However, nanoparticles are characterised by high cost and difficulty in the fabrication process, limiting their usage in composite synthesis. Generally, solid lubricants extracted from the coal mines cause drastic pollution, soil erosion, and environmental degradation. In addition to this, the extraction process releases large quantities of methane and CO_2 gasses, exacerbating climate change. Hence, in this study, graphite particles were extracted from corn stovers which were abundantly available in the Sambalpur area in the western region of Odisha. The pyrolysis process was used for graphite extraction and the extracted graphite was treated with hydrofluoric acid (HF) to enhance purity. The high-purity graphite was used as reinforcement in the A356 alloy to study its effect on the mechanical and tribological properties. The input process parameters in the tribology test such as applied load, sliding speed, and sliding distances were optimised against the output responses such as wear rate and COF.

2. Materials and Composite Synthesis

The A356 matrix powder ($\leq$20 μm) was procured from Nano Shell Enterprises, Chennai. The elemental constituents present in the A356 powder are shown in Table 1. The solid lubricant was extracted from the corn stover, and the extracted graphite was added as reinforcement in the A356 (Al-7Si) alloy. The elemental morphology through FESEM and the EDX spectra for the extracted Gr and A356 alloy are displayed in Figure 1. The SEM micrograph of Gr confirms the hexagonal planar structure (refer to Figure 1a) and no elements other than carbon were identified in the EDX spectra for Gr, as shown in Figure 1b.

Table 1. Elemental constituents in A356 alloy.

Element	Si	Ti	Mg	C	Fe	Remaining
Wt%	6.56	0.06	0.36	1.65	0.08	Aluminium

Figure 1. *Cont.*

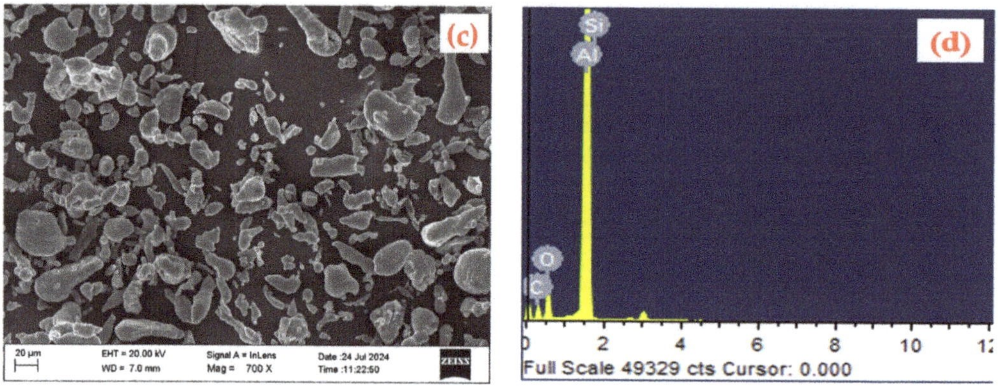

Figure 1. (**a**,**b**) FESEM and EDX spectra for Gr particles, and (**c**,**d**) FESEM and EDX for A356 alloy.

Table 2 shows the composites' compositions. The composites were manufactured using the bottom pouring stir casting technique. The 250 g Al-7Si alloy rods were initially heated and melted in the muffle furnace at 700 °C. The molten Al-7Si slurry was vigorously stirred for a duration of 5 min using a graphite stirrer rotating at a speed of 500 rpm [2,13]. The graphite particles were preheated at 130 °C and then added to the molten metal and stirred at 500 rpm for 5 min for the uniform dispersion of the Gr particles. Thereafter, the composite slurry was added to the mild steel die which was placed below the melting furnace. The schematic diagram for the bottom pouring stir casting and the fabricated composites for the tensile test are depicted in Figure 2a,b, respectively. For microstructural investigation, the casted specimens were polished with different grade emery papers ranging from 200 mesh size to 2300 mesh size, and then polished with the polishing cloth by applying diamond paste size varying from 6 µm to 12 µm, 3 µm to 4 µm, and 0.5 µm to 1 µm. The polished specimens were subjected to etching in Keller's for 30 s for the proper visualization of the microstructures.

Table 2. Synthesized bottom pouring stir-cast A356-Gr composites.

Composite Code	wt% of A356	wt% of Graphite Reinforcement
A356	100	0
C1	97.5	2.5
C2	95	5
C3	92.5	7.5

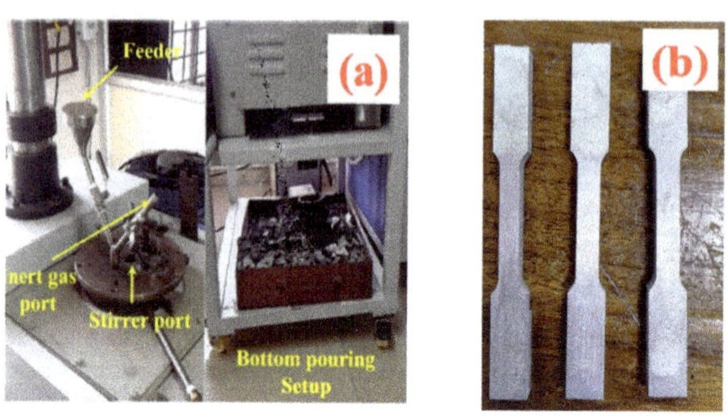

Figure 2. (**a**) Bottom pouring stir casting setup. (**b**) Fabricated tensile test specimens.

The compression strength and ultimate tensile strength (U.T.S) were measured on the M30 model universal testing machine as per ASTME9 [10] and ASTME8M standards, respectively [7,14]. The Brinell hardness was measured on an MD300 hardness tester according to ASTME10 specifications with the application of a 50 kg load by using a 2.5 mm dia ball indenter. The wear test at room temperature was performed on a DUCOM tribometer with an EN-31 hardened steel disc. The composite pins were prepared according to ASTM G99 standards [15]. The tribology tests were performed at three process parameter levels as tabulated in Table 3. The process parameters were optimised by using the ANOVA (Analysis of Variance) approach against the minimum wear and COF.

Table 3. Input tribological process parameters for ANOVA analysis.

Applied Load (AL in Newtons)	Sliding Distance (SL in Meters)	Sliding Velocity (SL in m/s)
20	1500	1
30	2000	2
30	2500	3

3. Results and Discussion

3.1. XRD Analysis and Microstructural Study

The optical microscope images for the cast A359—2.5 wt% Gr, A359—5 wt% Gr, and A359—7.5 wt% Gr composites are shown in Figure 3a–c. The images were captured under the same magnification for comparison. The graphite particles are identified as black-coloured spots in all composites, and the percentage of these particles is found to improve with Gr reinforcement percentage. The XRD patterns for the composite reveal graphite (Gr), silicon (Si), and aluminium particles at the Gr wt% of 2.5% and 5%. Despite this, oxygen (O), magnesium (Mg), and Al_2Mg elements were detected in the XRD pattern at 7.5% Gr reinforcement, as shown in Figure 4. These added magnesium elements for hydrogen absorption eliminate the blow holes during casting, react with the aluminium matrix, and form the Al_2Mg agglomeration at 7.5 wt% Gr addition. In addition to this, the added Gr reacts with the entrapped oxygen during the solidification process and forms graphene oxide (Gro).

The SEM images of the composites are depicted in Figures 5 and 6. The Gr reinforcements were dispersed uniformly in the A356 alloy up to the addition of 5 wt% Gr due to the uniform stirring and the proper interface bonds in the Gr and A356 alloy particles. However, Al_2Mg clusters were attained in the 7.5 wt% graphite composite, and the corresponding cluster was confirmed in the Al356—7.5 wt% Gr XRD pattern (refer to Figure 4). The disparity in the thermal properties of A356 and the generated Al_2Mg agglomeration causes weakening in the composite bonding strength and load-bearing capability. The EDX spectra for the composites reinforced with 2.5 wt% and 5 wt% Gr reveal the presence of C, Si, and aluminium, as shown in Figure 5a–d. The presence of magnesium (Mg) and oxygen (O) was confirmed at 7.5 wt% Gr addition, which shows the clusters of Al_2Mg and GrO in the A356—7.5%Gr composite, as depicted in Figure 6.

3.2. Hardness of the A356-Gr Composite

The Brinell microhardness of the casted samples was investigated by averaging the five readings taken at different locations on the surface, and the results are shown in Figure 7. It was found that the microhardness improved with the Gr reinforcements up to 5 wt%, and the maximum Brinell hardness of 92 BHN was noticed for A356—5 wt% Gr composite. The hardness value gradually decreased after the addition of 5 wt% Gr reinforcement. There was an improvement in the 6.1%, 13.5%, and 9.87% BHN for the addition of 2.5 wt%, 5 wt%, and 7.5 wt% reinforced Gr compared to the A356 alloy. The maximum hardness of the A356—5 wt% composite was attributed to the even distribution of the graphite particles and the strong cohesive interface bonding among the graphite and A356 particles [16]. The minimal porosity levels lead to strengthened load transfer along the grain boundaries of neighbouring particles, which makes hardness higher at 5 wt% graphite reinforcement. The reduction in the microhardness at 7.5 wt% Gr reinforcement can be attributed to two factors. The presence of brittle Al_2Mg clusters promotes different expansions than the A356 and Gr particles, which promotes the weak interfaces and micropores, and causes the composite to have a hardness lower than the 5 wt% Gr reinforcement.

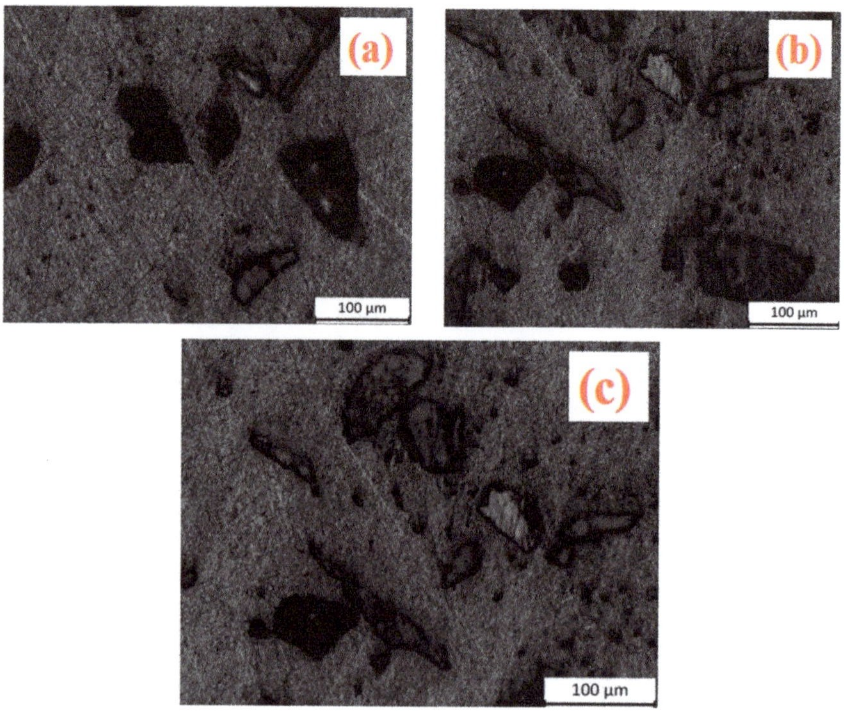

Figure 3. Optical microscope images for the (**a**) A359—2.5 wt% Gr, (**b**) A359—5 wt% Gr, and (**c**) A359—7.5 wt% Gr composites.

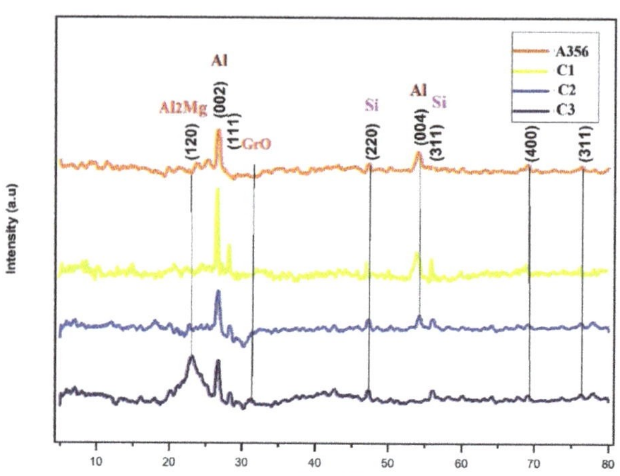

Figure 4. XRD pattern for the A356, C1, C2, and C3 composites.

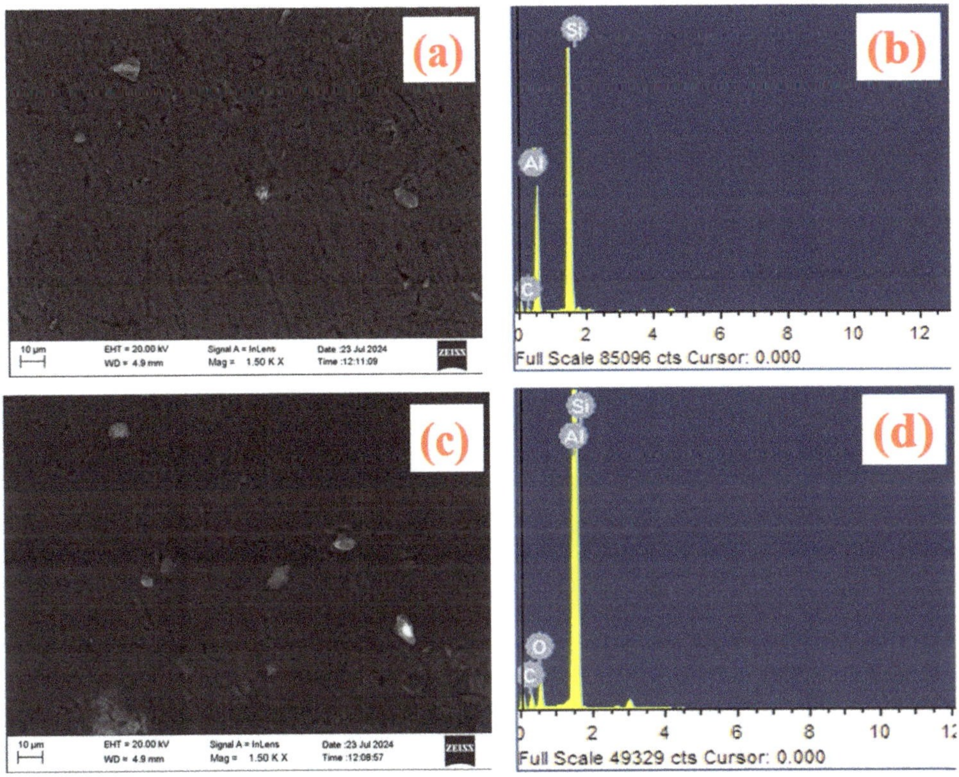

Figure 5. FESEM and EDS spectra for the (**a**,**b**) A356—2.5 wt% Gr and (**c**,**d**) A356-5 wt% Gr composites.

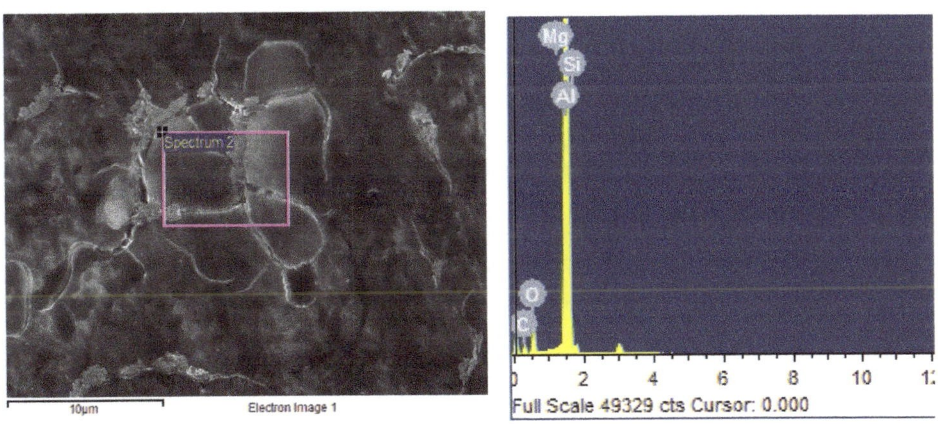

Figure 6. FESEM and EDX for the clusters in the A356—7.5 wt% Gr composite.

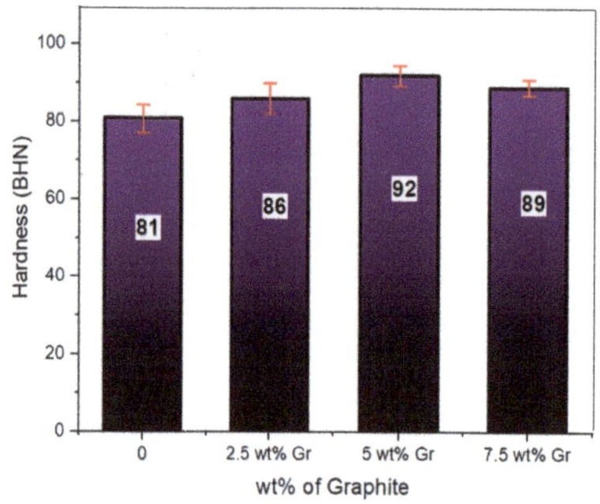

Figure 7. Hardness of the cast composites.

3.3. Tensile Strength and % Elongation of A356—Gr Composite

The U.T.S and percentage of elongation of the casted specimens are shown in Figure 8. The U.T.S and % elongation were considerably enhanced by 18.72% and 48.12% for the A356—5 wt% Gr composite compared to the A356 alloy. The highest values of 123 MPa compression strength and 7.11% percentage of elongation were attained at 5 wt% Gr reinforcement, and U.T.S and % elongation values were found to be decreased to 109 MPa and 6.8% for the A356—7.5wt% Gr composite. Experimentation on the incorporation of graphite particles strengthens the aluminium alloy. The decrement in the U.T.S and % elongation over the addition of 5 wt% Gr is due to the addition of the increased Gr weight percentage, accelerating the generation of a large number of slip planes at the grain boundaries [17,18]. During the application of tensile loads, the atoms find the easiest path to displace along the slip planes which leads to plastic deformation at the lower intensity of applied tensile loads. This slip phenomenon in the A356—7.5 wt% Gr composite tends to lower U.T.S and % elongation more than the A356 composite reinforced with 5 wt% Gr particles. The presence of various elastic characteristics of Al_2Mg, A356, and graphite generates a tri-axial state of stress, which accelerates crack initiation and propagation at the grain boundaries and leads to the early-stage failure of the composite with 7.5 wt% Gr [19,20].

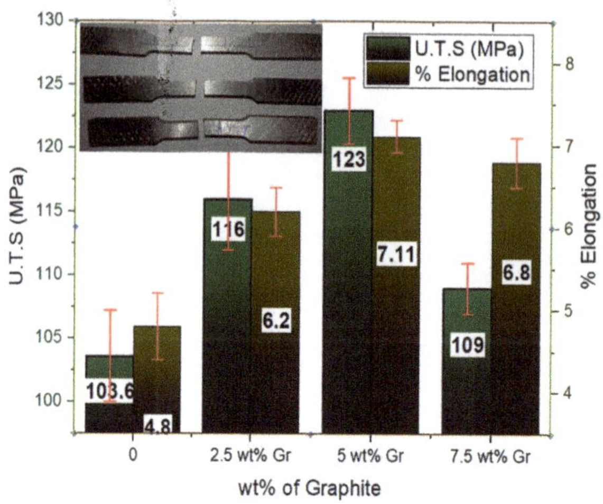

Figure 8. U.T.S and % elongation of the cast composites.

3.4. Tribological Behaviour of the A356-Gr Composite

The wear track used and the synthesized A356—Gr composite pins for the tribology test are depicted in Figure 9a,b. The wear test was conducted at 10 N, 20 N, 30 N, and 40 N applied loads, 1000 m sliding distance, and 1 m/s sliding velocity. The variation in the COF and wear rate at the constant sliding distance and loads is depicted in Figure 10. It was found that the COF and wear rate were increasing with an increase in the applied loads for all wt% of Gr reinforcements. The composite with 7.5 wt% Gr content exhibits minimal COF and wear loss compared to the composite reinforced with 2.5 wt% and 5 wt% Gr particles. The decreased wear rate and COF were attributed to the self-lubricating properties of the extracted graphite particles [9,21].

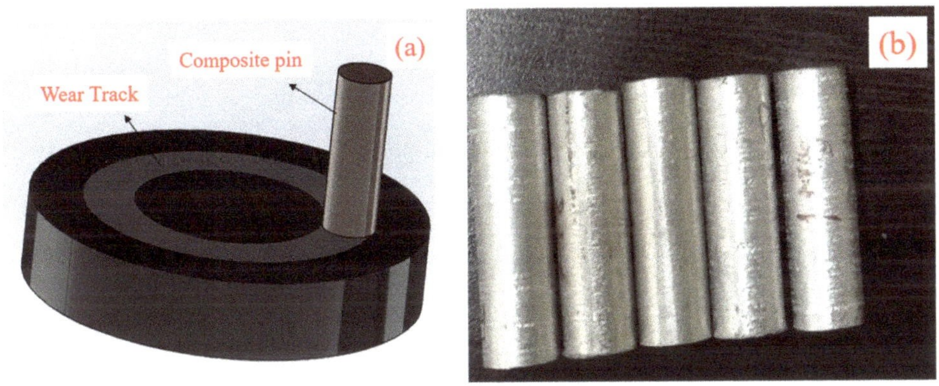

Figure 9. (**a**) Wear track for the tribology test and (**b**) fabricated composite pins.

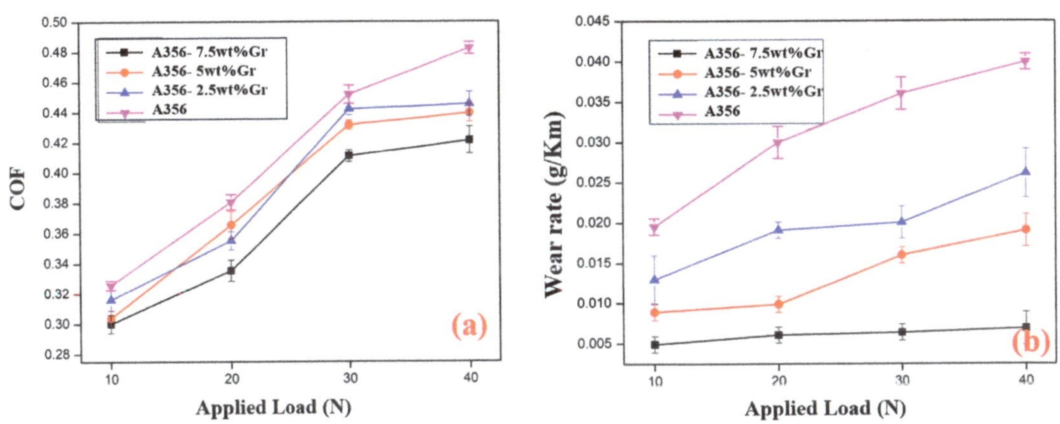

Figure 10. Variation in (**a**) COF and (**b**) wear rate for the cast composites.

The maximum wear rate of 0.040 g/km was observed for the unreinforced A356 alloy which was higher than all wt% Gr-reinforced A356 alloys. By increasing the Gr wt% from 2.5 wt% to 7.5 wt%, the wear rate was found to be decreased due to the existence of self-lubricated Gr particles. A wear rate of 0.041 g/km, 0.026 g/km, and 0.0056 g/km was attained for the A356—2.5 wt% Gr, A356—5 wt% Gr, and A356—7.5 wt% Gr composite, which was 16.3%, 19.8%, and 21.6% lower than the unreinforced A356 alloy. The decrease in the wear rate was attributed to the continuous rubbing of the protruding Gr particles against the disc surface, leading to the improved contact between the composite pin and disc which was prone to the generation of the lubricated tribo-layer, resulting in the minimal wear rate of the composite. The minimum wear rate for the 7.5 wt% Gr composite was due to the higher hardness of the Gr particles compared to the alloy at 5 wt% Gr particles. In addition to this, the generated frictional heat, due to the interaction between the pin's surface and disc, led to the oxidation of the pin surface. As a consequence, the reduced metal-to-metal interface contact led to a minimal wear rate at 7.5 wt% Gr reinforcement

It was inferred that the COF for all the samples was found to increase gradually with an increase in the applied loads. The minimal COF for the A356—7.5 wt% Gr resulted from the developed mechanically mixed layer (MML) at higher applied loads. The MML lowered the metal-to-metal contact and accelerates the shearing action of the composite pin surface. The generated heat during the shearing led to the detachment of the particles from the surfaces and generated MML, which contains elements from both the pin's and disc's surface [22,23]. The MML diminishes the COF during the rubbing of the composite pin against the disc.

The worn surface morphology for the pin surface at a higher applied load of 40 N, sliding velocity of 1 m/s, and sliding distance of 1000 m for the different Gr wt% is shown in Figure 11. The presence of deeper and longer grooves parallel to the sliding direction was observed in all the composites, which confirms the abrasive wear phenomenon. However, there were notable distinctions in the wear morphology of the base alloy and composite surface. In the A356 alloy, deeper grooves containing shallow craters were present, as depicted in Figure 11a. This was due to the continuous contact between the harder disc surface and the softer aluminium alloy [24]. The harder particles on the disc surface penetrated into the base alloy, leading to rigorous surface damage and generating deeper and longer grooves on the pin surface. Moreover, delamination wear and fine grooves were observed on the composites reinforced with 2.5 wt%, 5 wt%, and 7.5 wt% Gr particles. The presence of the Gr particles and the Si content in the A356 alloy led to the generation of microcracks at the higher magnitude of applied loads [25–27]. The developed subsurface microcracks propagated and met each other, eventually causing shearing action on the pin surface and resulting in the detachment of material from the pin surface and leading to delamination wear, as shown in Figure 11b–d.

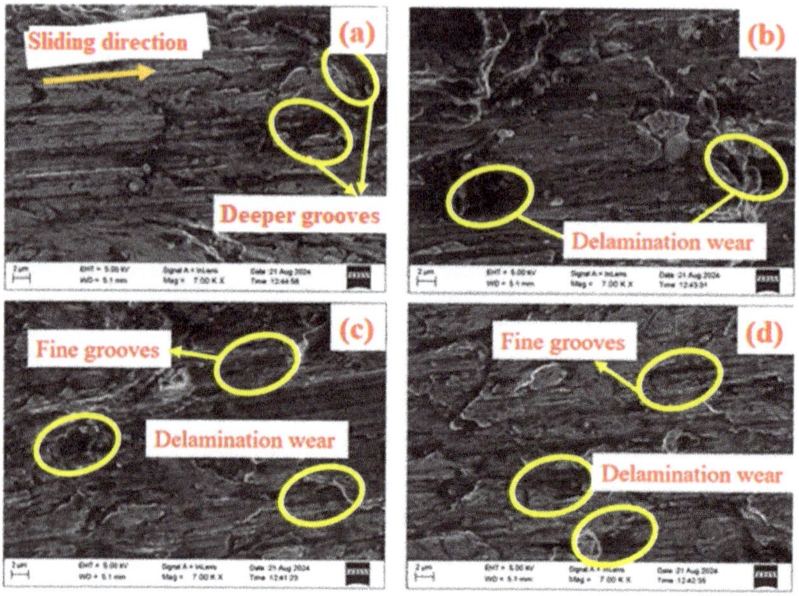

Figure 11. SEM micrographs for the worn surface at 1000 m sliding distance and 40 N applied load (**a**) sliding direction (**b**) Delamination wear (**c**,**d**) Fine grooves.

3.5. ANOVA Design of Experiments for Tribological Analysis

The maximum mechanical properties such as U.T.S and % of elongation were attained for the A356—5 wt% Gr composite. Hence, the composite specimen containing 5 wt% Gr was selected for the tribological analysis with a Taguchi L9 orthogonal array. The input parameters such as applied load, sliding velocity, and sliding distance were optimized against the COF and wear rate. The obtained output responses for the experimentation are shown in Table 4.

Table 4. Input process parameters and output responses.

Run Number	Applied Load (N)	Sliding Distance (m)	Sliding Velocity (m/s)	COF	Wear Rate (g/km)
1	20	1500	1	0.008	0.26
2	20	2000	2	0010	0.28
3	20	2500	3	0.012	0.29
4	30	1500	2	0.036	0.3
5	30	2000	3	0.039	0.31
6	30	2500	1	0.04	0.318
7	40	1500	3	0.042	0.322
8	40	2000	1	0.046	0.342
9	40	2500	2	0.052	0.36

The ANOVA (at 95% confidence level) for the wear rate response is shown in Table 5. It was concluded that applied load has attained the maximum sum of squares, and its effects are most influential, followed by sliding distance and sliding velocity. The *p*-value (allowable range 0 to 0.005) is used to determine the essential input factors. The Analysis of Variance table indicates that load and sliding distance are the two most influential variables that affect wear rate. Sliding velocity, with a *p*-value greater than 0.05, is considered a non-influential parameter in predicting wear rate. For the given selected output response (wear rate), the R^2 value of 0.9964 was closely aligned with the adjusted R^2 value of 0.9855. The R^2 predicted (0.9268) and R^2 adjusted (0.9855) are aligned closely and in favourable limits. Therefore, this model is capable of predicting the wear rate within the specified parameter ranges.

Table 5. ANOVA for wear rate.

Source	DOF	Adj SS	Adj MS	F-Value	*p*-Value
Applied Load (AL)	2	0.006273	0.003136	227.65	0.004
Sliding Distance (SD)	2	0.001244	0.000622	45.13	0.022
Sliding Velocity (SV)	2	0.000081	0.000040	2.94	0.254
Error	2	0.000028	0.000014		
Total	8	0.007625			

Model Summary: S R-sq R-sq(adj) R-sq(pred); 0.0037118, 99.64%, 98.55%, 92.68%.

The ANOVA for the COF (@ 95% confidence level) is illustrated in Table 6. It was clear that the applied load has the highest sum of square values and the most significant impact on the COF, surpassing sliding distance and velocity. Similar to the wear rate, the *p*-value (0 to 0.05) is used to determine the selection of the influential input parameters. According to the *p*-value in the Analysis of Variance, the most significant factors influencing the COF are applied load and sliding distance. The sliding velocity, which has a *p*-value larger than 0.05, is deemed a non-significant parameter in predicting the COF. The predicted R^2 value for COF (0.9332) is close to the R^2 adjusted (0.9848). Hence, this selected model is within the range of the parameters that can predict the COF.

Table 6. ANOVA for COF.

Source	DOF	Adj SS	Adj MS	F-Value	*p*-Value
Applied Load (AL)	2	0.002217	0.001108	255.77	0.004
Sliding Distance (SD)	2	0.000054	0.000027	6.23	0.038
Sliding Velocity (SV)	2	0.000005	0.000002	0.54	0.650
Error	2	0.000009	0.000004		
Total	8	0.002284			

Model Summary: S R-sq R-sq(adj) R-sq(pred); 0.0020817, 99.62%, 98.48%, 92.32%.

The obtained quadratic equations for the COF (1) and wear rate (2) are provided below.

$$COF = -0.0350 + 0.001833 \times AL + 0.000006 \times SD - 0.00017 \times SV \quad (1)$$

$$WR = 0.1541 + 0.003233 \times AL + 0.000029 \times SD + 0.00033 \times SV \quad (2)$$

The actual and predicted value graphs for the wear rate and COF are plotted in Figure 12a,b. The obtained points are very close to the straight line and the value falls under the acceptable range. In addition to this, the data points are well distributed, indicating an expected correlation among the predicted and actual responses.

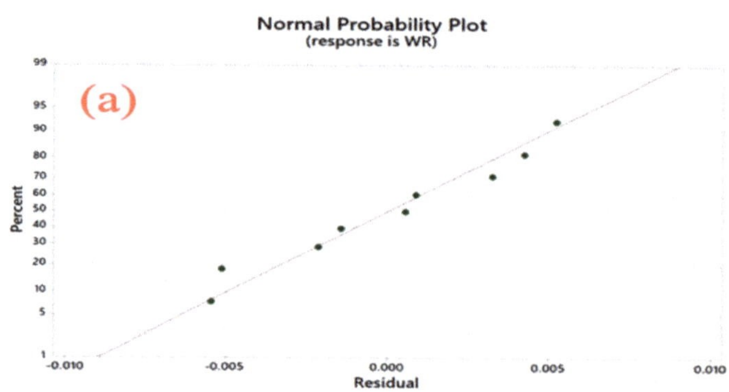

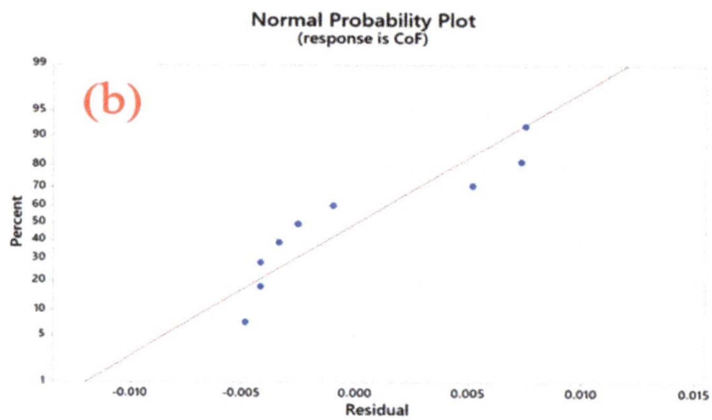

Figure 12. Actual and predicted values for the (**a**) wear rate and (**b**) COF.

The residual plots for the COF and wear rate are shown in Figure 13a,b. The residue plots demonstrated that all the residues ran between −0.0075 and +0.0075 for COF and between −0.0050 and +0.0050 for the wear rate, and there was no discernible pattern observed for both COF and wear rate. The three-dimensional graph for the effect of input process parameters on COF and wear rate under dry sliding conditions is shown in Figures 14 and 15. The selected process parameters in this experimentation were applied loads of 20 N, 30 N, and 40 N, sliding distances of 1500 m, 2000 m, and 2500 m, and sliding velocities of 1 m/s, 2 m/s, and 3 m/s.

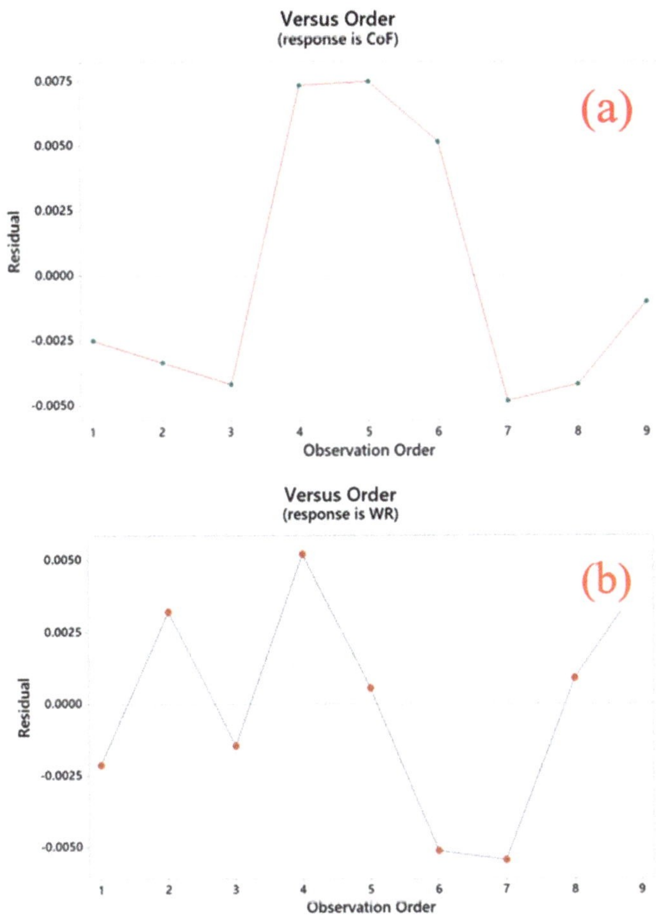

Figure 13. Residue vs. run for (**a**) COF and (**b**) wear rate.

The effect of input parameters on the COF for the dry sliding wear test is shown in Figure 14. Increased COF was observed with an increase in the applied load and sliding velocity. The minimum COF was attained at a sliding distance of 1 m/s and an applied load of 20 N. This was attributed to the lower sliding velocities. The composite pin surface tends to spend a prolonged period of time rubbing against the disc surface, which generates the self-lubricated tribofilm at the tribo pair. There was a 30% enhancement in the COF at 40 N applied load compared to the 20 N applied load. This phenomenon was due to the unstable MML, which was developed at the interfaces at higher applied loads. However, at a lower magnitude of applied loads, the MML was stable, which supports the generation of higher COF values.

When compared to the wear rate for A356-Gr composites, the wear rate for 40 N was significantly higher than that of 30 N and 20 N applied loads, as shown in Figure 15. This might be due to adhesive wear on the pin material, which is caused by a rise in temperature between the contact surfaces at the higher magnitude of applied load on the composite pin. The developed protective mechanically mixed layer on the tribo interfaces due to the detached Gr particles at higher loads (40 N) and sliding distances (2000 m) minimizes the wear rate at higher sliding distances. An increase in the sliding velocity from 1 m/s to 3 m/s leads to a decline in wear rate due to the presence of the protective layer consisting of self-lubricated Gr particles. It was concluded that with a minimum wear rate of 0.00563 g/km and a coefficient of friction (COF) of 0.3763, the best combination of input parameters were 20 N applied load, 1 m/s sliding speed, and 1000 m sliding distance for the A356—5 wt% Gr composite running against the EN-31 hardened steel disc.

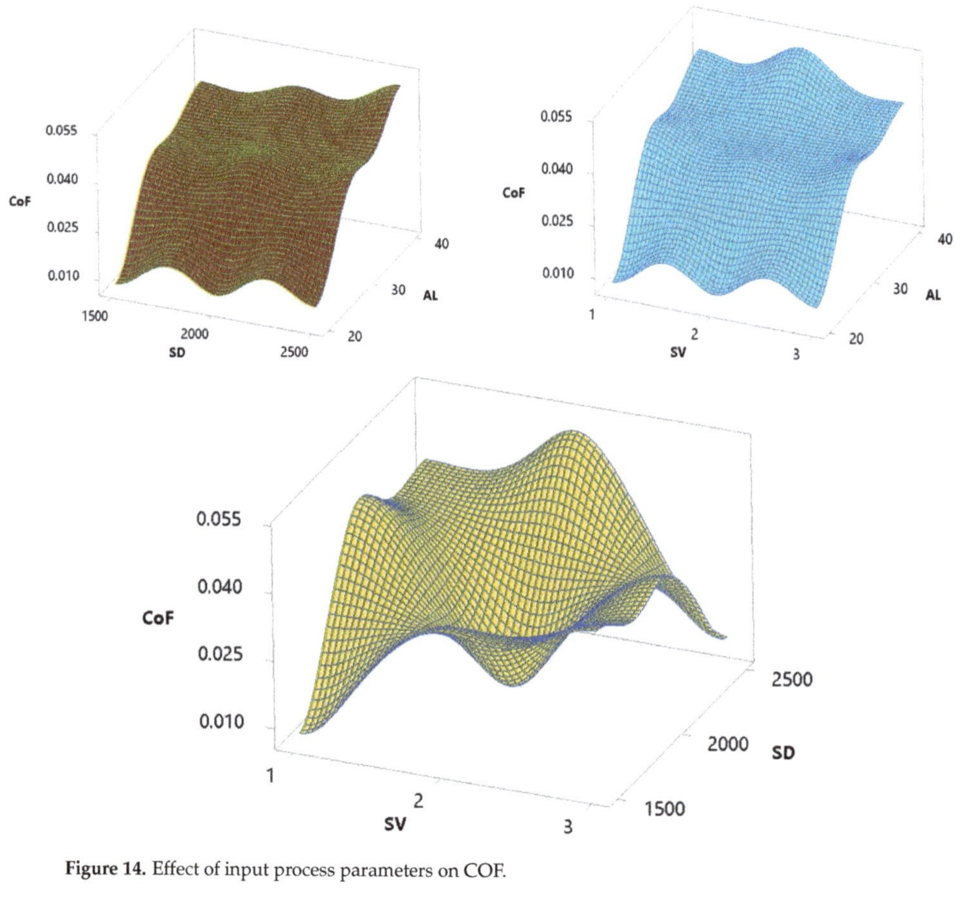

Figure 14. Effect of input process parameters on COF.

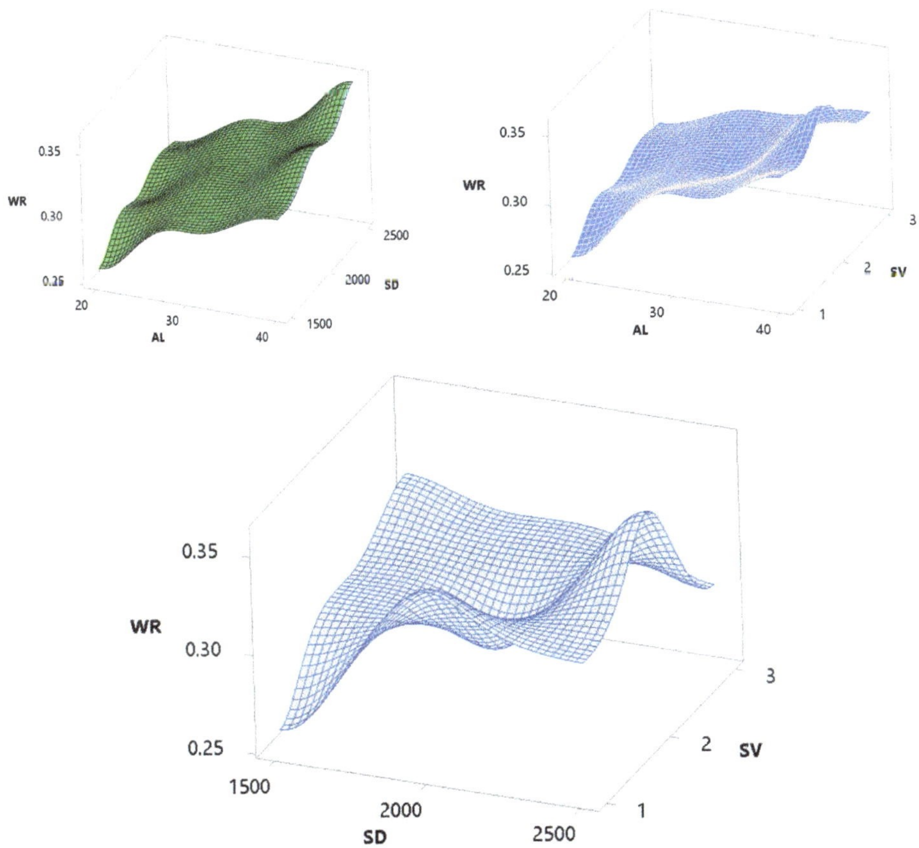

Figure 15. Effect of input process parameters on the wear rate.

4. Conclusions

The graphite lubricant was extracted from corn stover, and the extracted Gr was studied mechanically and tribologically by reinforcing the A356 alloy. The following results were obtained after the experiments.

1. The U.T.S and % elongation for the A356-Gr were found to be increased up to the 5 wt% Gr addition, and the maximum U.T.S of 123 MPa and 7.11% elongation were attained due to strong interface bonding between Gr and the A356 alloy.
1. The presence of an Al_2Mg cluster at 7.5 wt% Gr reinforcement led to a decrement in mechanical properties due to the uneven thermal expansions of the cluster and adjacent particles.
2. The obtained micro Brinell hardness at 5 wt% Gr reinforcement was 13.5%, 6.97%, and 3.37% higher than the A356 alloy, A356—2.5 wt% Gr, and A356—7.5 wt% composite.
2. The uniform dispersion of reinforcements was identified in the FESEM analysis of the A356—2.5 wt% Gr and A356—5 wt% Gr composite. However, Al_2Mg clusters were confirmed in the A356—7.5 wt% Gr composite due to the non-uniform sintering, and the corresponding Al_2Mg agglomeration was shown in the XRD pattern.
3. The wear rate and COF values were found to be decreased with an increase in % Gr addition due to the formation of the self-lubricated MML at the tribo interfaces. The minimum wear rate of 0.00563 g/km and COF of 0.3763 were found at 7.5 wt% Gr reinforcement with an applied load of 40 N.

4. The ANOVA results confirmed that the optimal process parameters for the minimum wear rate of 0.0052386 g/Km and 0.364 COF were 1 m/s sliding velocity, 1000 m sliding distance, and 20 N applied load conditions.

Author Contributions: Conceptualization, V.S.S.V.; methodology, V.S.S.V.; software V.S.S.V.; validation, P.R.V.; formal analysis, P.R.V.; investigation, V.S.S.V.; resources, V.S.S.V.; data curation, V.S.S.V.; writing—original draft preparation, V.S.S.V.; writing—review and editing, P.R.V.; visualization, V.S.S.V.; supervision, P.R.V.; project administration, P.R.V.; funding acquisition, P.R.V. All authors have read and agreed to the published version of the manuscript.

Funding: This research received no external funding.

Data Availability Statement: Data is contained within the article.

Conflicts of Interest: The authors declare no conflict of interest.

References

1. Wang, Y.Q.; Afsar, A.M.; Jang, J.H.; Han, K.S.; Song, J.I. Room temperature dry and lubricant wear behaviors of $Al_2O_{3f}/SiC_p/Al$ hybrid metal matrix composites. *Wear* **2010**, *268*, 863–870. [CrossRef]
2. Yadav, A.K.; Pandey, K.M.; Dey, A. Aluminium Metal Matrix Composite with Rice Husk as Reinforcement: A Review. *Mater. Today Proc.* **2018**, *5*, 20130–20137. [CrossRef]
3. Venkatesh, V.S.S.; Deoghare, A.B. Microstructural Characterization and Mechanical Behaviour of SiC and Kaoline Reinforced Aluminium Metal Matrix Composites Fabricated Through Powder Metallurgy Technique. *Silicon* **2021**, *14*, 3723–3737. [CrossRef]
4. Aydin, F.; Durgut, R.; Mustu, M.; Demir, B. Prediction of wear performance of ZK60/CeO_2 composites using machine learning models. *Tribol. Int.* **2023**, *177*, 107945. [CrossRef]
5. Reddy, A.P.; Krishna, P.V.; Rao, R.N. Tribological Behaviour of Al6061–2SiC-xGr Hybrid Metal Matrix Nanocomposites Fabricated through Ultrasonically Assisted Stir Casting Technique. *Silicon* **2019**, *11*, 2853–2871. [CrossRef]
6. Liu, S.; Wang, Y.; Muthuramalingam, T.; Anbuchezhiyan, G. Effect of B_4C and MOS_2 reinforcement on micro structure and wear properties of aluminum hybrid composite for automotive applications. *Compos. Part B Eng.* **2019**, *176*, 107329. [CrossRef]
7. Venkatesh, V.S.S.; Deoghare, A.B. Effect of Particulate Type Reinforcements on Mechanical and Tribological Behavior of Aluminium Metal Matrix Composites: A Review. In *Recent Advances in Mechanical Engineering*; Pandey, K.M., Misra, R.D., Patowari, P.K., Dixit, U.S., Eds.; Springer: Singapore, 2021; pp. 295–303.
8. Liu, J.; Li, Y.; Cheng, C.; Li, W.; Qin, X. Effect of temperature on the structure and mechanical properties of SiC–TiB_2 composite ceramics by solid-phase spark plasma sintering. *Ceram. Int.* **2022**, *48*, 23151–23158. [CrossRef]
9. Basavarajappa, S.; Chandramohan, G.; Mahadevan, A.; Thangavelu, M.; Subramanian, R.; Gopalakrishnan, P. Influence of sliding speed on the dry sliding wear behaviour and the subsurface deformation on hybrid metal matrix composite. *Wear* **2007**, *262*, 1007–1012. [CrossRef]
10. Singh, J. Fabrication characteristics and tribological behavior of Al/SiC/Gr hybrid aluminum matrix composites: A review. *Friction* **2016**, *4*, 191–207. [CrossRef]
11. Chou, S.-N.; Huang, J.-L.; Lii, D.-F.; Lu, H.-H. The mechanical properties of Al_2O_3/aluminum alloy A356 composite manufactured by squeeze casting. *J. Alloys Compd.* **2006**, *419*, 98–102. [CrossRef]
12. Maleki, K.; Alizadeh, A.; Hajizamani, M. Compressive strength and wear properties of SiC/Al6061 composites reinforced with high contents of SiC fabricated by pressure-assisted infiltration. *Ceram. Int.* **2021**, *47*, 2406–2413. [CrossRef]
13. Ajay, C.V.; Mohammed, K.T.; Hariharasakthisudhan, P.; Kumar, V.N.; Vishnu, R. Characteristics Study of Mechanical and Tribological Behaviour of Gr/Sn Dispersed Al-7Si Alloy Matrix Composite Processed Through Bottom Pouring Stir Casting Technique. *Silicon* **2023**, *15*, 5089–5104. [CrossRef]
14. Santhosh, N.; Praveena, B.A.; Jain, R.; Hasan, M.A.; Islam, S.; Khan, M.A.; Razak, A.; Daniyal, M. Analysis of friction and wear of aluminium AA 5083/WC composites for building applications using advanced machine learning models. *Ain Shams Eng. J.* **2023**, *14*, 102090. [CrossRef]
15. Huang, W.; Lin, Q.; Zhang, X. Investigation of Tribological Properties of Magnesium Alloys under Dry Sliding and Lubrication Condition. *Proc. Inst. Mech. Eng. Part J.* **2011**, *225*, 35–42. [CrossRef]
16. Manohar, G.; Pandey, K.M.; Maity, S.R. Effect of microwave sintering on the microstructure and mechanical properties of AA7075/B_4C/ZrC hybrid nano composite fabricated by powder metallurgy techniques. *Ceram. Int.* **2021**, *47*, 32610–32618. [CrossRef]
17. Manohar, G.; Pandey, K.M.; Maity, S.R. Effect of compaction pressure on mechanical properties of AA7075/B_4C/graphite hybrid composite fabricated by powder metallurgy techniques. *Mater. Today Proc.* **2020**, *38*, 2157–2161. [CrossRef]
18. Ramadoss, N.; Pazhanivel, K.; Anbuchezhiyan, G. Synthesis of B_4C and BN reinforced Al7075 hybrid composites using stir casting method. *J. Mater. Res. Technol.* **2020**, *9*, 6297–6304. [CrossRef]
19. Han, N.L.; Wang, Z.G.; Wang, W.L.; Zhang, G.D.; Shi, C.X. Low-cycle fatigue behavior of a particulate SiC/2024Al composite at ambient and elevated temperature. *Compos. Sci. Technol.* **1999**, *59*, 147–155. [CrossRef]

20. Pazhouhanfar, Y.; Eghbali, B. Microstructural characterization and mechanical properties of TiB_2 reinforced Al6061 matrix composites produced using stir casting process. *Mater. Sci. Eng. A* **2018**, *710*, 172–180. [CrossRef]
21. Hasan, M.S.; Kordijazi, A.; Rohatgi, P.K.; Nosonovsky, M. Triboinformatics Approach for Friction and Wear Prediction of Al Graphite Composites Using Machine Learning Methods. *J. Tribol.* **2022**, *144*, 011701. [CrossRef]
22. Radhika, N.; Karthik, R.; Gowtham, S.; Ramkumar, S. Synthesis of Cu-10Sn/SiC Metal Matrix Composites and Experimental Investigation of its Adhesive Wear Behaviour. *Silicon* **2019**, *11*, 345–354. [CrossRef]
23. Baradeswaran, A.; Perumal, A.E. Influence of B_4C on the tribological and mechanical properties of Al 7075–B_4C composites. *Compos. Part B Eng.* **2013**, *54*, 146–152. [CrossRef]
24. Kumar, H.G.P.; Xavior, M.A. Assessment of Mechanical and Tribological Properties of Al 2024-SiC-Graphene Hybrid Composites. *Procedia Eng.* **2017**, *174*, 992–999. [CrossRef]
25. Venkatesh, V.S.S.; Deoghare, A.B. Modelling and Optimisation of Wear Parameters for Spark Plasma Sintered Al-SiC-Kaoline Hybrid Composite. *Adv. Mater. Process. Technol.* **2022**, *8*, 1286–1304. [CrossRef]
26. Han, T.; Zhang, S.; Zhang, C. Unlocking the secrets behind liquid superlubricity: A state-of-the-art review on phenomena and mechanisms. *Friction* **2022**, *10*, 1137–1165. [CrossRef]
27. Han, T.; Zhang, C.; Luo, J. Macroscale Superlubricity Enabled by Hydrated Alkali Metal Ions. *Langmuir* **2018**, *34*, 11281–11291. [CrossRef]

lubricants

MDPI

Article

Wear Behavior Analysis and Gated Recurrent Unit Neural Network Prediction of Coefficient of Friction in Al10Cu-B_4C Composites

Mihail Kolev *, Ludmil Drenchev, Veselin Petkov, Rositza Dimitrova, Krasimir Kolev and Boris Yanachkov

Institute of Metal Science, Equipment and Technologies with Hydro- and Aerodynamics Centre "Acad. A. Balevski", Bulgarian Academy of Sciences, 1574 Sofia, Bulgaria; ljudmil.d@ims.bas.bg (L.D.); veselin.petkov@ims.bas.bg (V.P.); rossy@ims.bas.bg (R.D.); kkolev@ims.bas.bg (K.K.); byanachkov@ims.bas.bg (B.Y.)

* Correspondence: mihail1kolev@ims.bas.bg or mihail1kolev@gmail.com

Abstract: Aluminum-based metal matrix composites reinforced with B_4C are advanced materials recognized for their exceptional combination of lightweight properties, high hardness, and superior wear resistance. These characteristics make them perfectly suited for applications demanding exceptional performance in extreme mechanical and tribological environments. This study investigates the wear behavior, microstructural characteristics, and predictive modeling of Al10Cu-B_4C composites fabricated via powder metallurgy with varying B_4C contents (0, 2.5, 5, and 7.5 wt.%). The addition of B_4C microparticles to Al10Cu composites significantly influenced their tribological properties with 2.5 wt.% B_4C achieving a 21.74% reduction in the coefficient of friction (COF) and 7.5 wt.% B_4C providing a remarkable 65.00% improvement in wear resistance. Microstructural analysis using SEM and EDS was conducted on the unreinforced materials and the reinforced composites both before and after the wear tests. To further analyze and predict the tribological performance, a Gated Recurrent Unit neural network was developed to predict COF values. The need for this model arises from its potential to cost-effectively facilitate the prediction of COF, diminishing the need for extensive experimental testing while being noted for its simplicity and ease of implementation in practical applications. The model achieved excellent accuracy with an R^2 of 0.9965 for the test set and 0.9917 for the validation set. Additionally, feature importance analysis using Random Forest models identified reinforcement-related features as the dominant predictors for both COF and mass wear. These findings demonstrate the potential of Al10Cu-B_4C composites for emerging industrial applications, where enhanced wear resistance and controlled friction are critical for improving efficiency and durability under rigorous operating conditions. Furthermore, this study highlights the efficacy of neural network models in accurately predicting COF, providing a powerful tool for optimizing the performance of advanced composite materials.

Keywords: Al10Cu-B_4C composites; metal matrix composites; wear characteristics; dry-sliding friction; gated recurrent unit; neural network

1. Introduction

Aluminum-based alloys are widely valued in advanced material applications due to their light weight [1], high specific strength [2], and corrosion resistance [3], making them essential for automotive, aerospace, and structural applications. These alloys, especially those reinforced with ceramic particles, offer enhanced mechanical properties and wear resistance, enabling them to withstand high-stress environments while remaining cost-effective and easy to manufacture [4–6].

Aluminum and copper matrix composites reinforced with boron carbide (B_4C) have gained attention due to their superior mechanical properties, enhanced wear resistance, and potential for high-performance structural applications [7]. The hard, wear-resistant B_4C particles serve as a protective phase within the softer aluminum matrix, reducing material loss during sliding or abrasive wear processes [8]. This reinforcement effect is particularly beneficial in applications requiring high durability under frictional forces, as B_4C particles act as a barrier to surface deformation and material removal [9,10]. Additionally, the homogeneous distribution of B_4C within the matrix has been shown to improve load-bearing capacity, thereby extending the operational lifespan of these composites in wear-intensive environments [11]. Notably, B_4C-based composites, owing to their exceptionally high boron content, are particularly valuable for nuclear applications, where they can potentially replace traditional steel components due to their excellent neutron absorption capabilities while maintaining high strength and intercrystalline corrosion resistance [12,13].

The research by Alizadeh et al. [8] set the stage for understanding the mechanical enhancements achieved with B_4C reinforcement in Al–2 wt.% Cu composites. By employing mechanical milling and hot extrusion, their work demonstrated that incorporating B_4C led to a finer particle size distribution and reduced crystallite size in the matrix. These microstructural modifications enhanced both wear resistance and tensile strength, establishing a foundational link between B_4C reinforcement and improved mechanical performance. Their findings also emphasized the critical role of particle distribution and crystallite size in defining composite properties, guiding subsequent investigations into the effects of particle morphology on mechanical behaviors. Lokesh et al. [14] investigated Al-Cu reinforced with B_4C composites fabricated via powder metallurgy, focusing on the role of particle size distribution and mixing duration. Their results highlighted that an extended mixing time facilitated a more uniform distribution of B_4C particles, leading to improved tensile strength and fracture characteristics. This study underscored the importance of precise control over particle mixing in powder metallurgy to optimize the mechanical properties of Al-Cu/B_4C composites, offering a critical refinement to the insights from Alizadeh and Taheri-Nassaj's foundational work. In exploring wear behavior, Bujari et al. [10] conducted a comprehensive analysis on Al–4.5% Cu composites with varying B_4C content. Utilizing pin-on-disk wear testing, they observed that higher B_4C content provided enhanced wear resistance with B_4C particles acting as a protective layer against surface degradation. This study further illustrated the effective role of B_4C as a wear-resistant agent within the composite matrix, confirming that optimized B_4C content significantly contributes to the longevity and durability of Al-Cu-based composites under wear-intensive applications. Qadirkhan et al. [15] concentrated on the effects of prolonged milling time on the dispersion of B_4C particles. They discovered that extended milling promoted a more homogeneous dispersion of B_4C, which, in turn, resulted in higher composite hardness. Their findings emphasized the relationship between milling duration and particle distribution uniformity, which is critical for achieving enhanced hardness and strength in Al-Cu/B_4C composites. In related work, Ozkaya et al. [16] studied the effects of B_4C content and milling times on the mechanical properties of AlCuMg-B_4C nanocomposites. Their results demonstrated that although higher B_4C content and extended milling reduced the density of hot-pressed composites, these factors also contributed to significant improvements in tensile strength and hardness. This balance between density reduction and mechanical enhancement highlights an important optimization consideration for achieving superior strength in AlCuMg-B_4C nanocomposites. Expanding on the benefits of B_4C reinforcement, Uthayakumar et al. [17] explored hybrid Al–SiC–B_4C composites. Through dry sliding wear tests and focused ion beam (FIB) analysis, they showed that the combined reinforcement of SiC and B_4C particles yielded superior wear resistance compared to single-reinforced composites. This synergistic reinforcement approach demonstrated that integrating multiple hard particles could provide cumulative benefits in terms of wear behavior, opening a pathway for more durable composite materials. Akçamlı and Şenyurt [18] examined the effects of both milling duration and B_4C content on Al–Si–Cu composites. Using powder metallurgy techniques and heat treatments, they observed that increased B_4C content and optimized milling duration significantly enhanced the microhardness and wear resistance of the composites. Their findings reinforced the role of B_4C in providing hardness and wear benefits while also identifying optimal processing parameters for maximizing these properties. Bharathiraja et al. [19] investigated the effect of hybrid reinforcement with B_4C and graphene in Al6061 composites. They found that the addition of graphene alongside B_4C, combined with T6 heat treatment, led to marked increases in hardness and compressive strength. These enhancements were attributed to the synergistic effects of the reinforcements, which contributed to a composite structure with improved integrity and load-bearing capacity that was particularly suitable for high-strength applications. In another study by Qadirkhan and Siddiqui [15], a similar approach was applied to synthesize Al–Cu–B_4C composites with 12 wt.% B_4C reinforcement through ball milling under an argon atmosphere, utilizing ethanol as a process control agent. They varied milling times and maintained a ball-to-powder weight ratio of 10:1, observing through SEM that the addition of B_4C particles accelerated the work-hardening rate of the Al matrix. Their results showed that the full homogeneity of B_4C distribution within the Al matrix was achieved at a steady-state milling time, and EDX confirmed minimal contamination during the process. An increase in milling time corresponded with greater hardness in the composites, reinforcing the effectiveness of mechanical alloying for improving material properties in B_4C-reinforced systems. Alizadeh et al. [7] examined nanostructured Al2Cu composites reinforced with B_4C, achieving substantial compressive strength improvements and reduced wear rates through hot pressing. Their study highlighted the Orowan strengthening mechanism as a key factor, where B_4C particles at the nanoscale impeded dislocation movement, thus contributing to enhanced mechanical stability and wear resistance.

The application of machine learning (ML) in the design and analysis of Al-based composites is emerging as a transformative tool, offering new ways to optimize and predict material properties with precision [20,21]. An important step was achieved by Banerjee et al. [22] who successfully employed artificial neural networks (ANNs) integrated with genetic algorithms to design alumina-reinforced aluminum matrix composites. This approach led to the multi-objective optimization of tribo-mechanical properties, facilitating the customized development of composites with superior wear resistance and mechanical performance. It highlights the potential of integrating computational intelligence into composite design. In another study, Thankachan et al. [23] further illustrated the potential of ML by applying ANN models to accurately predict the surface roughness and material removal rates in the wire electrical discharge machining of Al-based alloys/composites. Their findings emphasized the efficiency of ML in refining machining processes to achieve the desired surface qualities and material properties. Focusing on wear analysis, Idrisi et al. [24] utilized ANN models to evaluate the wear performance of AA5083-SiC micro- and nanocomposites. Their study not only highlighted the ability of ML to accurately predict wear behavior but also demonstrated its capability to model complex interactions in tribological systems, ensuring improved material design and application. Further advancements in ML applications are illustrated by Deng et al. [25], who employed ML algorithms, including sequential minimal optimization algorithm for support vector regression with a PUK kernel, to predict the mechanical properties of Cu-Al alloys. This effort enabled the development of alloys with targeted mechanical characteristics, achieving superior tensile strength and hardness with minimal error, which validates the predictive power and practical utility of ML models in alloy design.

Despite substantial progress in the study of Al-Cu/B_4C composites, there remain critical gaps in understanding the specific behavior of Al10Cu-B_4C systems, especially regarding tribological performance, microstructural characterization, and predictive modeling. While previous studies have documented the general benefits of B_4C reinforcement in Al-based composites—such as enhanced wear resistance and hardness—there is a notable lack of comprehensive investigations specifically focused on Al10Cu-B_4C composites. Key aspects such as the influence of varying B_4C content on wear behavior and the development of predictive models for tribological properties have not been thoroughly addressed, leaving a gap in the optimization and application of these advanced materials.

This study aims to investigate the wear behavior, microstructural characteristics, and predictive modeling of Al10Cu composites reinforced with varying B_4C content (0, 2.5, 5, and 7.5 wt.%) fabricated via powder metallurgy. The primary objectives are to analyze the microstructural evolution and particle dispersion using SEM and EDS, evaluate the tribological performance of the composites through pin-on-disk testing, measure their microhardness, and determine the influence of reinforcement concentrations on wear mechanisms. Additionally, the study aims to develop and validate a Gated Recurrent Unit (GRU) neural network to predict the COF and mass wear, leveraging machine learning for precise performance modeling. By integrating experimental analysis with computational methods, this research seeks to optimize reinforcement content and provide insights into the design of high-performance Al10Cu-B_4C composites for tribological applications.

2. Materials and Methods

2.1. Fabrication

The Al10Cu-B_4C composites were fabricated using a powder metallurgy approach, which is a method selected for its effectiveness in producing particulate-reinforced Al-based matrix composites with uniform particle distribution and enhanced mechanical properties. The sintered composites were analyzed to investigate the effects of B_4C reinforcement on their microstructure, wear resistance, and hardness. This fabrication approach is based on methods used in prior studies, such as the work by Akçamlı et al. [18], which employed powder metallurgy to fabricate B_4C-reinforced Al-Si-Cu matrix composites with varying B_4C concentrations. In that study, powder mixing was conducted using a planetary ball mill at 500 rpm for various durations, which was followed by sintering at 530 °C for 3 h. This approach aligns well with the current study's process, which involves ball milling at 500 rpm for 5 h and sintering at 530 °C for 3 h, ensuring complete densification and consistent microstructural characteristics. The primary materials used in this study include Al powder ($\leq$44 μm), Cu powder (<44 μm), and B_4C particles ($\leq$10 μm). The solid densities of the powders are as follows: Al: 2.70 g/cm^3, Cu: 8.94 g/cm^3, and B_4C: 2.52 g/cm^3. The composite compositions were prepared with an aluminum matrix containing 10 wt.% copper (Al–10 wt.% Cu) and varying concentrations of B_4C (0 wt.%, 2.5 wt.%, 5 wt.%, and 7.5 wt.%). The Al, Cu, and B_4C powders were homogenized using a high-energy ball milling technique. The high-energy ball milling process, conducted at 500 rpm for 5 h with a ball-to-powder ratio of 6:1, was optimized to ensure the uniform dispersion of Cu and B_4C particles within the Al matrix. This process prevents particle segregation by facilitating mechanical interlocking and even distribution. Following milling, the mixed powders were compacted into cylindrical samples using a hydraulic press and a cylindrical steel die with a diameter of 16 mm. A uniaxial compaction pressure of 160 MPa was applied to form green compacts suitable for sintering, further consolidating the uniformity of the powders. The green compacts were sintered in an argon atmosphere to prevent oxidation. An initial debinding phase was performed at 400 °C for 2 h. Subsequently, the sintering temperature was raised to 530 °C and maintained for a duration of 3 h. The sintering process was carried out in a furnace to facilitate complete densification and interparticle bonding. After sintering, the samples were allowed to cool in the furnace until they reached ambient temperature. The fabrication process is presented in the flowchart in Figure 1.

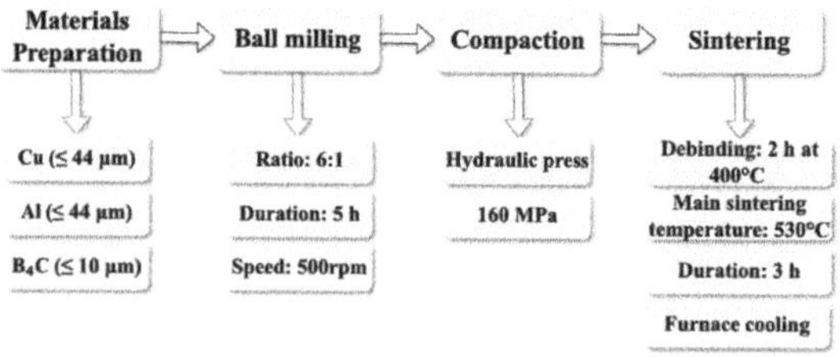

Figure 1. Flowchart of the fabrication process for producing Al10Cu-B_4C composites.

2.2. Characterization

Microstructural characterization of the Al10Cu-B_4C composites was conducted using SEM and EDX. The SEM analysis was performed with a Hirox SEM (Hirox Japan Co., Ltd., Tokyo, Japan), while elemental analysis was carried out using an EDX system (BRUKER Co., Frankfurt, Germany). These techniques enabled a detailed examination of the distribution and morphology of B_4C particles in the Al10Cu matrix.

XRD was employed to identify the phases present in the composite samples. Diffraction measurements were conducted using a TDM-10 X-ray diffractometer (Dandong Tongda Science and Technology Co., Ltd., Dandong, China) equipped with a scintillation detector and Ni-filtered Cu Kα radiation. The parameters used were as follows: operating voltage: 40 kV, current: 35 mA, scan speed: 0.24°/min, step size: 0.02°, and a 2θ interval from 15° to 100°.

Dry wear testing of the composites was performed using a TR-20 model Ducom Rotary tribometer (Ducom Instruments, Bangalore, India) in a pin-on-disk configuration. The wear test parameters included a load of 50 N, a sliding speed of 2.4 m/s, and a total sliding distance of 1000 m. The load of 50 N was specifically selected based on its use in related studies, such as one involving Al–2 wt.% Cu reinforced with B_4C nanoparticles fabricated via hot extrusion [8], which also employed a steel counter-disk with similar hardness. For the dry friction test in this study, the counter-disk material was EN-31 steel with a hardness of 62 HRC. Both the composite samples and the counter-disk were prepared with a uniform surface roughness of Ra = 0.4 µm.

The microhardness of the composite cross-sectional areas was analyzed using a Polyvar Met light microscope (Reichert Jung, Vienna, Austria) equipped with a semi-automated micro-Vickers hardness measurement system (Micro-Duromat 5000 computer control, Reichert Jung, Vienna, Austria). Microhardness measurements were performed with a standardized Vickers indenter—a diamond pyramid with a 136° angle at the vertex using a 0.05 kgf load applied for a duration of 10 s with pressure maintained for an additional 10 s to ensure consistency in indentation depth. Five samples were tested for each composition, and the average HV values were reported.

2.3. Predictive Modeling and Feature Importance Analysis

In this study, a data-driven ML model based on a GRU neural network was developed to predict the COF over time for the Al10Cu-B_4C composite. The GRU neural network was selected due to its capability to efficiently capture temporal dependencies in sequential data, such as the COF measurements over time. Unlike traditional recurrent neural networks, GRUs use a gating mechanism to regulate information flow, which reduces the risk of vanishing gradients and enables faster training with fewer computational resources [26,27]. This makes GRUs particularly advantageous for time-series prediction tasks where data patterns evolve over successive time steps. The need for this model arises from its potential to cost-effectively facilitate the prediction of COF, diminishing the need for extensive experimental testing while being noted for its simplicity and ease of implementation in practical applications.

The model utilized sequential time-series data and was trained, validated, and tested on experimentally obtained COF values to evaluate its predictive accuracy. The dataset, loaded from an XLSX file, consists of COF values measured over time, with TIME as the index variable and COF as the target variable for prediction. The prediction model was trained using 428 data points for time (in seconds, consistent across all materials) and 428 unique data points for the COF for each material. COF values were normalized to a [0, 1] range using MinMax scaling. This normalized dataset was then transformed into a supervised learning format, where each sample consists of a sequence of three consecutive time steps, which were chosen as a look-back period. The dataset was split into training (70%), validation (15%), and testing (15%) subsets. This splitting ensures robust model evaluation while avoiding overfitting. Each subset was reshaped into a 3D array format required for GRU model input with dimensions [samples, time steps, features]. A sequential model was constructed with the GRU layer as the primary component, which was designed to capture temporal dependencies in the COF data. The model architecture consisted of a GRU layer with four units, which was followed by a dense layer with one unit to predict the COF. Model training was conducted over 150 epochs with a batch size of 1, using mean_squared_error as the loss function and the Adam optimizer. Validation data were used to monitor training performance, ensuring generalization. The trained model's predictive performance was evaluated using the root mean squared error (RMSE), mean absolute error (MAE), and coefficient of determination (R^2) metrics on both the validation and test datasets. Validation metrics were used to fine-tune the model, while test metrics assessed final model performance. For interpretability and validation, actual and predicted COF values were plotted over time. Additionally, regression plots were generated for validation and test datasets to assess predictive accuracy visually. The evaluation metrics (RMSE, MAE, R^2) for each dataset were saved to a text file for documentation.

In addition to the predictive modeling for COF, feature importance analysis was conducted to evaluate the influence of various input variables on the mass wear and COF of all materials. A Random Forest Regressor with 50 estimators and a fixed random state (42) was employed to compute the relative importance of input features. The features considered included reinforcement content and microhardness. The importance values, normalized to a range of [0, 1], were extracted and visualized. This approach provides valuable insights into the contribution of each factor with bar plots highlighting their relative influence [28,29]. These technical analyses are pivotal for interpreting experimental results and guiding the optimization of material properties.

3. Results and Discussion

3.1. Microstructural Analysis

The microstructural analysis of the Al10Cu alloy and its composites reinforced with varying B_4C contents (0, 2.5, 5, and 7.5 wt.%) was conducted using SEM and EDS to investigate the matrix zones, intermetallic Al_2Cu regions, and B_4C reinforcement particles where applicable. Figure 2a–d and Table 1 provide a detailed summary of the microstructural features and the corresponding elemental compositions of key regions within the analyzed specimens. In the unreinforced Al10Cu alloy, the microstructure consisted primarily of an Al-rich matrix and intermetallic Al_2Cu phases (Figure 2a). The matrix zone was confirmed to be predominantly aluminum, as EDS analysis of region 3 indicated a composition of 89.1 wt.% Al. This Al-rich matrix served as the primary material phase. Intermetallic Al_2Cu regions were evident with region 1 exhibiting a high copper concentration of 58.8 wt.%. These phases are known to contribute to hardness and wear resistance. The addition of 2.5 wt.% B_4C to the Al10Cu alloy introduced significant microstructural changes with the appearance of three distinct regions corresponding to the matrix, intermetallic Al_2Cu phases, and dispersed B_4C particles (Figure 2b). The SEM image reveals a small amount of B_4C particles, which is expected due to the low reinforcement content used. The matrix zone remained aluminum-rich, with EDS results for region 3 showing a high Al content of 92.0 wt.%, indicating minimal interference from the reinforcement or intermetallic phases. The intermetallic Al_2Cu regions persisted, as observed in region 1, which retained a Cu concentration of 51.5 wt.%, contributing to the material's hardness. Importantly, the incorporation of B_4C particles was confirmed in region 2, where EDS detected 40.4 wt.% B and 59.6 wt.% C. The well-dispersed B_4C particles enhanced the wear resistance of the composite through their load-bearing capability and abrasive effects. Increasing the B_4C content to 5 wt.% further improved the integration of reinforcement particles alongside the matrix and intermetallic phases (Figure 2c). The Al matrix remained stable, with region 3 exhibiting a composition of 94.4 wt.% Al, suggesting that the increased reinforcement did not compromise the primary matrix material. The Cu-rich intermetallic phases were also maintained, as evidenced by a Cu concentration of 49.7 wt.% in region 2, which continued to strengthening the composite. The B_4C particles, represented in region 1, showed a high boron concentration of 62.4 wt.%, confirming their effective incorporation. The optimized distribution of B_4C particles at this reinforcement level enhanced the composite's structural integrity and wear resistance. At the highest B_4C content of 7.5 wt.%, the microstructure exhibited a uniform dispersion of reinforcement particles without noticeable agglomeration (Figure 2d). The Al-rich matrix composition in region 2 remained consistent, with 92.4 wt.% Al, while the intermetallic Al_2Cu phases persisted, with region 1 exhibiting a Cu content of 50.3 wt.%. Region 3, corresponding to the B_4C particles, showed concentrated boron and carbon levels of 60.2 wt.% and 21.9 wt.%, respectively, confirming effective incorporation of the reinforcement. The uniform distribution of B_4C particles at 7.5 wt.% contributed to enhanced structural integrity and wear resistance, countering the typical challenges associated with higher reinforcement levels. The EDS analysis demonstrates a clear progression in microstructural complexity with increasing B_4C content. The transition from a simple Al-Cu matrix in the unreinforced alloy to a composite reinforced with uniformly dispersed B_4C particles illustrates the benefits of reinforcement in enhancing wear resistance and mechanical properties. These observations underscore the importance of balancing reinforcement content to achieve optimal composite performance without compromising structural integrity.

The wear mechanisms affecting the Al10Cu alloy and its composites reinforced with varying B_4C contents (0, 2.5, 5, and 7.5 wt.%) were evaluated through the SEM analysis of worn surfaces, as presented in Figure 3. The results reveal a distinct trend in the severity of wear, which decreases progressively with increasing B_4C reinforcement. The unreinforced Al10Cu alloy exhibited the most pronounced wear damage, which was characterized by deep grooves and significant material removal, as presented in Figure 3a. The worn surface displayed evidence of severe plastic deformation, which was indicative of adhesive wear as the primary mechanism. The absence of reinforcing particles left the softer aluminum matrix and intermetallic Al_2Cu phases vulnerable to surface shearing and material transfer during sliding contact. This lack of reinforcement contributed to a rapid degradation of the material under the applied mechanical load. Incorporating 2.5 wt.% B_4C into the Al10Cu matrix resulted in a noticeable reduction in wear severity, as presented in Figure 3b. The worn surface of the composite showed shallower grooves, although evidence of abrasive wear was still apparent. The small volume fraction of B_4C particles, as evident in the SEM image, improved wear resistance by acting as load-bearing elements and reducing stress concentration in the Al matrix. However, the relatively low reinforcement content was insufficient to fully mitigate surface deformation and material removal. A further increase in B_4C content to 5 wt.% resulted in significant improvements in wear resistance, as presented in Figure 3c. The SEM images of the worn surface revealed fine, evenly spaced grooves with minimal material loss, which were consistent with a dominant abrasive wear mechanism. The optimal distribution of B_4C particles within the matrix at this reinforcement level provided effective load transfer and enhanced resistance to surface deformation, reducing wear damage. The combination of hard B_4C particles and intermetallic Al_2Cu phases contributed synergistically to the material's ability to resist mechanical wear, making the 5 wt.% B_4C composite one of the best-performing materials in this study. At 7.5 wt.% B_4C, the worn surface behavior was comparable to that observed at 5 wt.% reinforcement, as shown in Figure 3d. The SEM image revealed similarly fine grooves with minimal material removal, indicating that increasing the reinforcement beyond 5 wt.% did not yield further significant improvements in wear resistance. The well-dispersed B_4C particles continued to provide effective load-bearing capacity and surface protection, maintaining high wear resistance without noticeable drawbacks such as particle agglomeration. The progression in wear behavior across the tested materials highlights the critical influence of B_4C content on the wear mechanisms. The unreinforced Al10Cu alloy was dominated by adhesive wear, leading to severe surface damage. With the introduction of B_4C reinforcement, the wear mechanisms shifted toward abrasive wear with increasing reinforcement content progressively reducing the severity of surface damage up to 5 wt.%. Notably, the composites with 5 wt.% and 7.5 wt.% B_4C demonstrated the best overall performance, achieving high wear resistance due to the protective and load-bearing effects of the B_4C particles. This trend underscores the effectiveness of B_4C reinforcement in mitigating wear with diminishing returns observed at higher reinforcement levels.

Table 1. EDS analysis results in mass norm. % of Al10Cu specimens (with and without B_4C reinforcement) before wear tests, focusing on regions 1–3 as indicated in Figure 2a–d.

Specimen	Analysis No.	Al	Cu	B	C
Al10Cu (Figure 2a)	1	41.2	58.8	–	–
	2	57.1	42.9	–	–
	3	89.1	10.9	–	–
Al10Cu + 2.5 wt.% B_4C (Figure 2b)	1	48.5	51.5	–	–
	2	–	–	40.4	59.6
	3	92.0	8.0	–	–
Al10Cu + 5.0 wt.% B_4C (Figure 2c)	1	18.8	–	62.4	18.8
	2	50.3	49.7	–	–
	3	94.4	5.6	–	–
Al10Cu + 7.5 wt.% B_4C (Figure 2d)	1	49.7	50.3	–	–
	2	92.4	7.6	–	–
	3	17.8	–	60.3	21.9

Figure 2. SEM images of specimen before subjected to pin-on-disk tests: (**a**) Al10Cu material; (**b**) composites reinforced with 2.5 wt.% B_4C; (**c**) composites reinforced with 5 wt.% B_4C; (**d**) composites reinforced with 7.5 wt.% B_4C.

The XRD analysis of the Al10Cu alloy and its composites with varying B_4C contents (0, 2.5, 5, and 7.5 wt.%) was conducted to identify the crystalline phases present and to confirm the incorporation of B_4C into the matrix. The XRD patterns, shown in Figure 4a–d, provide a comprehensive insight into the phase composition and structural evolution of the materials as the B_4C reinforcement level increases. The XRD pattern for the unreinforced Al10Cu alloy (Figure 4a) revealed the dominant presence of Al peaks, corresponding to the FCC structure of Al, alongside peaks attributed to the Al_2Cu intermetallic phase. The Al_2Cu phase, known for its contribution to mechanical strength and hardness, was prominently observed, confirming the natural segregation of copper in the alloy. However, no peaks associated with B_4C were detected in this sample, which was expected due to the absence of reinforcement particles. In the composites reinforced with B_4C (Figure 4b–d), the XRD patterns demonstrate the continued presence of the dominant Al and Al_2Cu phases with B_4C peaks appearing in all reinforced samples. Notably, the intensity of the B_4C peaks remained relatively stable across all reinforcement levels, suggesting that the B_4C particles were successfully incorporated into the matrix albeit without significant variation in their concentration or distribution detectable via XRD. The XRD results confirm the successful incorporation of B_4C particles into the Al10Cu matrix while maintaining the structural stability of the Al and Al_2Cu phases. These findings support the observations in microstructural and wear analyses, highlighting the compatibility of B_4C with the matrix and its role in enhancing the composite's properties.

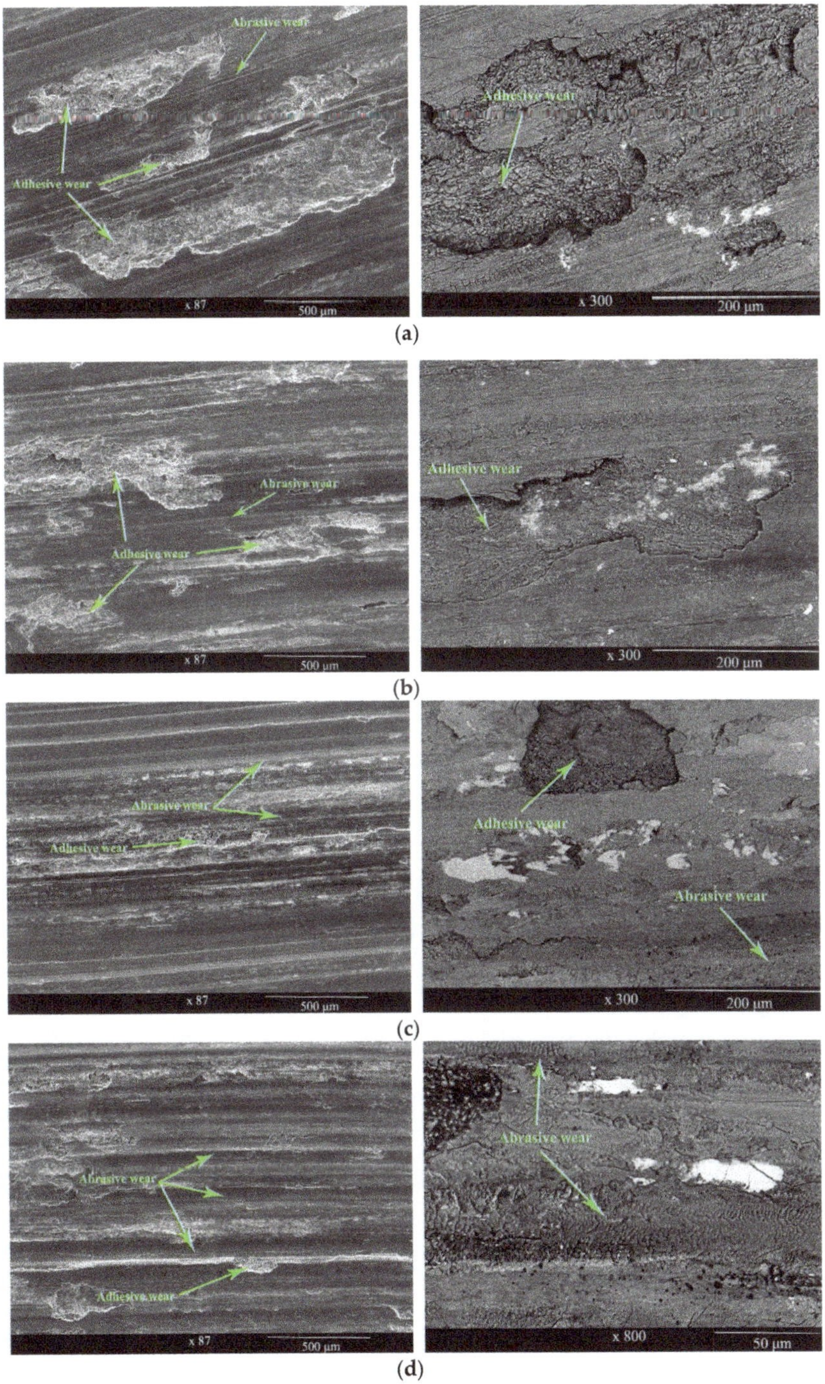

Figure 3. SEM images of specimen subjected to pin-on-disk: (**a**) Al10Cu material; (**b**) composites reinforced with 2.5 wt.% B_4C; (**c**) composites reinforced with 5 wt.% B_4C; (**d**) composites reinforced with 7.5 wt.% B_4C.

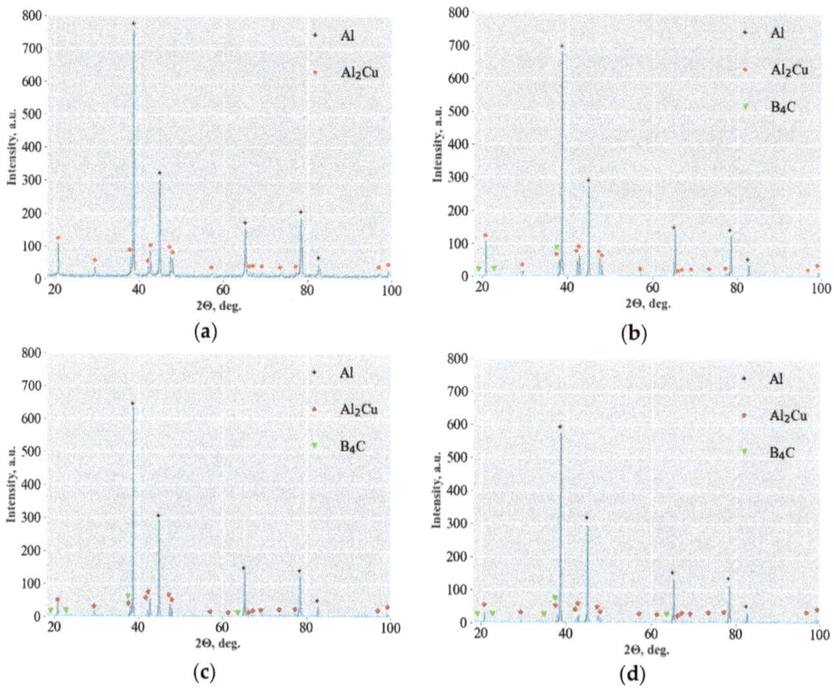

Figure 4. XRD pattern of Al10Cu alloy and its composites reinforced with varying B_4C contents: (**a**) unreinforced Al10Cu alloy, showing dominant Al and Al_2Cu intermetallic phases; (**b**) Al10Cu + 2.5 wt.% B_4C with traces of B_4C reinforcement alongside Al and Al_2Cu phases; (**c**) Al10Cu + 5 wt.% B_4C, displaying enhanced B_4C peaks with stable Al and Al_2Cu phases; (**d**) Al10Cu + 7.5 wt.% B_4C, exhibiting intensified B_4C peaks and retained Al and Al_2Cu phases, indicating successful reinforcement integration.

3.2. Microhardness and Wear Behavior

The microhardness (HV) of the Al10Cu alloy and its composites reinforced with varying B_4C contents (0, 2.5, 5, and 7.5 wt.%) was evaluated to assess the influence of reinforcement on the mechanical properties (Figure 5). The microhardness of Al10Cu composites reinforced with varying percentages of B_4C microparticles showed a consistent increase compared to the unreinforced baseline (32 HV). Adding 2.5 wt.% B_4C resulted in a microhardness of 40 HV, marking a 25.00% improvement. With 5 wt.% B_4C, the hardness further increased to 50 HV, representing a 56% rise. The highest hardness, 53 HV, was observed at 7.5 wt.% B_4C, reflecting a 65.37% enhancement. These results demonstrate the significant hardening effect of B_4C microparticles, with incremental gains becoming more gradual at higher reinforcement levels, emphasizing the material's improved mechanical properties with increasing B_4C content.

The wear characteristics of the Al10Cu alloy and its composites reinforced with varying B_4C contents (0, 2.5, 5, and 7.5 wt.%) were evaluated using pin-on-disk tests at dry sliding conditions. The results, focusing on the COF (Figure 6a) and mass wear (Figure 6b), provide critical insights into the tribological performance of these materials under dry sliding conditions. The COF of the unreinforced material was measured at 0.46. Introducing 2.5 wt.% B_4C significantly reduced the COF by 21.74%, reaching a minimum value of 0.36. This reduction suggests that a moderate amount of B_4C effectively lowers surface friction due to the improved surface hardness and load-bearing capacity imparted by the reinforcement. However, further increasing the B_4C content to 5 wt.% led to a smaller reduction of 4.35% compared to the baseline, indicating a diminishing rate of improvement in COF reduction. At 7.5 wt.% B_4C, the COF increased by 19.57% above the baseline, which was likely due to excessive reinforcement creating abrasive interactions with the steel counterbody, leading to higher friction.

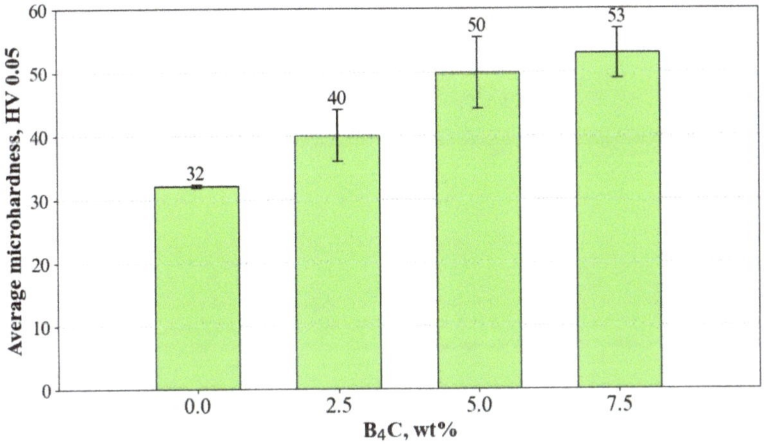

Figure 5. Microhardness values of Al10Cu alloy and its composites reinforced with varying B_4C contents (0, 2.5, 5, and 7.5 wt.%).

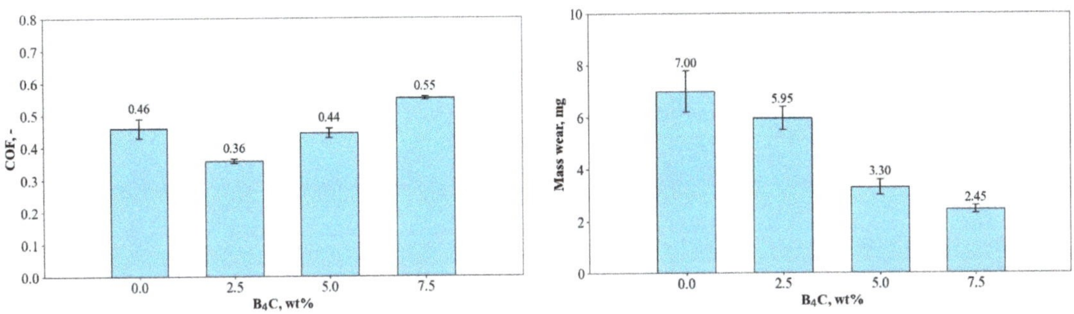

(**a**) (**b**)

Figure 6. Wear behavior of Al10Cu alloy and its composites reinforced with varying B_4C contents (0, 2.5, 5, and 7.5 wt.%): (**a**) COF values; (**b**) mass wear values.

Mass wear exhibited a consistent decrease with the addition of B_4C, showcasing the positive impact of reinforcement on wear resistance. The unreinforced composite had the highest mass wear at 7.00 mg. Incorporating 2.5 wt.% B_4C reduced wear by 15.00%, which was followed by a significant reduction of 52.86% at 5 wt.% B_4C. The lowest wear value of 2.45 mg was recorded at 7.5 wt.% B_4C, reflecting a 65.00% improvement over the baseline. The SEM images (Figure 3) confirm this trend, showing finer grooves and minimal material removal at higher reinforcement levels, which was attributed to the uniform dispersion of B_4C particles and their role in enhancing surface protection and load-bearing capacity.

3.3. COF Prediction and Model Performance Evaluation

The COF prediction and model performance evaluation for the Al10Cu alloy and its composites with varying B_4C contents (0, 2.5, 5, and 7.5 wt.%) were conducted using a

GRU-based neural network, as presented in Figure 7. The predictive performance of the model was assessed using evaluation metrics and compared against experimental COF values. Additionally, descriptive statistics provided further insight into the distribution of COF values for each material composition. The GRU model demonstrated high accuracy in predicting the COF for all materials, as evidenced by the evaluation metrics summarized in Table 2.

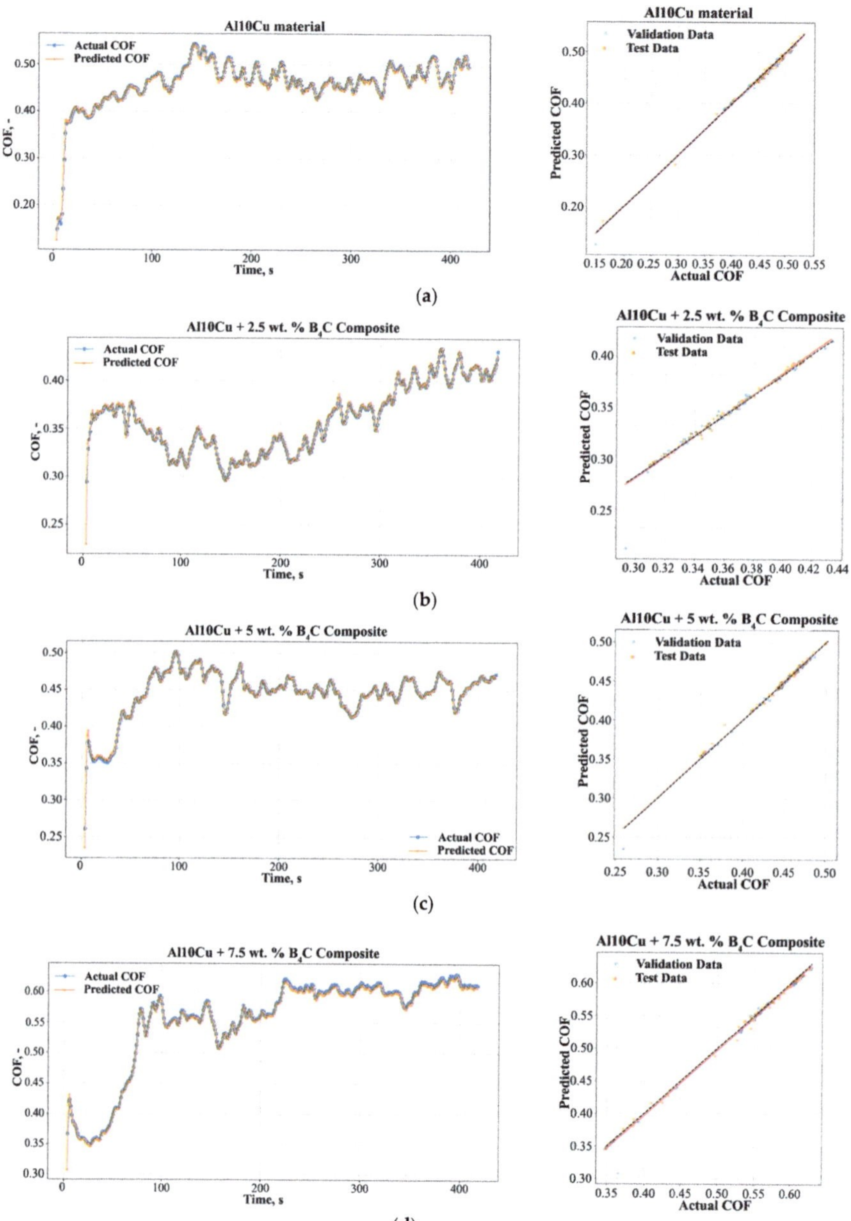

Figure 7. COF prediction and model performance evaluation of Al10Cu materials: (**a**) without reinforcement; (**b**) with 2.5 wt.% B_4C; (**c**) with 5 wt.% B_4C; (**d**) with 7.5 wt.% B_4C.

Table 2. Evaluation metrics for GRU model on Al10Cu composites with different B_4C concentrations.

Material	Validation RMSE	Validation MAE	Validation R^2	Test RMSE	Test MAE	Test R^2
Al10Cu	0.0054	0.0040	0.9892	0.0047	0.0036	0.9946
Al10Cu + 2.5 wt.% B_4C	0.0063	0.0031	0.9625	0.0028	0.0022	0.9934
Al10Cu + 5 wt.% B_4C	0.0039	0.0022	0.9916	0.0029	0.0021	0.9939
Al10Cu + 7.5 wt.% B_4C	0.0085	0.0044	0.9917	0.0051	0.0041	0.9965

For the unreinforced Al10Cu alloy, the model achieved a validation RMSE of 0.0054 and an R^2 of 0.9892, indicating an excellent correlation between predicted and experimental values. The COF predictions for the unreinforced material showed higher mean values (0.4577) and a wider range of variation, as reflected in the descriptive statistics from Table 3. These findings are consistent with the higher friction observed experimentally, which was attributed to the adhesive wear mechanisms dominating in the absence of reinforcement. The introduction of 2.5 wt.% B_4C reinforcement improved the predictive performance of the GRU model with a validation RMSE of 0.0063 and an R^2 of 0.9625. The mean COF value for this composite was reduced to 0.3582, representing a 21.75% decrease compared to the unreinforced material. The descriptive statistics revealed a narrower distribution of COF values (Std. Dev. = 0.0407), highlighting the role of B_4C particles in stabilizing frictional behavior by reducing surface interaction between the counterface and the aluminum matrix. For the composite with 5 wt.% B_4C, the GRU model achieved the best overall predictive performance with a validation RMSE of 0.0039 and an R^2 of 0.9916. The mean COF value for this material was recorded at 0.4436, representing only a 3.08% reduction compared to the unreinforced alloy. The narrower COF distribution observed in the descriptive statistics (Std. Dev. = 0.0448) further supports the enhanced frictional stability of this composition. The optimal dispersion of B_4C particles at this reinforcement level contributed to a more uniform load-bearing capacity, reducing the overall friction and enhancing the tribological performance. For the composite with 7.5 wt.% B_4C, the model maintained high predictive accuracy with a validation RMSE of 0.0085 and an R^2 of 0.9917. However, the mean COF value slightly increased to 0.5509 compared to the 5 wt.% B_4C composite, representing a 24.23% increase. Despite this, the mean COF for the 7.5 wt.% B_4C composite still reflected a significant improvement over the unreinforced material. The descriptive statistics showed a broader distribution of COF values (Std. Dev. = 0.0860).

Table 3. Descriptive statistics for COF of Al10Cu composites with different B_4C concentrations.

Material	Mean	Std. Dev.	Min	25%	50%	75%	Max
Al10Cu	0.4577	0.0612	0.0093	0.4463	0.4667	0.4865	0.5447
Al10Cu + 2.5 wt.% B_4C	0.3582	0.0407	0.0291	0.3297	0.3580	0.3864	0.4376
Al10Cu + 5 wt.% B_4C	0.4436	0.0448	0.0234	0.4408	0.4510	0.4659	0.5022
Al10Cu + 7.5 wt.% B_4C	0.5509	0.0860	0.0318	0.5456	0.5816	0.6082	0.6306

The GRU model's predictive performance across all materials underscores its robustness in capturing the complex relationships between material composition, microstructure, and tribological behavior. The results further illustrate the effectiveness of B_4C reinforcement in reducing friction and improving wear resistance with the optimal performance observed for the 5 wt.% B_4C composite. The descriptive statistics and evaluation metrics collectively highlight the balance between reinforcement content and particle dispersion, emphasizing the importance of achieving uniform reinforcement distribution to maximize the benefits of B_4C in Al10Cu composites.

3.4. Feature Importance Analysis for Mass Wear and COF Prediction

The feature importance analysis for predicting the COF and mass wear in Al10Cu composites reinforced with varying B_4C contents (0, 2.5, 5, and 7.5 wt.%) was conducted using Random Forest models. The results, presented in Figure 8, reveal the relative contributions of key features—reinforcement content and microhardness—to the predictive accuracy of the models for COF (Figure 8a) and mass wear (Figure 8b). These insights underscore the dominant role of material properties, particularly reinforcement content and microhardness, in determining tribological performance.

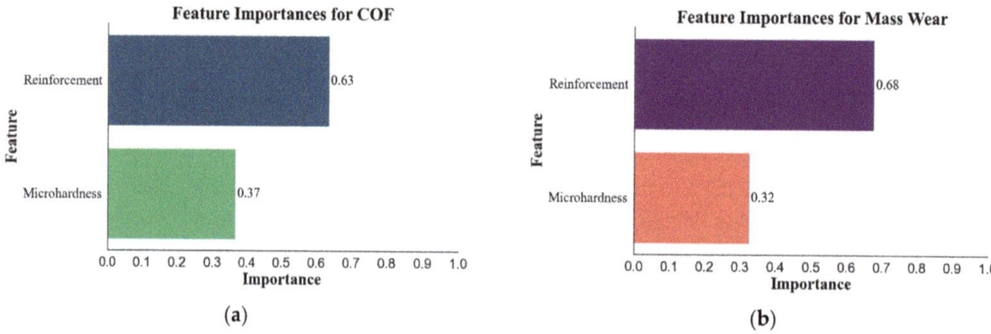

Figure 8. Feature importance analysis for Random Forest models used to predict outcomes in Al10Cu composites with varying B_4C reinforcement levels: (**a**) COF; (**b**) mass wear.

For COF prediction (Figure 8a), reinforcement content was identified as the most critical feature, with an importance value of 0.63, indicating its dominant influence on frictional behavior. The introduction of B_4C particles altered the surface interaction between the matrix and the counterbody, reducing COF by mitigating adhesion and enabling more efficient load transfer. Microhardness followed closely as the second most significant feature with an importance value of 0.37. This reflects the role of increased surface hardness, achieved through reinforcement, in minimizing surface deformation and reducing COF. The combined influence of reinforcement and microhardness indicates that material properties are the primary drivers of COF changes in Al10Cu-B_4C composites.

For mass wear prediction (Figure 8b), the importance of reinforcement content was even more pronounced with the highest importance value of 0.68. This highlights the critical role of B_4C particles in enhancing wear resistance by acting as hard, load-bearing reinforcements that reduce material loss during sliding. Microhardness was the second most influential feature with an importance value of 0.32, demonstrating the synergy between reinforcement and surface hardness in resisting wear.

Comparing the two plots, reinforcement content has a slightly higher importance for predicting mass wear (0.68) compared to COF (0.63), reflecting the stronger correlation between wear resistance and reinforcement content. In contrast, microhardness plays a more balanced role across both outcomes with similar importance values for COF (0.37) and mass wear (0.32). These findings suggest that while reinforcement content directly governs the wear mechanisms, microhardness complements this effect by mitigating surface damage, influencing both friction and wear. The feature importance trends align well with the experimental observations. Increasing the B_4C content progressively reduced wear severity, as observed in the SEM images, and enhanced the tribological performance of Al10Cu-B_4C composites under dry sliding conditions.

4. Conclusions

This study investigated the wear behavior, microstructural characteristics, and tribological performance of Al10Cu composites reinforced with varying B_4C concentrations (0, 2.5, 5, and 7.5 wt.%) fabricated via powder metallurgy in the as-sintered condition. Additionally, a GRU neural network was developed to predict the COF, providing a novel integration of ML with advanced composite analysis. The results offer valuable insights into the interplay between reinforcement content, microstructure, mechanical properties, and tribological performance, highlighting the potential of these composites for advanced applications.

1. The incorporation of B_4C particles into the $Al_{10}Cu$ matrix was successful with uniform dispersion observed at 5 wt.% and 7.5 wt.% reinforcement. This uniformity enhanced the composites' structural integrity and wear resistance by reducing particle agglomeration.
2. Pin-on-disk testing demonstrated that the addition of B_4C reinforcement led to substantial reductions in COF and mass wear. The lowest COF (0.36) was achieved with 2.5 wt.% B_4C, representing a 21.74% reduction compared to the unreinforced alloy. However, further increases in B_4C content resulted in diminishing returns with an eventual increase in COF. On the other hand, wear resistance showed consistent improvement with increasing B_4C content, beginning with a 15.00% reduction in mass wear at 2.5 wt.% B_4C and culminating in a remarkable 65.00% reduction at 7.5 wt.%, illustrating the protective effects of the hard B_4C particles.
3. Microhardness results revealed a consistent improvement in hardness with increasing B_4C content. The microhardness of the unreinforced Al10Cu alloy (32 HV) increased to 40 HV (25.00%) at 2.5 wt.% B_4C, further rising to 50 HV (56.00%) at 5 wt.% B_4C, and achieving a maximum of 53 HV (65.37%) at 7.5 wt.% B_4C. These findings highlight the significant hardening effect of B_4C microparticles with incremental gains becoming more gradual at higher reinforcement levels.
4. The GRU neural network developed in this study demonstrated exceptional accuracy in predicting the COF with R^2 values of 0.9965 and 0.9917 for the test and validation sets, respectively.
5. Feature importance analysis identified reinforcement-related parameters as the dominant predictors of COF and wear behavior, underscoring the critical role of B_4C particles in enhancing tribological performance.

These findings underscore the potential of Al10Cu-B_4C composites for high-performance applications requiring superior wear resistance and mechanical properties. Furthermore, the integration of ML demonstrates a powerful approach to predicting and optimizing the tribological performance of these materials, paving the way for advancements in composite design and manufacturing.

Author Contributions: Conceptualization, M.K. and L.D.; methodology, M.K., L.D., V.P., R.D. and K.K.; software, M.K., L.D., V.P., R.D. and K.K.; validation, M.K., L.D., V.P., R.D. and K.K.; formal analysis, M.K.; investigation, M.K., V.P., R.D., K.K. and B.Y.; resources, M.K., V.P., R.D., K.K. and B.Y.; data curation, M.K., V.P., R.D., K.K. and B.Y.; writing—original draft preparation, M.K.; writing—review and editing, M.K. and L.D.; visualization, M.K., V.P., R.D. and K.K.; supervision, M.K. and L.D.; project administration, M.K.; funding acquisition, M.K. All authors have read and agreed to the published version of the manuscript.

Funding: This research was funded by the Bulgarian National Science Fund, Project КП-06-Н77/5 "Self-lubricating hybrid aluminum metal matrix composites: synthesis, experimental and computer modeling of mechanical and tribological properties".

Data Availability Statement: Data will be made available on request.

Conflicts of Interest: The authors declare no conflicts of interest.

References

1. Sharath, B.N.; Pradeep, D.G.; Madhu, K.S. A Review on the Potential Impact of Age Hardening on Aluminium Alloys and Hybrid Composites for Engineering Applications. *Prog. Eng. Sci.* **2024**, *1*, 100013. [CrossRef]
2. Annamalai, A.R.; James, J.; Kuppan, P.; Hemakumar, S.; Jen, C.-P. Studies on Mechanical Property of Squeeze Cast and Heat Treated AA6061-ZrO2 Composite. *Mater. Res. Express* **2020**, *7*, 086506. [CrossRef]
3. Dyakova, V.; Kostova, Y.; Tzaneva, B.; Spasova, H.; Kovacheva, D. Effect of Cu as a Minority Alloying Element on the Corrosion Behaviour of Amorphous and Crystalline Al-Ni-Si Alloy. *Materials* **2023**, *16*, 5446. [CrossRef] [PubMed]
4. Godbole, K.; Bhushan, B.; Narayana Murty, S.V.S.; Mondal, K. Al-Si Controlled Expansion Alloys for Electronic Packaging Applications. *Prog. Mater. Sci.* **2024**, *144*, 101268. [CrossRef]
5. Dyakova, V.; Spasova, H.; Kostova, Y.; Mourdjeva, Y.; Stefanov, G. Effect of Cu as Minority Alloying Element on Glass Forming Ability and Crystallization Behavior of Rapidly Solidified Al-Si-Ni Ribbons. Environment. Technology. Resources. In Proceedings of the 14th International Scientific and Practical Conference, Rezekne, Latvia, 15–16 June 2023; Volume 3, pp. 69–73. [CrossRef]
6. Dumitraschkewitz, P.; Gerstl, S.S.A.; Stephenson, L.T.; Uggowitzer, P.J.; Pogatscher, S. Clustering in Age-hardenable Aluminum Alloys. *Adv. Eng. Mater.* **2018**, *20*, 1800255. [CrossRef]
7. Alizadeh, A.; Maleki, M.; Abdollahi, A. Preparation of Super-High Strength Nanostructured B 4 C Reinforced Al-2Cu Aluminum Alloy Matrix Composites by Mechanical Milling and Hot Press Method: Microstructural, Mechanical and Tribological Characterization. *Adv. Powder Technol.* **2017**, *28*, 3274–3287. [CrossRef]
8. Alizadeh, A.; Taheri-Nassaj, E. Mechanical Properties and Wear Behavior of Al–2wt.% Cu Alloy Composites Reinforced by B4C Nanoparticles and Fabricated by Mechanical Milling and Hot Extrusion. *Mater. Charact.* **2012**, *67*, 119–128. [CrossRef]
9. Asghar, Z.; Latif, M.A.; Rafi-ud-Din; Nazar, Z.; Ali, F.; Basit, A.; Badshah, S.; Subhani, T. Effect of Distribution of B4C on the Mechanical Behaviour of Al-6061/B4C Composite. *Powder Metall.* **2018**, *61*, 293–300. [CrossRef]
10. Bujari, S.S.; Kurahatti, R.V. Microstructure and Wear Behavior of B4C Particulates Reinforced Al-4.5%cu Alloy Composites. *IOSR J. Mech. Civ. Eng.* **2017**, *14*, 4–9. [CrossRef]
11. Manohar, G.; Pandey, K.M.; Maity, S.R. Characterization of Boron Carbide (B4C) Particle Reinforced Aluminium Metal Matrix Composites Fabricated by Powder Metallurgy Techniques—A Review. *Mater. Today* **2021**, *45*, 6882–6888. [CrossRef]
12. Pozdniakov, A.V.; Zolotorevskiy, V.S.; Barkov, R.Y.; Lotfy, A.; Bazlov, A.I. Microstructure and Material Characterization of 6063/B4C and 1545K/B4C Composites Produced by Two Stir Casting Techniques for Nuclear Applications. *J. Alloys Compd.* **2016**, *664*, 317–320. [CrossRef]
13. Zhang, P.; Li, Y.; Wang, W.; Gao, Z.; Wang, B. The Design, Fabrication and Properties of B4C/Al Neutron Absorbers. *J. Nucl. Mater.* **2013**, *437*, 350–358. [CrossRef]
14. Lokesh, G.N.; Karunakara, S. Impact of Particle Size Distribution for Variable Mixing Time on Mechanical Properties and Microstructural Evaluation of Al-Cu/B4C Composite. *Mater. Today* **2020**, *22*, 1715–1722. [CrossRef]
15. Qadirkhan, M.A.; Arif Siddiqui, M. Fabrication and Testing of Al-Cu-B4C Composite by Mechanical Alloying. *Int. J. Innov. Res. Sci. Eng. Technol.* **2015**, *4*, 8362–8368. Available online: https://www.ijirset.com/upload/2015/september/50_Fabrication.pdf (accessed on 25 December 2024).
16. Ozkaya, S.; Canakci, A. Effect of the B4C Content and the Milling Time on the Synthesis, Consolidation and Mechanical Properties of AlCuMg-B4C Nanocomposites Synthesized by Mechanical Milling. *Powder Technol.* **2016**, *297*, 8–16. [CrossRef]
17. Uthayakumar, M.; Aravindan, S.; Rajkumar, K. Wear Performance of Al–SiC–B4C Hybrid Composites under Dry Sliding Conditions. *Mater. Des.* **2013**, *47*, 456–464. [CrossRef]
18. Akçamlı, N.; Şenyurt, B. B4C Particulate-Reinforced Al-8.5 Wt% Si-3.5 Wt% Cu Matrix Composites: Powder Metallurgical Fabrication, Age Hardening, and Characterization. *Ceram. Int.* **2021**, *47*, 6813–6826. [CrossRef]
19. Bharathiraja, P.; Xavior, A. Effect of B4C and Graphene on the Microstructural and Mechanical Properties of Al6061 Matrix Composites. *J. Mater. Res. Technol.* **2024**, *31*, 496–505. [CrossRef]
20. Liu, X.; Tian, S.; Tao, F.; Du, H.; Yu, W. Machine Learning-Assisted Modeling of Composite Materials and Structures: A Review. In Proceedings of the AIAA Scitech 2021 Forum, Virtual, 11 January 2021; American Institute of Aeronautics and Astronautics: Reston, VA, USA, 2021.
21. Kordijazi, A.; Zhao, T.; Zhang, J.; Alrfou, K.; Rohatgi, P. A Review of Application of Machine Learning in Design, Synthesis, and Characterization of Metal Matrix Composites: Current Status and Emerging Applications. *JOM* **2021**, *73*, 2060–2074. [CrossRef]
22. Banerjee, T.; Dey, S.; Sekhar, A.P.; Datta, S.; Das, D. Design of Alumina Reinforced Aluminium Alloy Composites with Improved Tribo-Mechanical Properties: A Machine Learning Approach. *Trans. Indian Inst. Met.* **2020**, *73*, 3059–3069. [CrossRef]
23. Thankachan, T.; Soorya Prakash, K.; Malini, R.; Ramu, S.; Sundararaj, P.; Rajandran, S.; Rammasamy, D.; Jothi, S. Prediction of Surface Roughness and Material Removal Rate in Wire Electrical Discharge Machining on Aluminum Based Alloys/Composites Using Taguchi Coupled Grey Relational Analysis and Artificial Neural Networks. *Appl. Surf. Sci.* **2019**, *472*, 22–35. [CrossRef]

4. Idrisi, A.H.; Hamid Ismail Mourad, A. Wear Performance Analysis of Aluminum Matrix Composites Using Artificial Neural Network. In Proceedings of the 2019 Advances in Science and Engineering Technology International Conferences (ASET), Dubai, United Arab Emirates, 26 March–10 April 2019; IEEE: Piscataway, NJ, USA, 2019.
5. Deng, Z.-H.; Yin, H.-Q.; Jiang, X.; Zhang, C.; Zhang, G.-F.; Xu, B.; Yang, G.-Q.; Zhang, T.; Wu, M.; Qu, X.-H. Machine-Learning-Assisted Prediction of the Mechanical Properties of Cu-Al Alloy. *Int. J. Miner. Metall. Mater.* **2020**, *27*, 362–373. [CrossRef]
6. Hou, B.; Wang, Z.; Hu, C.; Zhu, Y. Gate Recurrent Unit Neural Network Based High-Precision Feedforward Control for Piezoelectric Nanopositioning Stage. *Eng. Appl. Artif. Intell.* **2024**, *133*, 108126. [CrossRef]
7. Alakbari, F.S.; Mohyaldinn, M.E.; Ayoub, M.A.; Hussein, I.A.; Muhsan, A.S.; Ridha, S.; Salih, A.A. A Gated Recurrent Unit Model to Predict Poisson's Ratio Using Deep Learning. *J. Rock Mech. Geotech. Eng.* **2023**, *16*, 123–135. [CrossRef]
8. Hasan, M.S.; Kordijazi, A.; Rohatgi, P.K.; Nosonovsky, M. Triboinformatic Modeling of Dry Friction and Wear of Aluminum Base Alloys Using Machine Learning Algorithms. *Tribol. Int.* **2021**, *161*, 107065. [CrossRef]
9. Kolev, M. A Novel Approach to Predict the Effect of Chemical Composition and Thermo-Mechanical Processing Parameters on Cu–Ni–Si Alloys Using a Hybrid Deep Learning and Ensemble Learning Model. *Compos. Commun.* **2024**, *48*, 101903. [CrossRef]

lubricants

MDPI

Article

Influence of Titanium Oxide and Calcium Carbonate Powder as Additives on Tribological Characteristics of Lubricants

Abhinit Shrivastava [1], Surinder Kumar [1], Sumit Kumar [1], Hemant Kumar [2], Sumika Chauhan [3,4], Govind Vashishtha [3,5,*] and Rajesh Kumar [1]

1 Precision Metrology Laboratory, Department of Mechanical Engineering, Sant Longowal Institute of Engineering and Technology, Longowal 148106, India; abhinitindia@gmail.com (A.S.); surinderthakur38@gmail.com (S.K.); sumitkumar@sliet.ac.in (S.K.); rajesh_krs@sliet.ac.in (R.K.)
2 Department of Chemistry, Sant Longowal Institute of Engineering and Technology, Longowal 148106, India; hemantk1331@gmail.com
3 Faculty of Geoengineering, Mining and Geology, Wroclaw University of Science and Technology, Na Grobli 15, 50-421 Wroclaw, Poland; sumi.chauhan2@gmail.com
4 Division of Research and Development, Lovely Professional University, Phagwara 144411, India
5 Department of Mechanical Engineering, Graphic Era Deemed to be University, Dehradun 248002, India
* Correspondence: govindyudivashishtha@gmail.com

Abstract: The properties of lubricating oils are greatly enhanced by the incorporation of additives. With technological advancements, numerous additives have been developed and proven effective for this purpose. Some additives enhance anti-wear and anti-friction characteristics, while others improve the oil's viscosity index. It has been noted that certain additives influence more than one property of the lubricating oil. In this study, a mixture of TiO_2 at 0.2 wt.%, 0.3 wt.%, and 0.4 wt.% and $CaCO_3$ at 0.4 wt.%, 0.6 wt.%, and 0.8 wt.% was used as an additive in gear oil EP140 to prepare the samples. A pin-on-disc test was conducted for the tribological characterization of the various samples. A combination of 0.2 wt.% TiO_2 and 0.4 wt.% $CaCO_3$ particles in the gear oil resulted in a remarkable 88.23% reduction in wear compared to the base gear lubricating oil (EP140). The combination of 0.3 wt.% TiO_2 and 0.6 wt.% $CaCO_3$ particles in the gear oil led to a significant 36.84% reduction in the coefficient of friction. Field Emission Scanning Electron Microscopy (FESEM) revealed that the pin tested with sample S1 (gear oil containing 0.2 wt.% TiO_2 and 0.4 wt.% $CaCO_3$) exhibited a smoother wear surface than the base lubricating oil.

Keywords: $CaCO_3$; TiO_2; base oil

1. Introduction

Lubrication, a vital component of modern engineering, involves strategically introducing substances between moving surfaces to reduce wear and tear. These substances, known as lubricants, include a diverse array of materials from solid forms like graphite to semi-solid greases and liquid lubricating oils, each designed for specific applications and operating conditions [1]. The primary role of a lubricant is to prevent direct metal to metal contact, thereby significantly reducing friction, minimizing wear rates, and averting catastrophic failures, such as seizing. The effectiveness of lubricants, especially liquid lubricating oils, goes beyond merely reducing friction; they also actively maintain surface cleanliness, remove debris, and prevent the buildup of harmful contaminants that could compromise system integrity [2–5].

With technological progress, the development of high-performance lubricants has advanced significantly. While base oils, traditionally derived from crude oil refining, provide a fundamental lubricating capability, they often fail to meet the demands of modern machinery operating under diverse and often extreme conditions [6,7]. The inherent limitations of base oils necessitate the incorporation of additives that are carefully selected and precisely formulated to enhance their performance characteristics [8,9]. These additives, which include both organic and inorganic compounds, are introduced at concentrations typically ranging from 0.1% to 5% of the total oil volume and meticulously adjusted to meet the specific requirements of the intended application. This customization is crucial because the optimal additive package for a high-speed engine differs substantially from that required for a heavy-duty industrial gear system [9,10]. Additives may exist in solution, fully dissolved within the base oil, or as finely dispersed solid particles, ensuring optimal distribution and performance throughout the lubricated system [11].

The versatility of additive technology lies in its capacity to address numerous performance deficiencies in base oils. Additives are not just enhancements; they are essential components that enable lubricants to achieve performance levels far beyond the capabilities of the base oil [12]. These enhancements span a broad spectrum of crucial properties. For instance, anti-wear additives minimize surface damage and extend the lifespan of components. Anti-friction additives further reduce energy lost to friction, improve efficiency, and lower operating temperatures [13]. Viscosity modifiers ensure that the oil maintains optimal viscosity across a wide range of temperatures, which is vital for maintaining lubrication under both extreme heat and cold conditions. Pour point depressants counteract the tendency of oils to thicken and solidify at low temperatures, ensuring proper lubrication even in subzero environments. Other additives address issues like oxidation stability, corrosion protection, and foam prevention [14]. This complex interplay between the base oil and additives results in a synergistic effect where the combined properties far exceed the sum of their individual components, producing a sophisticated and highly effective lubricating fluid. This introduction lays the groundwork for a deeper exploration of specific types of additives, their mechanisms of action, and their impact on the overall performance of lubricating oils [15,16]. In general, additives perform three basic lubricating roles [15]:

- Enhance the existing properties of the base oil, such as the viscosity index, corrosion resistance, wear, and friction resistance.
- Suppress the undesirable properties of the base oil, such as reducing the pour point of the oil using pour point depressants.
- Impart new properties to base oils, such as high load-carrying capacity, using extreme pressure (EP) additives.

The integration of nanoparticles into lubricants has proven to be a highly effective strategy for enhancing tribological performance. Numerous studies have demonstrated the significant potential of nanoparticles in reducing friction and wear, thereby improving efficiency and extending the lifespan of components across various applications [17]. Research has underscored the positive impact of different types of nanoparticles on lubricant properties. For instance, adding copper nanoparticles to lubricating oil has been shown to effectively reduce both the coefficient of friction and the wear rate on contacting surfaces [18]. Moreover, incorporating 2 wt.% copper oxide (CuO) nanoparticles into polyalphaolefin (POA) base oil resulted in a remarkable 273% increase in load-carrying capacity [19], highlighting the potential of nanoparticles to significantly enhance lubricant performance under heavy loads.

The advantages of these nanoparticles extend beyond those of their metallic counterparts. Incorporating 0.4 vol.% silicon dioxide (SiO_2) nanoparticles into R134a refrigerant not only enhanced the refrigerant's heat transfer properties but also improved the tribological characteristics of polyalkylene glycol (PAG) lubricants used alongside it [11]. Similarly, hybrid nanoparticles made from zinc oxide (ZnO) and multiwalled carbon nanotubes

(MWCNTs) yielded promising outcomes. An optimal concentration of 0.25 wt.% of these hybrid nanoparticles, in a 3:1 ZnO:MWCNT ratio, mixed into 10W40 engine oil resulted in significant reductions in both the coefficient of friction (32.30%) and the wear volume (74.48%) compared to the base oil alone [11]. Further studies confirmed the positive effects of ZnO nanoparticles, showing enhancements in the thermophysical and tribological properties of diesel oil. A 0.25 wt.% concentration of ZnO nanoparticles dispersed in oleic acid (ZnO-OA) and added to POA40 base oil led to a 25% reduction in the friction coefficient and an 82% reduction in wear (measured by cross-sectional area) compared to the neat base oil [20–24].

Titanium dioxide (TiO_2) nanoparticles have been demonstrated to be effective as lubricant additives. A concentration of 0.3 wt.% of TiO_2 nanoparticles in multi-grade engine oil led to an impressive 86% reduction in the coefficient of friction [25]. Moreover, varying concentrations of TiO_2 nanoparticles (ranging from 0.005% to 0.025% by volume) in the base oil has been shown to enhance the load-carrying capacity of journal bearings by up to 40% [26]. Further research has investigated the effects of TiO_2 nanoparticles at concentrations of 0.1%, 0.5%, and 1% wt. in SAE 10 engine oil, highlighting the range of potential benefits [25]. In some instances, the mechanism behind these improvements is attributed to the formation of boundary lubrication films and adsorption films of oleic acid on rubbing surfaces, as observed with oleic-acid-modified lanthanum titanate (OA/$LaTiO_2$) particles in rapeseed oil [27,28]. These OA/$LaTiO_2$ particles, with an average diameter of 20 nm, enhance both the anti-wear and friction-reducing properties of rapeseed oil.

Calcium carbonate ($CaCO_3$) nanoparticles have demonstrated potential in enhancing the load-carrying capacity, wear resistance, and friction characteristics of polyalphaolefin (PAO) base oils [17]. When combined with cerium oxide (CeO_2) nanoparticles, these properties are further improved, particularly in terms of anti-wear and anti-friction capabilities. The optimal anti-wear performance was observed at a 5 wt.% concentration of $CaCO_3$ in an amphiphilic copolymer [17]. Additionally, studies examining the relationship between lubricant chemical structure and friction/wear behavior under boundary conditions have revealed a correlation between increased molecular size and reduced wear and friction [25]. Collectively, these findings highlight the significant potential of nanoparticles to enhance lubricant performance and address critical tribological challenges. All of these references are summarized in Table 1.

In recent years, extensive research has concentrated on enhancing tribological properties, including, specifically, the reduction of friction and wear of lubricating oils. A substantial portion of this research highlights the effectiveness of incorporating solid particles, particularly metallic oxides and carbonates, to boost lubrication performance. While numerous studies have individually examined the effects of metallic oxides or carbonates, there has been relatively limited exploration of the combined impact of these two additive groups on lubricating oil properties. This study aims to fill this gap by investigating the tribological behavior of gear-lubricating oil modified with a blend of metallic oxides and carbonates. The primary objective is to analyze how these two types of additives collectively influence the key tribological characteristics of the resulting lubricant. This approach facilitates a more comprehensive understanding of the interaction between these additive types and their impact on overall lubrication performance, potentially leading to the development of even more effective lubricant formulations.

Table 1. Summary of the literature review.

S.No.	Lubricant	Additive	Observation
1	General lubricating oils	Anti-wear additives	Minimize surface damage and prolong the lifespan of components.
2	General lubricating oils	Anti-friction additives	Reduce energy lost to friction, improve efficiency, and reduce operating temperatures.
3	General lubricating oils	Viscosity modifiers	Maintain optimal viscosity across a wide range of temperatures.
4	General lubricating oils	Pour point depressants	Counteract the tendency of oils to thicken and solidify at low temperatures.
5	General lubricating oils	Metallic oxides and carbonates	Improve friction and wear reduction.
6	Polyalphaolefin (POA) base oil	Copper oxide (CuO) nanoparticles (2 wt.%)	A 273% increase in load-carrying capacity.
7	R134a refrigerant	Silicon dioxide (SiO_2) nanoparticles (0.4 vol.%)	Positive influence on heat transfer properties of the refrigerant and tribological characteristics of polyalkylene glycol (PAG) lubricants.
8	10W40 engine oil	Zinc oxide (ZnO) and multiwalled carbon nanotube (MWCNT) hybrid nanoparticles (0.25 wt.%, 3:1 ratio)	A 32.30% reduction in coefficient of friction and a 74.48% reduction in wear volume compared to the base oil alone.
9	POA40 base oil	Zinc oxide dispersed in oleic acid (ZnO-OA) (0.25 wt.%)	A 25% reduction in the friction coefficient and an 82% reduction in wear (measured by cross-sectional area) compared to the neat base oil.
10	Multi-grade engine oil	Titanium dioxide (TiO_2) nanoparticles (0.3 wt.%)	An 86% reduction in the coefficient of friction.
11	Base oil	Titanium dioxide (TiO_2) nanoparticles (0.005% to 0.025% by volume)	Increased load-carrying capacity of journal bearings by as much as 40%.
12	SAE 10 engine oil	Titanium dioxide (TiO_2) nanoparticles (0.1%, 0.5%, and 1% wt.)	Illustrates the range of potential benefits.
13	Rapeseed oil	Oleic-acid-modified lanthanum titanate (OA/LaTiO_2) particles (20 nm diameter)	Enhances both the anti-wear and friction-reducing properties. Formation of boundary lubrication films and adsorption films of oleic acid on rubbing surfaces.
14	Polyalphaolefin (PAO) base oils	Calcium carbonate ($CaCO_3$) nanoparticles	Improves load-carrying capacity, wear resistance, and friction characteristics.
15	Amphiphilic copolymer	Calcium carbonate ($CaCO_3$) nanoparticles (5 wt.%) and cerium oxide (CeO_2) nanoparticles	Optimal anti-wear characteristics.

2. Materials and Methods

This section provides a detailed account of the materials and methods employed in preparing the lubricating oils for this study. It describes the properties of the chosen base oil and outlines the procedures used to create the various lubricant samples tested in the experiments.

2.1. Base Oil

The developed lubricating oils were intended for use in a gearbox. Consequently, a readily available market oil specifically designed for gearbox lubrication was chosen as the base oil. EP 140 is one such oil, and it has been utilized as the base oil; its characteristics are detailed in Table 2.

Table 2. Characteristics of the base oil.

S. No.	Property	Value
1	Kinematic viscosity (c-St) at 100 °C	28–33
2	Viscosity index, min	90
3	Flash point (°C)	190
4	Pour point (°C), max	−3

2.2. Additives

To enhance the properties of lubricants, additives are commonly employed. Among these, nanoparticles have garnered significant interest from researchers investigating their potential as lubricant additives. In this study, titanium oxide (TiO_2) and calcium carbonate ($CaCO_3$) powders, in the form of microparticles, were utilized to improve the tribological characteristics of EP 140 gear oil. These powders were procured from the market, and their size, shape, and composition were assessed using X-Ray Diffraction (XRD), Field Emission Scanning Electron Microscopy (FESEM), and Energy Dispersive X-Ray Spectroscopy (EDS), as depicted in Figures 1 and 2, respectively. Additionally, the average size of the additives was determined with the aid of SEM, as illustrated in Figure 3.

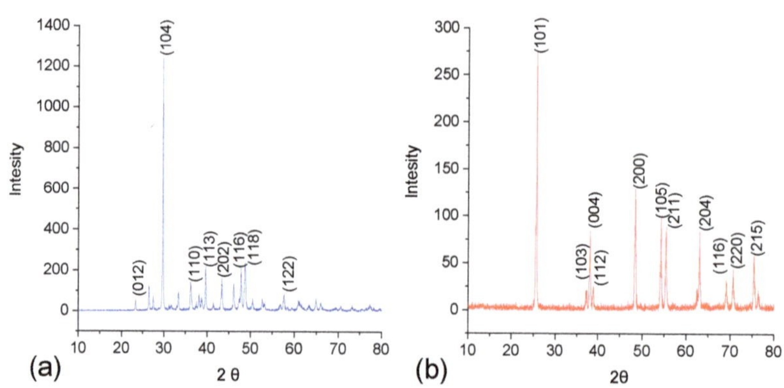

Figure 1. XRD results of (**a**) $CaCO_3$ powder and (**b**) TiO_2 powder.

2.3. Determination of Concentration of Additives

Previous research has shown that titanium dioxide (TiO_2) and calcium carbonate ($CaCO_3$) each independently reduce friction and wear in lubricating oils. This study explores the combined effects of these additives on the lubrication characteristics of base oil. Building on prior research that suggests optimal concentrations of 0.3 wt.% TiO_2 and 0.6 wt.% $CaCO_3$, samples were prepared with concentrations both above and below these optimal values, as detailed in Table 3.

Table 3. Composition of additives in base oil (BS) used for sample preparation.

S. No.	Sample Oil	TiO_2 (wt.%)	$CaCO_3$ (wt.%)
1	Sample 1 (S1)	0.2	0.4
2	Sample 2 (S2)	0.3	0.6
3	Sample 3 (S3)	0.4	0.8

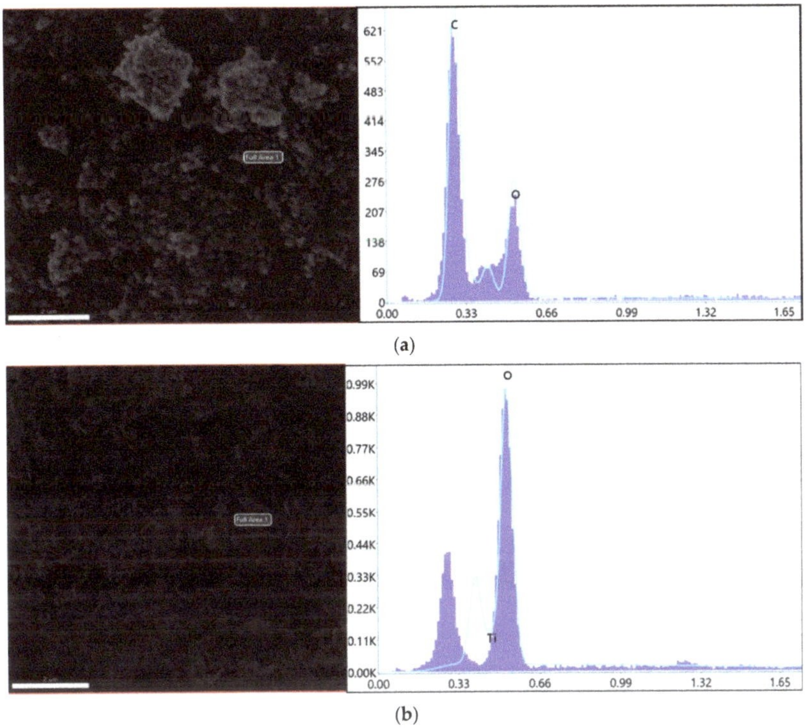

Figure 2. EDX results of (**a**) $CaCO_3$ powder and (**b**) TiO_2 powder.

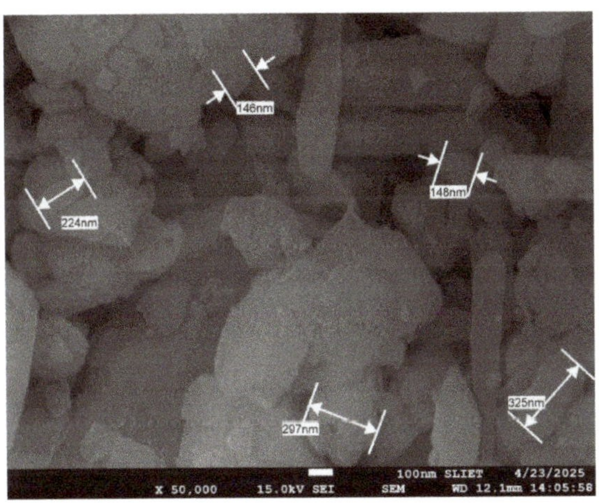

Figure 3. Size of the additives.

2.4. Methods for Lubricant Synthesis

To prepare lubricant samples, one liter of the base oil was initially weighed. The necessary amounts of titanium dioxide (TiO_2) and calcium carbonate ($CaCO_3$) particles, determined as a weight percentage of the base oil, were then measured. These additives were incorporated into the base oil and mixed with a magnetic stirrer for 30 min. To ensure thorough dispersion and minimize particle agglomeration, the mixture was sonicated overnight. The resulting homogeneous lubricant samples are detailed in Table 2. The preparation process is outlined in the flowchart shown in Figure 4.

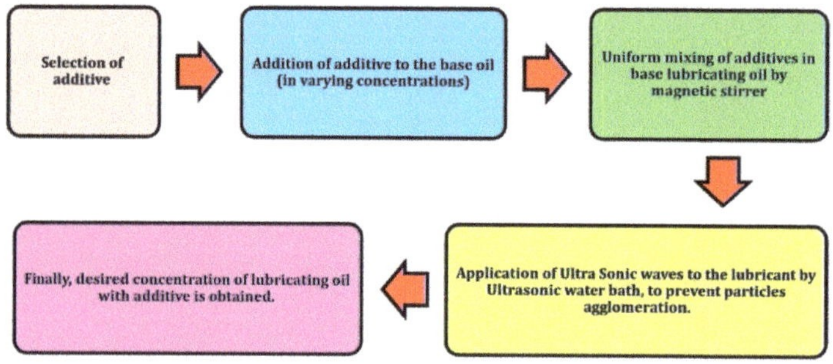

Figure 4. Procedure for lubricant synthesis.

2.5. Viscosity Index Determination

Each 400 mL sample of lubricating oil was placed in a separate beaker, which was then immersed in a temperature-controlled water bath and heated to 70 °C. Viscosity measurements were conducted at various temperatures using a digital Brookfield viscometer. Figure 5 presents a flowchart that illustrates the viscosity–temperature characterization process.

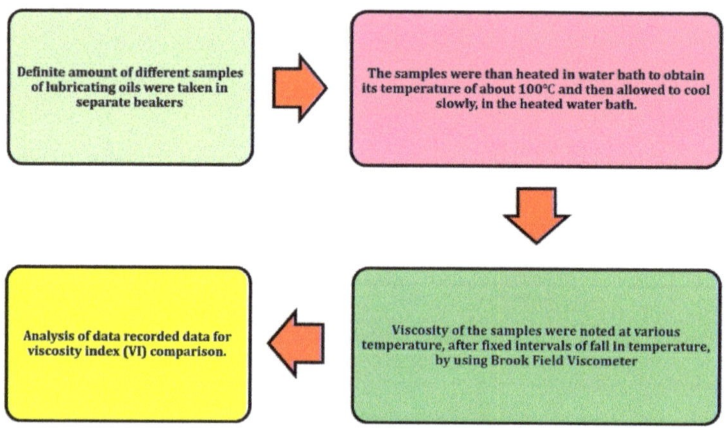

Figure 5. Procedure for viscosity–temperature relationship evaluation.

3. Tribological Testing of Synthesized Lubricants

3.1. Test Pin Fabrication

The test specimens, specifically the pins, were crafted from an MS round bar of grade 43A. These pins were machined to meet the specific dimensions required by the tribometer. The pin's diameter was consistently maintained at 8 mm, with a length of 30 mm, as illustrated in Figure 6.

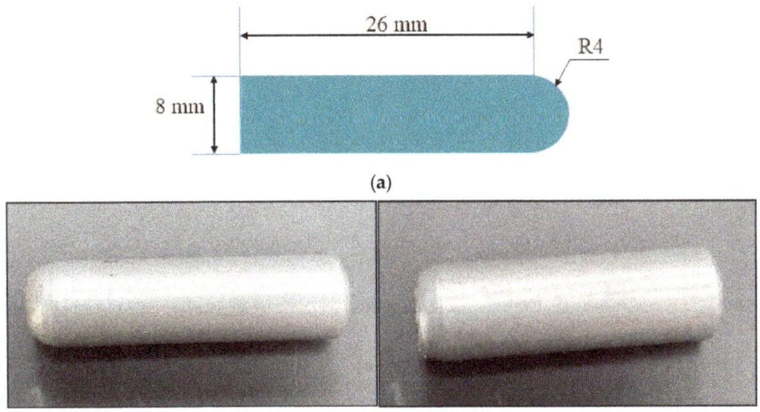

(b) (c)

Figure 6. (**a**) Sketch of friction pin, (**b**) pictorial view of pin before test, and (**c**) pictorial view of pin after test.

3.2. Pin Disc on Tribometer

The friction and wear characteristics of the prepared lubricant samples were assessed using a pin-on-disc tribometer integrated with WIN-DUCOM 2010 software. This tribometer featured a motorized spindle and chuck to secure the rotating disc, a lever arm mechanism to hold the test pin, and a loading system that allowed for controlled force application between the pin and the disc, as illustrated in Figure 7. The wear track formed on the disc was circular, accommodating multiple wear passes along the same path. Equipped with a force transducer, the tribometer precisely measured the frictional force, enabling the determination of the coefficient of friction. The complete tribometer setup is depicted in Figure 8a,b. WIN-DUCOM software facilitated data acquisition and analysis of the friction and wear data obtained during testing.

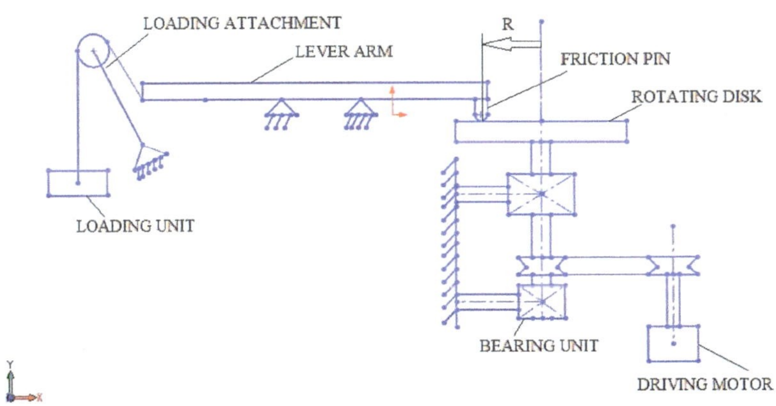

Figure 7. Schematic sketch of pin-on-disc tribometer.

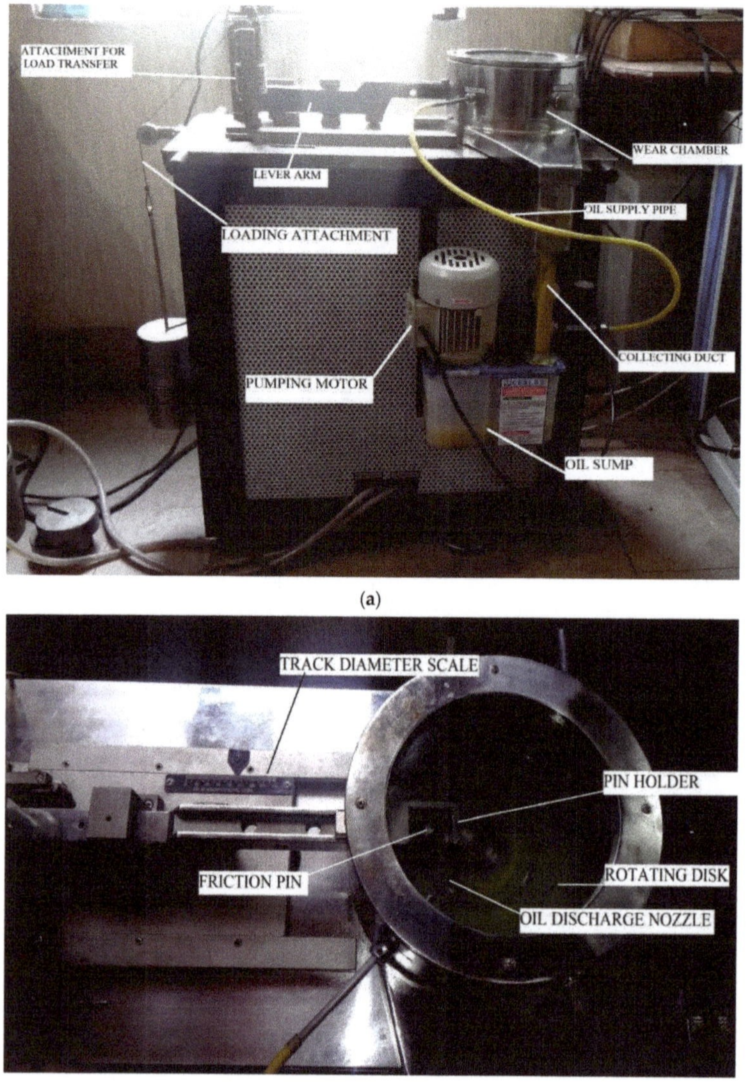

Figure 8. (**a**) Pin-on-disc tribometer; (**b**) top view of tribometer.

The operating principle of the tribometer is depicted in Figure 7. The test pin was firmly secured in a lever arm assembly and pressed against a rotating disc. This disc was mounted on a rotating chuck, which was directly connected to a constant-speed motor to ensure a uniform rotational speed throughout the test. The opposite end of the lever arm was linked to a loading unit via a loading attachment. A specified load was applied to this loading unit, and it was transferred to the pin through the lever arm, maintaining consistent contact pressure between the pin and the rotating disc for the duration of the experiment. This setup ensured that a constant load was maintained throughout the tests.

Tribological testing using a pin-on-disc apparatus involved the following steps:

1. The test pin and disc were thoroughly cleaned and dried to eliminate contaminants or foreign materials that could have affected the results.

2. The test pin was securely mounted in its holder to ensure perpendicular alignment with the disc surface to maintain consistent contact conditions.
3. The tribometer motor was started, and the rotational speed was adjusted to a predetermined value. At this stage, the pin was not in contact with the rotating disc.
4. The lever arm was adjusted to set the desired diameter for the wear track on the rotating disc.
5. The required number of disc rotations and total test duration were programmed into the tribometer.
6. An appropriate amount of mass was then added to the loading unit. The loading attachment then transferred this force to the test pin via the lever arm, applying the desired normal force to the pin–disc contact zone.
7. The lubricant pump was activated to ensure a continuous supply of lubricant to the pin–disc contact zone.
8. With the pin in contact with the rotating disc under load, the test commenced and ran until the preset time was reached, automatically stopping the apparatus.
9. The test pin and disc were carefully removed from the tribometer, and any wear debris was meticulously removed from the surfaces. Subsequent analysis of the pin and the disc allowed for the determination of the wear volume and the coefficient of friction.

4. Results and Discussion

The viscosity of lubricating oils, like most oils, is known to inversely correlate with temperature; as the temperature rises, the viscosity decreases. This reduction in viscosity lessens the oil's lubricating ability, thereby adversely impacting the overall efficiency of the system. Enhancing the viscosity index (VI) of the oil is a crucial approach to counteracting this temperature-dependent viscosity decline; a higher VI signifies a smaller change in viscosity across a given temperature range. This study examined the effect of titanium dioxide (TiO_2) and calcium carbonate ($CaCO_3$) particles on the viscosity and stability of the base oil. Three lubricant samples were prepared, each with different concentrations of additives. The viscosity of each sample, along with that of the base oil, was measured at various temperatures. The resulting viscosity values (in mPa·s) are presented in Table 4, facilitating a comparison of the viscosity–temperature behavior among the different lubricant formulations. This comparison underscores the effectiveness of additives in maintaining viscosity at elevated temperatures.

Table 4. Viscosity of oils at different temperatures.

Oil Sample	Temperature						
	30 °C	35 °C	40 °C	45 °C	50 °C	55 °C	60 °C
Base oil	431.5	280.0	244.8	121.2	108.1	96.4	87.5
Sample 1	471.6	325.0	206.0	144.0	117.9	104.0	93.4
Sample 2	484.8	324.6	240.0	179.0	132.9	115.0	107.0
Sample 3	528.0	363.9	275.0	195.3	146.4	142.0	132.3

The experimental results demonstrated a positive correlation between additive concentration and oil viscosity; as the amount of TiO_2 and $CaCO_3$ additives increased, so did the viscosity of the lubricant. This rise in viscosity can be attributed to the formation of agglomerates of additive particles within the oil, which effectively enhances the bulk viscosity. As illustrated in Figure 9, the viscosity of all lubricant samples, including the base oil, decreased with rising temperature, aligning with the typical behavior of lubricating oils. However, at elevated temperatures, samples with higher additive concentrations exhibited greater viscosities than those with lower concentrations. Moreover, the rate at which viscosity decreased with temperature was more pronounced with higher additive concentrations. Lubricants with greater additive concentrations showed a steeper decline in viscosity as the temperature rose. This indicates a complex interplay between additive concentration, temperature, and agglomeration in influencing the overall viscosity of the formulated lubricating oils.

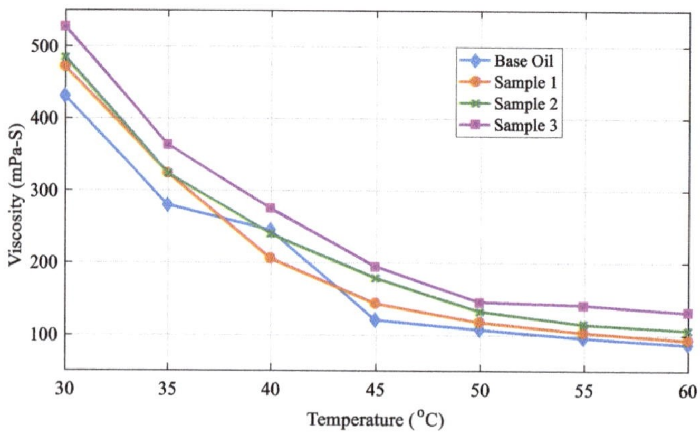

Figure 9. Viscosity variation of oils with temperature.

4.1. Anti-Friction Test of Lubricating Oils

To assess the impact of various TiO_2 and $CaCO_3$ combinations on the anti-friction properties of the base oil, a series of tribological tests were conducted using a pin-on-disc tribometer. Multiple samples of the formulated lubricating oils were tested under different operating conditions to evaluate their anti-friction performance. Specifically, tests were carried out at two rotational speeds and two distinct load levels to thoroughly examine how these parameters influence the lubricating properties of the formulated oils, as detailed in Table 5. The following parameters were kept constant for all tests: a pin revolution track diameter of 90 mm and a total test duration of 10 min. This consistent experimental design ensured that any variations in the measured friction coefficients could be directly attributed to differences in lubricant composition rather than other factors. The results of these tests provided valuable insights into the effectiveness of various additive combinations in reducing friction.

Table 5 presents the coefficient of friction (COF) results obtained under various test conditions. At low speed and low load, sample S2, which contains 0.3 wt.% TiO_2 and 0.6 wt.% $CaCO_3$, exhibited the lowest COF. A general trend emerged, showing that the COF initially decreased to a minimum at intermediate additive concentrations before rising again with further increases in concentration. This suggests the existence of an optimal additive concentration range for minimizing friction under these specific conditions. However, under high load and low speed conditions, S3 (0.4 wt.% TiO_2 and 0.8 wt.% $CaCO_3$) demonstrated the lowest COF, as shown in Table 5. This indicates that the optimal additive concentration for reducing friction can be influenced by the applied load. The observed nonlinear relationship between additive concentration and COF suggests a complex interplay between the additives and their interactions within the base oil under various load and speed conditions.

Table 5. Coefficient of friction of oils at different load and speed conditions.

Condition	Oil Sample	Load (kg)	Speed (rpm)	Frictional Force (N)	Coefficient of Friction	% Reduction in COF
Low load, low speed	No lubricant	4	350	2.9	0.068	-
	Base oil	4	350	1.4	0.035	48.53
	S1	4	350	1.4	0.035	48.53
	S2	4	350	1.3	0.032	52.94
	S3	4	350	2.1	0.05	26.47
High load, low speed	No lubricant	10	350	12	0.105	-
	Base oil	10	350	6.4	0.062	40.95
	S1	10	350	5.8	0.057	45.71
	S2	10	350	6.3	0.061	41.90
	S3	10	350	5.4	0.052	50.47
Low load, high speed	No lubricant	4	500	9.2	0.212	-
	Base oil	4	500	1.5	0.038	82.07
	S1	4	500	1.8	0.044	79.24
	S2	4	500	1.1	0.024	88.67
	S3	4	500	2.1	0.054	74.58
High load, high speed	No lubricant	10	500	33	0.31	-
	Base oil	10	500	5.4	0.054	82.58
	S1	10	500	5.8	0.057	81.61
	S2	10	500	7.1	0.07	77.41
	S3	10	500	7.5	0.075	75.80

Under conditions of low load and high speed, the lubricant sample with an intermediate additive concentration demonstrated the lowest coefficient of friction (COF), as shown in Table 5. However, a notable change in behavior occurred when the load increased from 4 kg to 10 kg while maintaining high speed. In this high-load scenario, the COF decreased as the additive concentration increased, with the base oil (lacking additives) showing the lowest COF. This unexpected result suggests a breakdown in the tribofilm formation mechanism at higher speeds and loads. Although higher additive concentrations are effective at lower loads and speeds, they may fail to sustain an effective lubricating tribofilm under high loads and speeds, resulting in a higher COF compared to the base oil, which may rely on different lubrication mechanisms under these extreme conditions. The inability to form a stable and effective tribofilm with high additive concentrations at high speeds and loads likely contributed to increased friction.

As shown in Table 5, the reduction in the COF becomes significantly more pronounced at higher speeds, irrespective of the applied load. Among all of the samples, S2 exhibited the greatest reduction in COF, particularly under conditions of low load and high speed, achieving nearly a 90% reduction, as highlighted in Table 5. The base oil and S1 samples performed moderately, with improved results under high-speed conditions. In contrast, S3 demonstrated poor performance under low load and low speed but showed improvement at higher speeds. This pattern indicates that speed exerts a more substantial influence on friction reduction than load. High-speed conditions likely facilitate the formation of a lubricating film or decrease contact resistance, leading to a more effective reduction in COF.

4.2. Anti-Wear Test of Lubricating Oils

To evaluate the anti-wear properties of the formulated lubricating oils, wear tests were conducted alongside friction tests using the same pin-on-disc tribometer. This simultaneous testing ensured consistent experimental conditions, minimizing variations that could result from changes in environmental factors or machine settings between tests. The anti-wear performance of each lubricant sample was assessed, and the results are presented in Table 6. This combined approach facilitates a direct comparison between the anti-friction and anti-wear characteristics of various lubricant formulations.

Table 6. Wear at different load and speed conditions.

Condition	Oil Sample	Load (kg)	Speed (rpm)	Wear (Micron)	Reduction in Wear (%)
Low load, low speed	No lubricant	4	350	0.013	-
	Base oil	4	350	0.004	69.2
	S1	4	350	0.001	92.23
	S2	4	350	0.006	53.84
	S3	4	350	0.008	38.46
High load, low speed	No lubricant	10	350	0.021	-
	Base oil	10	350	0.012	42.85
	S1	10	350	0.006	71.42
	S2	10	350	0.008	61.9
	S3	10	350	0.01	52.38
Low load, high speed	No lubricant	4	500	0.056	-
	Base oil	4	500	0.038	32.14
	S1	4	500	0.008	85.71
	S2	4	500	0.026	53.57
	S3	4	500	0.042	25.00
High load, high speed	No lubricant	10	500	0.262	-
	Base oil	10	500	0.153	41.6
	S1	10	500	0.018	93.13
	S2	10	500	0.096	63.35
	S3	10	500	0.171	34.73

The analysis of the wear test results presented in Table 6 indicates that under conditions of low load and low speed, the lubricant sample containing 0.2 wt.% TiO_2 and 0.4 wt.% $CaCO_3$ achieved the lowest wear rate. This formulation demonstrated an impressive 92.23% reduction in wear compared to a no lubrication control test, underscoring the significant anti-wear performance of this specific additive combination under these operating conditions. This finding highlights the potential of this optimized additive mixture to enhance the wear resistance of lubricating oils in low-stress environments.

Under conditions of high load and low speed, a lubricant sample with 0.2 wt.% TiO_2 and 0.4 wt.% $CaCO_3$ achieved the lowest wear rate. This formulation resulted in a 50% reduction in wear compared to the base oil without additives. A clear trend was observed, as wear increased as the concentrations of TiO_2 and $CaCO_3$ additives rose. Similar findings were noted under low-load, high-speed conditions, where the 0.2 wt.% TiO_2 and 0.4 wt.% $CaCO_3$ mixture again achieved the lowest wear rate. In this scenario, wear also increased with higher additive concentrations, as detailed in Table 6. The consistent observation of minimal wear at this specific additive concentration across varying load and speed conditions suggests that this combination provides optimal anti-wear protection, regardless of the operating parameters, within the tested range.

Table 6 presents the percentage of wear reduction (%) across various lubricant samples tested under four distinct combinations of loads and speeds. Sample S1 demonstrates the highest wear reduction under conditions of low load, low speed, high load, and high speed (highlighted with red circles), indicating its exceptional performance in both mild and severe operating conditions. The base oil sample shows moderate performance, but it exhibits lower wear reduction at low loads and high speeds, suggesting reduced effectiveness when only speed is increased. Sample S2 displays a balanced performance across all four conditions, maintaining relatively consistent wear reduction. Sample S4 shows the least wear reduction, particularly under high-speed conditions, indicating its unsuitability for high-speed applications. Figure 10 illustrates that both load and speed influence the wear-reducing performance of the lubricants. However, sample S1 proves to be the most effective across a wide range of conditions, making it a strong candidate for applications involving varying mechanical stresses.

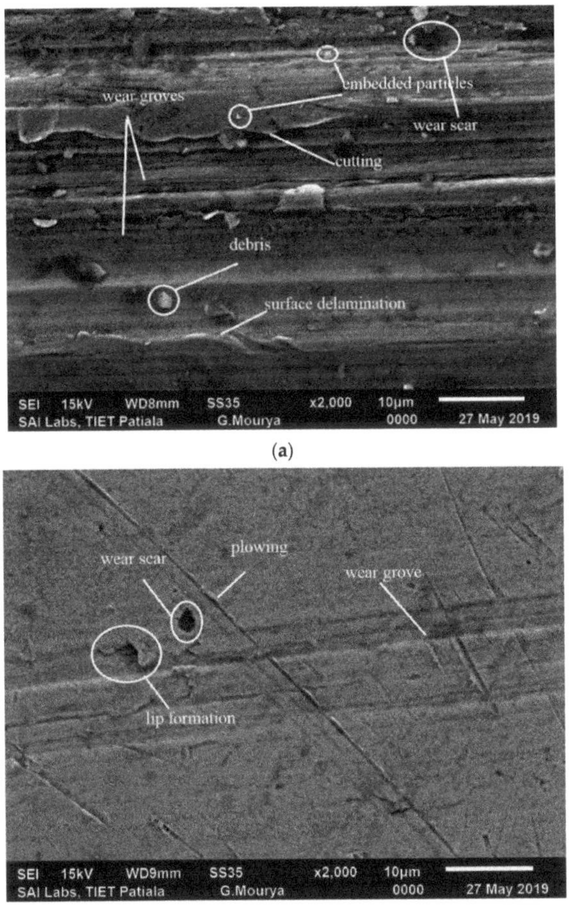

(a)

(b)

Figure 10. (**a**) SEM image of specimen base oil (BO) at 10 kg load and 500 rpm speed; (**b**) SEM image of specimen (S1) tested at 10 kg load and 500 rpm speed.

4.3. SEM Testing of Specimen

To explore how rotational speed affects wear patterns, scanning electron microscopy (SEM) was utilized to examine the morphology of the worn test pins. SEM images were captured for two samples tested under a 10 kg load and a speed of 500 rpm, one lubricated with sample S1 (containing additives) and the other with the base oil, as depicted in Figure 10a and Figure 10b, respectively. The specimen lubricated with base oil, shown in Figure 10b, exhibited significant wear damage, including deep grooves, signs of material cutting, and considerable debris accumulation on the worn surface. This severe wear aligned with the higher wear rates observed for the base oil, as indicated in Table 6. Conversely, the inclusion of nanomaterials (TiO_2 and $CaCO_3$) in sample S1, as seen in Figure 10a, led to a distinctly different wear pattern. The SEM images showed that the nanomaterials effectively filled the micro-grooves and valleys on the worn surface, forming a protective layer that minimized direct metal to metal contact. This enhanced surface protection mechanism is a crucial factor in the significant reduction in wear observed with additive-containing lubricants, as shown in Table 5. The difference in surface morphology highlights the improved tribological performance resulting from the incorporation of nanomaterials into the base lubricant.

Comparing the SEM images revealed clear differences in wear patterns at various speeds. At low speeds, the wear surface displayed a relatively smooth morphology. However, as the speed increased, the wear pattern shifted to one marked by more material cutting and the formation of lip-like features at the edges of the wear track. This increase in cutting and lip formation resulted directly from the higher shear stresses that developed at elevated speeds. Furthermore, at higher speeds, the presence of embedded particles within the worn surface indicates a more aggressive interaction between the lubricant and the test pin. The presence of wear debris and signs of plowing further contributed to the degradation of the surface texture and an increase in overall wear at higher speeds. The transition from smooth wear at lower speeds to more severe damage characterized by cutting, lip formation, particle embedding, and plowing at higher speeds highlights the significant impact of rotational speed on wear mechanisms in the pin-on-disc test.

The physiochemical interactions between $CaCO_3$, TiO_2, and the base oil (EP 140) achieved homogeneous dispersion. This is because the surface energy of the nanoparticles is compatible with the surface tension of the base oil. Also, the nanoparticles naturally attract each other through van der Waals forces, leading to agglomeration. The combination of these factors provides a more robust and durable protective film, improving wear resistance and load-carrying capacity. It can lead to a further reduction in the coefficient of friction, resulting in even better energy efficiency. $CaCO_3$ can help mitigate any potential abrasive effects of TiO_2, while TiO_2 can compensate for any potential thickening effects of $CaCO_3$, leading to a more balanced and optimized formulation.

The effect of the developed nanoparticles on the MS round bar (pin) is that it reduces the friction between the pin and the counter surface, which leads to less removal of the material from the pin, extending its lifespan. The surface asperities of the pin have been filled by the developed nanoparticles, providing more contact points and distributing the load more evenly. To validate these findings, the SEM and EDS analyses of the worn-out surface were carried out, as shown in Figure 11. The composition of the elements of the worn-out surface obtained through EDS is tabulated in Table 7. From the results obtained through EDS, it can be easily observed that calcium (Ca) is present on the worn-out surface.

Table 7. Composition of the worn-out surface in the case of sample S1.

Element	At. No.	Netto	Mass Norm. [%]	Atom [%]	Abs. Error [%] (3 Sigma)
C	6	30,245	52.25	78.89	6.89
O	8	3145	5.70	6.46	0.80
Na	11	1660	1.29	1.02	0.28
Cl	17	6023	1.20	0.61	0.11
K	19	2355	0.59	0.27	0.07
Ca	20	2024	0.56	0.25	0.07
Cr	24	2253	0.76	0.26	0.09
Fe	26	68,739	37.66	12.23	2.66
		Sum	100.00	100.00	

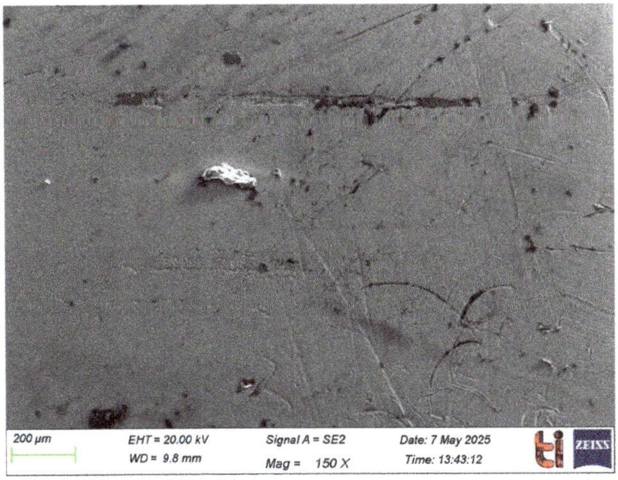

(a)

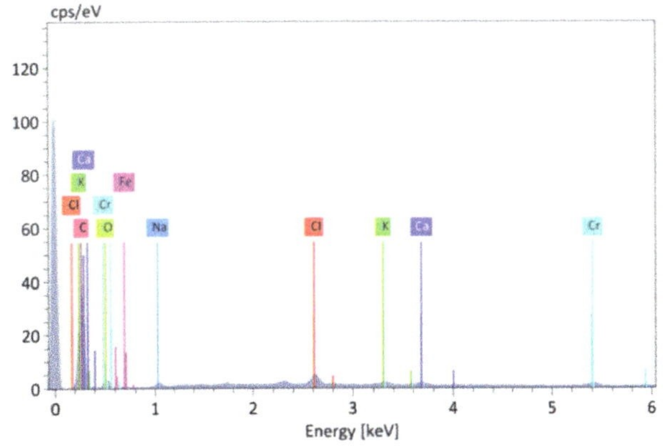

(b)

Figure 11. (**a**) SEM image of the worn-out surface; (**b**) EDS of the worn-out surface in the case of sample S1.

5. Conclusions

Viscosity–temperature tests demonstrated that adding titanium dioxide (TiO_2) and calcium carbonate ($CaCO_3$) particles increased the base oil's viscosity without adversely affecting its viscosity index. Moreover, higher concentrations of these additives led to increased viscosities, likely due to particle agglomeration. The combined use of these additives effectively reduced both the coefficient of friction and the wear rate of the gear oil. Although the reduction in the coefficient of friction varied with test conditions, a mixture of 0.2 wt.% TiO_2 and 0.4 wt.% $CaCO_3$ notably decreased the wear rate. This improvement is attributed to the formation of a protective tribofilm in the contact zone, which effectively separates the contacting surfaces. However, excessively high additive concentrations may increase the sliding friction and, consequently, the wear rates due to potential interference with effective tribofilm formation. Future research could explore incorporating polymers alongside the TiO_2 and $CaCO_3$ additives to further enhance the viscosity index and investigate the impact of compatible extreme-pressure additives on load-carrying capacity. This would broaden the understanding of the synergistic effects of these additive types on the overall tribological performance of gear oil.

Author Contributions: A.S.: data curation, software, writing—original draft, methodology; S.K. (Surinder Kumar): data curation, software, writing—original draft, methodology; S.K. (Sumit Kumar): editing and methodology; H.K.: writing—review and editing, supervision; S.C.: writing—review and editing; G.V.: writing—review and editing, supervision; R.K.: writing—review and editing, supervision. All authors have read and agreed to the published version of the manuscript.

Funding: This research received no external funding.

Data Availability Statement: Data will be made available upon reasonable request.

Conflicts of Interest: The authors declare no conflicts of interest.

References

1. Xu, N.; Zhang, M.; Li, W.; Zhao, G.; Wang, X.; Liu, W. Study on the selectivity of calcium carbonate nanoparticles under the boundary lubrication condition. *Wear* **2013**, *307*, 35–43. [CrossRef]
2. Bartz, W.J. Tribology, lubricants and lubrication engineering—A review. *Wear* **1978**, *49*, 1–18. [CrossRef]
3. Blau, P.J. *Friction Science and Technology: From Concepts to Applications, Second Edition*, 2nd ed.; CRC Press: Boca Raton, FL, USA, 2008. [CrossRef]
4. Shenoy, B.S.; Binu, K.G.; Pai, R.; Rao, D.S.; Pai, R.S. Effect of nanoparticles additives on the performance of an externally adjustable fluid film bearing. *Tribol. Int.* **2012**, *45*, 38–42. [CrossRef]
5. Wan, S.; Tieu, A.K.; Xia, Y.; Zhu, H.; Tran, B.H.; Cui, S. An overview of inorganic polymer as potential lubricant additive for high temperature tribology. *Tribol. Int.* **2016**, *102*, 620–635. [CrossRef]
6. Chen, L.; Zhu, D. Preparation and tribological properties of unmodified and oleic acid-modified CuS nanorods as lubricating oil additives. *Ceram. Int.* **2017**, *43*, 4246–4251. [CrossRef]
7. Sharma, V.; Timmons, R.B.; Erdemir, A.; Aswath, P.B. Interaction of plasma functionalized TiO_2 nanoparticles and ZDDP on friction and wear under boundary lubrication. *Appl. Surf. Sci.* **2019**, *489*, 372–383. [CrossRef]
8. Miao, W.J.; Wang, S.Q.; Wang, Z.H.; Wu, F.B.; Zhang, Y.Z.; Ouyang, J.H.; Wang, Y.M.; Zou, Y.C. Additive Manufacturing of Advanced Structural Ceramics for Tribological Applications: Principles, Techniques, Microstructure and Properties. *Lubricants* **2025**, *13*, 112. [CrossRef]
9. Gupta, R.N.; Harsha, A.P. Synthesis, Characterization, and Tribological Studies of Calcium–Copper–Titanate Nanoparticles as a Biolubricant Additive. *J. Tribol.* **2016**, *139*, 021801. [CrossRef]
10. Abdul Basir, N.F.; Iliya Nor Za'im, S.; Abdul Wahab, Y.; Johan, M.R.; Ghaffari Khaligh, N. Impact Symmetry on Tribology Performance of 2,5-Bis-Cycloalkylsulfenyl[1,3,4]Thiadiazole Derivatives. *Tribol. Trans.* **2025**, *68*, 206–223. [CrossRef]
11. Pendbhaje, G.; Ali, M.; Bajaj, D. Bio-lubricants as viable replacements for mineral oils in automotive engines: A review. *Tribol.-Mater. Surf. Interfaces* **2025**, *19*, 24–38. [CrossRef]
12. Kamel, B.M.; Arafa, E.L.; Mohamed, A. Tribological and rheological properties of the lubricant containing hybrid graphene nanosheets (GNs)/titanium dioxide (TiO_2) nanoparticles as an additive on calcium grease. *J. Dispers. Sci. Technol.* **2023**, *44*, 2675–2682. [CrossRef]
13. Alghani, W.; Ab Karim, M.S.; Bagheri, S.; Amran, N.A.M.; Gulzar, M. Enhancing the Tribological Behavior of Lubricating Oil by Adding TiO_2, Graphene, and TiO_2/Graphene Nanoparticles. *Tribol. Trans.* **2019**, *62*, 452–463. [CrossRef]
14. Sivakumar, B.; Ranjan, N.; Ramaprabhu, S.; Kamaraj, M. Tribological properties of graphite oxide derivative as nano-additive: Synthesized from the waster carbon source. *Tribol. Int.* **2020**, *142*, 105990. [CrossRef]
15. Xia, W.; Zhao, J.; Wu, H.; Zhao, X.; Zhang, X.; Xu, J.; Jiao, S.; Wang, X.; Zhou, C.; Jiang, Z. Effects of oil-in-water based nanolubricant containing TiO_2 nanoparticles in hot rolling of 304 stainless steel. *J. Mater. Process. Technol.* **2018**, *262*, 149–156. [CrossRef]
16. Razavi, S.; Sabbaghi, S.; Rasouli, K. Comparative investigation of the influence of CaCO3 and SiO2 nanoparticles on lithium-based grease: Physical, tribological, and rheological properties. *Inorg. Chem. Commun.* **2022**, *142*, 109601. [CrossRef]
17. Tóth, Á.D.; Hargitai, H.; Szabó, Á.I. Tribological Investigation of the Effect of Nanosized Transition Metal Oxides on a Base Oil Containing Overbased Calcium Sulfonate. *Lubricants* **2023**, *11*, 337. [CrossRef]
18. Choi, Y.; Lee, C.; Hwang, Y.; Park, M.; Lee, J.; Choi, C.; Jung, M. Tribological behavior of copper nanoparticles as additives in oil. *Curr. Appl. Phys.* **2009**, *9*, e124–e127. [CrossRef]
19. Peña-Parás, L.; Taha-Tijerina, J.; Garza, L.; Maldonado-Cortés, D.; Michalczewski, R.; Lapray, C. Effect of CuO and Al2O3 nanoparticle additives on the tribological behavior of fully formulated oils. *Wear* **2015**, *332–333*, 1256–1261. [CrossRef]

20. Vardhaman, B.S.A.; Amarnath, M.; Ramkumar, J.; Mondal, K. Enhanced tribological performances of zinc oxide/MWCNTs hybrid nanomaterials as the effective lubricant additive in engine oil. *Mater. Chem. Phys.* **2020**, *253*, 123447. [CrossRef]
21. Mousavi, S.B.; Zeinali Heris, S. Experimental investigation of ZnO nanoparticles effects on thermophysical and tribological properties of diesel oil. *Int. J. Hydrog. Energy* **2020**, *45*, 23603–23614. [CrossRef]
22. Han, X.; Thrush, S.J.; Zhang, Z.; Barber, G.C.; Qu, H. Tribological characterization of ZnO nanofluids as fastener lubricants. *Wear* **2021**, *468–469*, 203592. [CrossRef]
23. Mariño, F.; López, E.R.; Arnosa, Á.; Gómez, M.A.G.; Piñeiro, Y.; Rivas, J.; Alvarez-Lorenzo, C.; Fernández, J. ZnO nanoparticles coated with oleic acid as additives for a polyalphaolefin lubricant. *J. Mol. Liq.* **2022**, *348*, 118401. [CrossRef]
24. Gara, L.; Zou, Q. Friction and Wear Characteristics of Oil-Based ZnO Nanofluids. *Tribol. Trans.* **2013**, *56*, 236–244. [CrossRef]
25. Laad, M.; Jatti, V.K.S. Titanium oxide nanoparticles as additives in engine oil. *J. King Saud Univ.-Eng. Sci.* **2018**, *30*, 116–122. [CrossRef]
26. Binu, K.G.; Shenoy, B.S.; Rao, D.S.; Pai, R. A Variable Viscosity Approach for the Evaluation of Load Carrying Capacity of Oil Lubricated Journal Bearing with TiO_2 Nanoparticles as Lubricant Additives. *Procedia Mater. Sci.* **2014**, *6*, 1051–1067. [CrossRef]
27. Ghasemi, R.; Fazlali, A.; Mohammadi, A.H. Effects of TiO_2 nanoparticles and oleic acid surfactant on the rheological behavior of engine lubricant oil. *J. Mol. Liq.* **2018**, *268*, 925–930. [CrossRef]
28. Gu, K.; Chen, B.; Chen, Y. Preparation and tribological properties of lanthanum-doped TiO_2 nanoparticles in rapeseed oil. *J. Rare Earths* **2013**, *31*, 589–594. [CrossRef]

lubricants

MDPI

Article

Mechanical and Wear Behaviour of Microwave Sintered Copper Composites Reinforced with Tungsten Carbide and Graphite Under Different Lubrication Conditions

V. S. S. Venkatesh [1], B. Surekha [2], Pandu Ranga Vundavilli [1,*] and Manas Mohan Mahapatra [1]

[1] School of Mechanical Sciences, Indian Institute of Technology Bhubaneswar, Khordha 752050, India; vssvenkateshnits@gmail.com (V.S.S.V.); mmmahapatra@iitbbs.ac.in (M.M.M.)

[2] School of Mechanical Engineering, KIIT University, Khordha 751024, India; surekhafme@kiit.ac.in

* Correspondence: pandu@iitbbs.ac.in

Abstract

This present study epitomises the fabrication of Cu-15%WC-X%Gr (X = 0, 3, 6, 9, 12) hybrid composites through a microwave sintering process. The synthesised composites were evaluated for hardness and compression strength as per ASTM standards. The composite corresponding to Cu-15%WC-9%Gr shows the optimal compression strength of 395 MPa. Based on this, the composite corresponding to the maximum compression strength was selected for subsequent wear investigations under dry, oil, and SiC nanofluid lubrication conditions. The SiC nanofluids were prepared by dispersing 1 wt% SiC, 1.5 wt% SiC, and 2 wt% SiC nanoparticles in soluble oils. Increasing the nanoparticle content enhanced both the thermal conductivity and zeta potential, indicating an improved heat transfer and dispersion stability. The wear test under different lubricating regimes demonstrates that the lubricating type had a pronounced influence on the wear rate and C.O.F. The minimum rate of wear of 0.0235 mm^3/m and C.O.F. of 0.28 were achieved for the 2 wt% SiC nanofluid lubrication. The worn surfaces under dry and oil-lubricated regimes revealed prominent microcracks and delamination wear. In contrast, surfaces tested under nanofluid lubrication exhibited smoother grooves with minimal surface damage and an absence of microcracking.

Keywords: copper composite; lubrication; tribology; worn surface morphology

1. Introduction

Copper (Cu) is extensively employed in industrial and engineering applications owing to its outstanding electrical and thermal conductivity, excellent corrosion resistance, and superior formability. These attributes make copper a critical material in sectors such as electronics, automotive systems, electrical power transmission, and heat exchange technologies [1–3]. Its wide availability and cost-effectiveness further establish copper as a preferred base material for both structural and functional components. Despite these benefits, pure copper's lack of mechanical properties and poor wear characteristics limit its applicability in situations involving prolonged mechanical loading or high temperatures. [4,5]. Considerable research has focused on the manufacturing of copper matrix composites (CMCs) to overcome these intrinsic constraints. The incorporation of suitable reinforcements aims to preserve copper's excellent thermal and electrical properties while substantially enhancing its mechanical characteristics, including hardness, strength, thermal stability, and wear resistance. Such enhancements are particularly critical in applications where structural reliability must coexist with high conductivity. To improve the properties of copper, a variety of reinforcement materials have been investigated, such as titanium carbide (TiC), graphene nanoplatelets (GNPs), silicon carbide (SiC), and titanium dioxide (TiO_2). These reinforcements exhibit distinct strengthening mechanisms and impart varying influences on the overall composite behaviour [6,7]. For instance, in Tugba et al. [8], the Cu–8 wt% TiO_2 composite fabricated via spark plasma sintering followed by hot extrusion exhibited a superior 290 MPa yield strength, representing a 72% improvement over pure copper while retaining appreciable ductility. El-Zaidia et al. [9] investigated Cu/graphene/flyash composites and found that, at an 8 wt% graphite reinforcement and 850 °C sintering temperature, the optimal tensile and hardness values of 323.3 MPa and 735.3 MPa were attained. Samal et al. [10] reported that the incorporation of Gr phases into the copper matrix enhanced the densification and hardness, resulting in a relative density of up to 96%. Additionally, the rupture strength improved with the addition of graphite reinforcement up to 5 vol.%, beyond which it decreased at higher graphite concentrations.

Based on previous studies, the present work focuses on the development of a copper metal matrix composite reinforced with tungsten carbide and graphite particles. The composite was synthesised using microwave sintering, and its mechanical and tribological behaviours were evaluated under dry, oil, and nanofluid lubrication regimes.

2. Materials and Methods

2.1. Materials

A copper powder with an exceptional purity level of 99.9% and an average particle size of 30 micrometres was selected to serve as the matrix phase. The copper powder used in the present study was procured from Vedayukt Company, East Singhbhum, India. The chemical composition of the matrix and reinforcement phases is shown in Table 1. Tungsten carbide (WC) powder, having a similar mean particle size (~30 μm) and a purity of 99.5%, together with graphite (Gr) particulates having an APS less than 30 μm, was incorporated as a reinforcement phase, as shown in Figure 1.

Table 1. Chemical composition of the matrix and reinforcement particles.

Material	Composition (wt%)/Purity
Copper matrix	99.9%
Tungsten carbide	WC $\geq$ 99%
Graphite	C $\geq$ 99%
Composite material	Cu–15%WC–9%Gr
Nanofluid lubricant	Soluble oil + 2 SiC

2.2. Fabrication of the Composite

Cu-WC-Gr hybrid composites were synthesised by incorporating the 15 wt% WC particles and Gr (3 wt%, 6 wt%,9 wt%, 12 wt%) into the copper matrix. The composite powders were homogenised through dry mixing in a planetary ball mill for a milling duration of 2 h under an argon atmosphere, employing WC balls having a 5 mm diameter as milling media and a WC vial with a charge ratio of 10:1 [11,12]. The blended powders were subsequently cold compacted using a 13 mm diameter steel die under an applied uniaxial pressure of 350 MPa [13,14]. To identify the optimum sintering conditions to achieve the maximum densification of copper, the preliminary sintering trials were performed on pure copper at 950 °C with holding times of 30, 60, and 120 min. Among the investigated conditions, a 60 min holding time yielded the highest density. The green compacts were heated using a microwave furnace (900 W and frequency of 2.35 GHz) at a sintering

https://doi.org/10.3390/lubricants14020054

temperature of 950 °C at an 80 °C/min sintering rate, with the samples embedded in a graphite bed to facilitate uniform heating [15,16]. The diagrammatic representation of the complete process is shown in Figure 2.

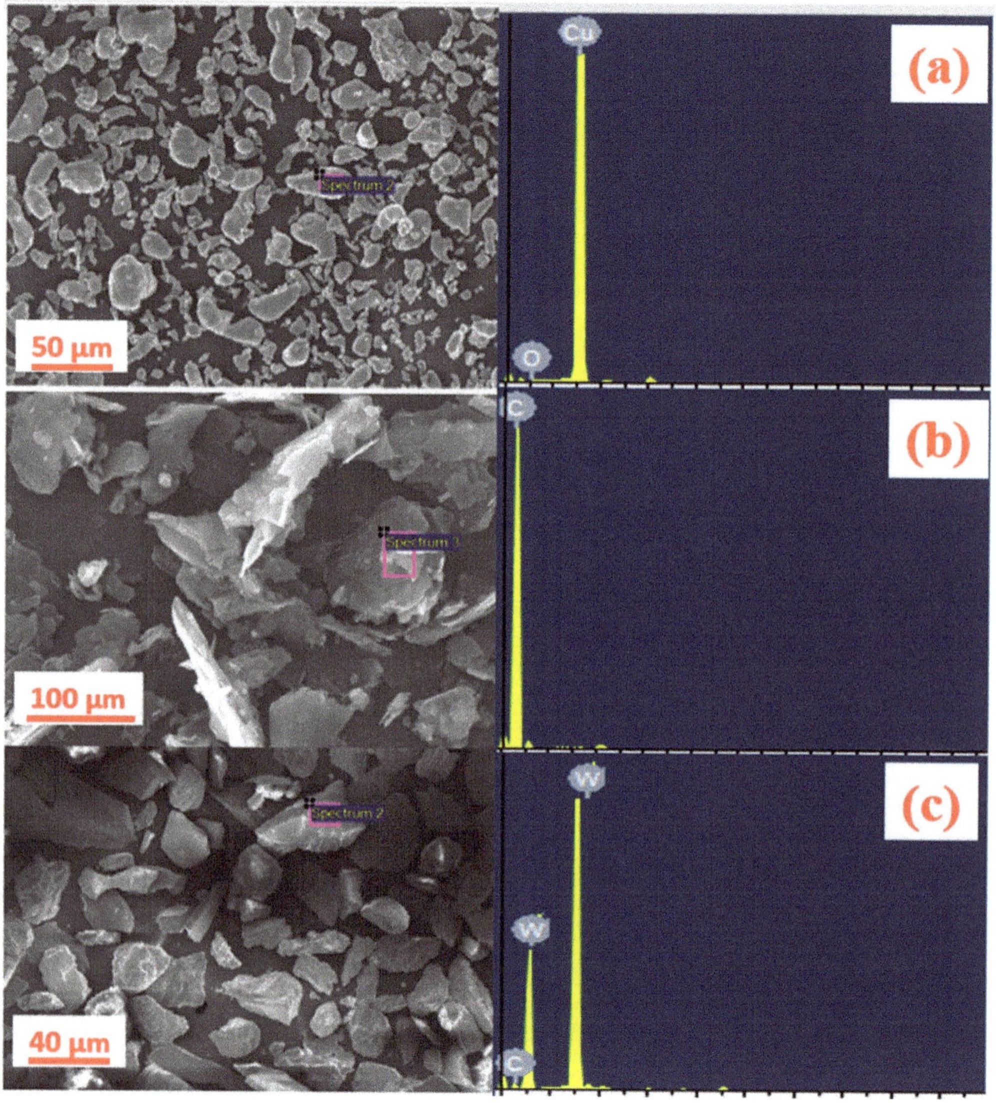

Figure 1. SEM and EDX mapping of as-received (**a**) copper, (**b**) graphite, and (**c**) WC particles.

https://doi.org/10.3390/lubricants14020054

Figure 2. Schematic of composite Ffbrication and characterizations performed.

3. Characterization Techniques

Microhardness evaluation was carried out on the polished sample surface using a Vickers hardness tester (Koopa universal hardness tester, UV1, Sari, Iran). Ten successive indentations were performed for each sample under an applied load of 98.1 N with a dwell time of 10 s at room temperature [17–19]. Compression tests were performed according to ASTM E9 standards [20]. The wear rate and C.O.F. were evaluated on the pin-on-disc wear tester in accordance with the ASTM-G99 standard. The composite pins were pressed against the EN 31 steel disc under the dry, oil, and nanofluid lubricating conditions. The tribology test was performed at a constant applied load of 40 N at the tract radius of 50 mm for 30 min time, which corresponds to a sliding distance of 3000 m [21]. The morphological features and elemental distribution of the synthesised composites and the wear track after the wear test were examined using Scanning Electron Microscopy (SEM, Zesis Supra 55, Oberkochen, Germany) coupled with an EDS analyser. XRD (SmartLab, Tokyo, Japan) was performed to reveal the phases present in the composite [22,23].

4. Preparation of SiC Nanofluids and Thermal Conductivity Measurement

SiC nanoparticles (<100 nm) were purchased from the nanoshell company, Bhubaneswar. Nanofluids with SiC nanoparticles were synthesised through a two-step approach by dispersing in ethylene glycol at concentrations of 1 wt% SiC, 1.5 wt% SiC, and 2 wt% SiC. The mixture was initially mixed with a magnetic stirrer at 600 rpm, followed by probe ultrasonication (FS-300 N, 24 Hz, 8 mm probe dia) to ensure homogeneous dispersion of SiC nanoparticles [24,25]. Sonication was performed for a 15 h duration in a 100 mL container at ambient temperature, as depicted in Figure 3.

https://doi.org/10.3390/lubricants14020054

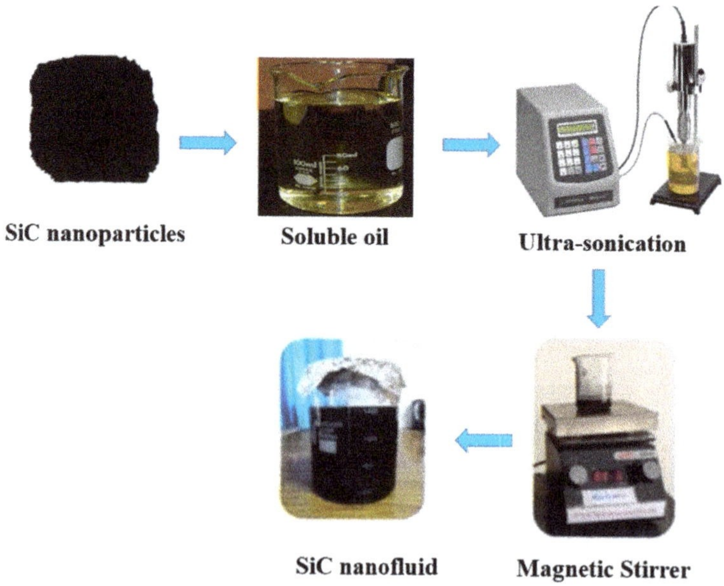

Figure 3. Schematic of the SiC nanofluid preparation.

The thermal conductivity (K) plays a decisive role in the energy dissipation capability of the nanofluids. In the present work, the "K" of the SiC nanofluid was determined using a hot disc thermal analyser at 25 °C for the 1 wt% SiC, 1.5 wt% SiC, and 2 wt% SiC nanofluids [26,27]. The results indicate that the "K" was increased from 0.761 W/m·K to 0.792 W/m·K with the addition of SiC nanoparticles from 1 wt% to 2 wt%. This improvement in "K" was due to the increased presence of the highly conductive SiC nanoparticles within the ethylene glycol matrix. The addition of SiC nanoparticles beyond 2 wt% leads to an agglomeration of SiC nanoparticles, leading to a reduction in the "K" of the nanofluid, as shown in Figure 4.

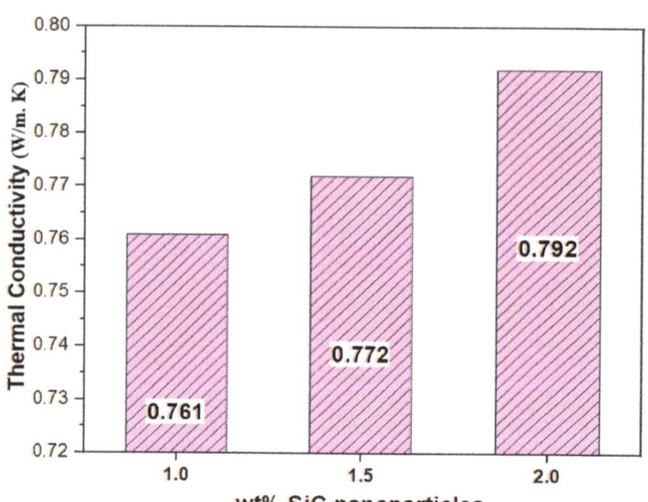

Figure 4. Thermal conductivity (K) of nanofluids with wt% of SiC particles.

https://doi.org/10.3390/lubricants14020054

5. Stability Analysis of the SiC/Ethylene Glycol Nanofluids

The dispersion stability of the SiC/ethylene glycol nanofluids was evaluated using Zetasizer Nano ZS integrated with a He-Ne laser and a fixed disposable folded capillary cell containing integrated electrodes connected to the instrument's voltage terminals. Under the applied electric field, the SiC nanoparticles migrated through the ethylene glycol medium with velocities proportional to their zeta potential [28,29]. Nanoparticles exhibiting higher zeta potentials indicate a higher electrophoretic mobility. The SiC nanoparticle movement was quantified through laser Doppler velocimetry as the laser beam passed through the charged suspension.

The zeta potential experiments were performed at 28 °C for SiC wt% of 1 wt% SiC, 1.5 wt% SiC, and 2 wt% SiC, and the results are illustrated in Figure 5. The zeta potential values in this present study exceeded 45 mV, with a maximum value of 58 mV attained for the 2 wt% SiC nanofluid, signifying a uniform dispersion and minimal agglomeration. Consequently, the 2 wt% SiC nanofluids were selected for the tribological analysis owing to their superior thermal conductivity and electrostatic stability.

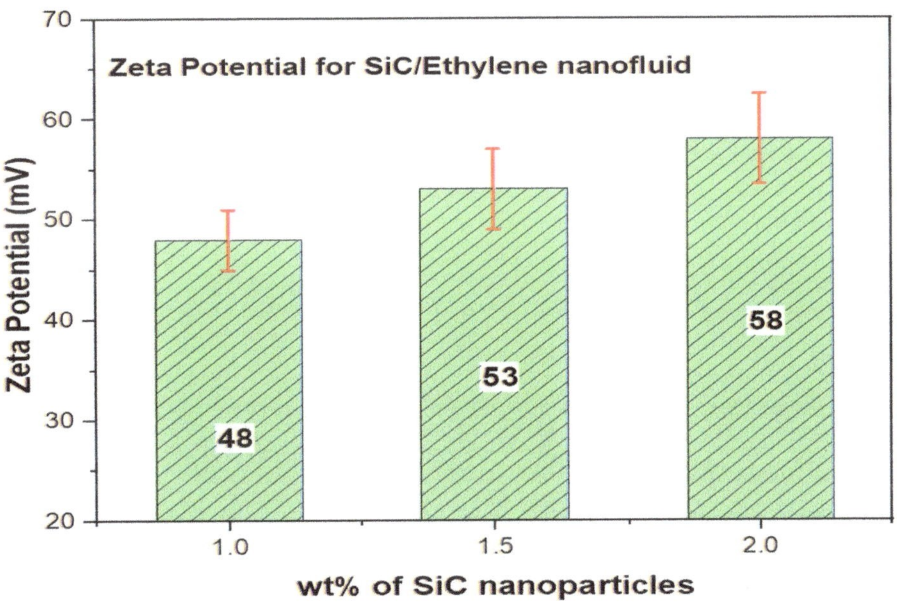

Figure 5. Zeta potential values for the prepared SiC nanofluids.

6. Results and Discussion

6.1. Morphological Study of the Cu-WC-Gr Composite

The synthesised composites were investigated to reveal the reinforcement dispersion through SEM analysis. The addition of the graphite (Gr) reinforcement up to 9 wt% leads to a uniform dispersion inside the copper matrix, as shown in Figure 6a,b for 6 wt% and 9 wt% Gr particles. However, the addition of Gr phases at 12 wt% creates brittle agglomerations by reacting with the softer copper matrix at the reinforcement and matrix interfaces, as depicted in Figure 6c. The peak related to the agglomerated peak was confirmed through the XRD analysis, and it was concluded that the peak at 43° matches with the Cu_2C_2 intermetallic, confirmed by ICDD card number 25-0015, as shown in Figure 7 [30,31]. The uniform dispersion of the tungsten, carbon, and graphite phases was clearly identified through the EDS mapping of the corresponding composite, as shown in Figure 6a,b. The

https://doi.org/10.3390/lubricants14020054

formation of the secondary agglomerations acted as an efficient interlocking among the copper matrix and reinforcement phases, thereby enhancing the load transfer among the interface particles. A controlled extent of interlocking is treated as beneficial. However, the existence of higher activation energies at the agglomeration regions leads to brittle and impurity-rich intermetallic layers, which create pores at the interfaces and decline the mechanical properties of the composites [32,33].

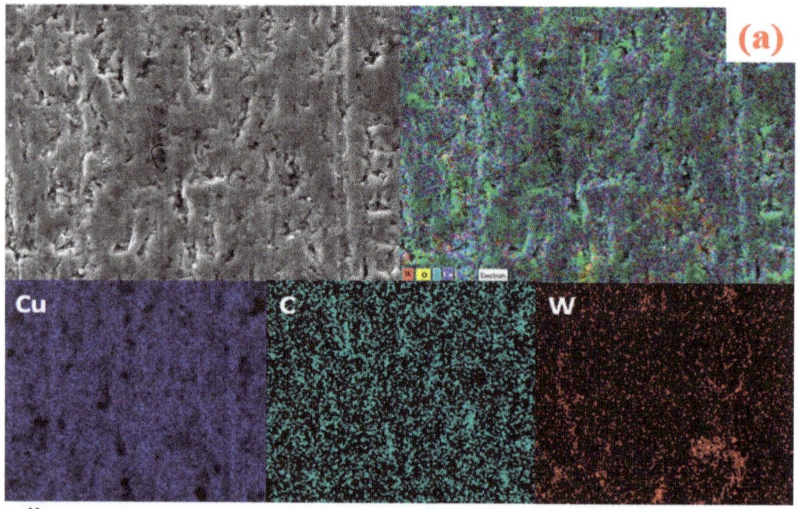

coeff

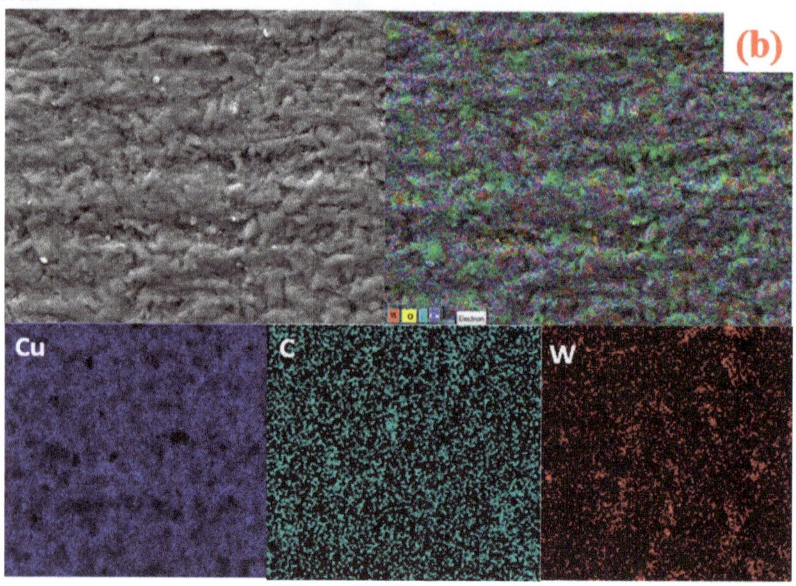

Figure 6. *Cont.*

https://doi.org/10.3390/lubricants14020054

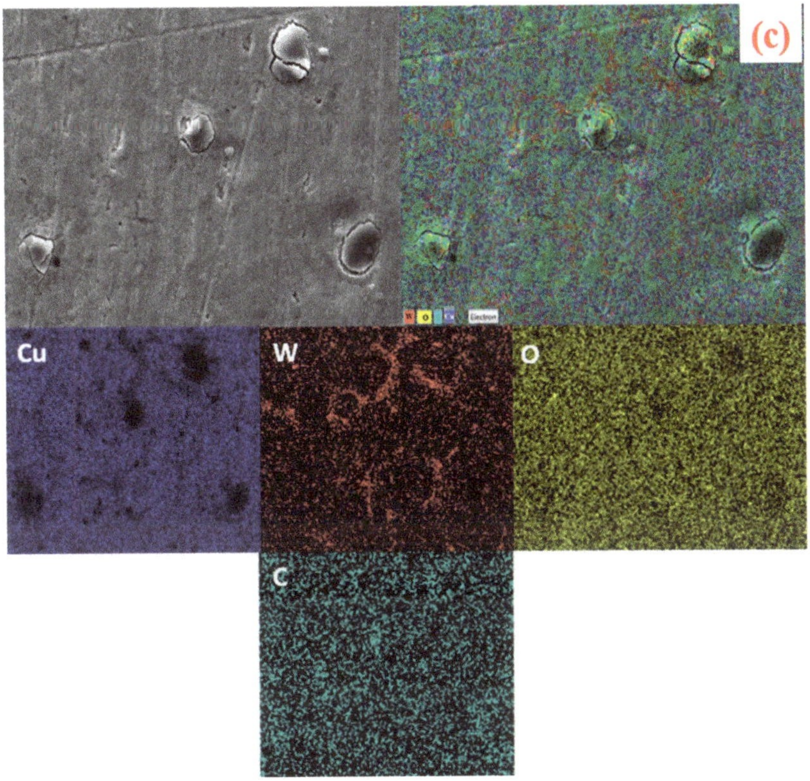

Figure 6. SEM and EDS images for (**a**) Cu-15%WC-6 wt% Gr, (**b**) Cu-15%WC-9 wt% Gr, and (**c**) Cu-15%WC-12 wt% Gr composites.

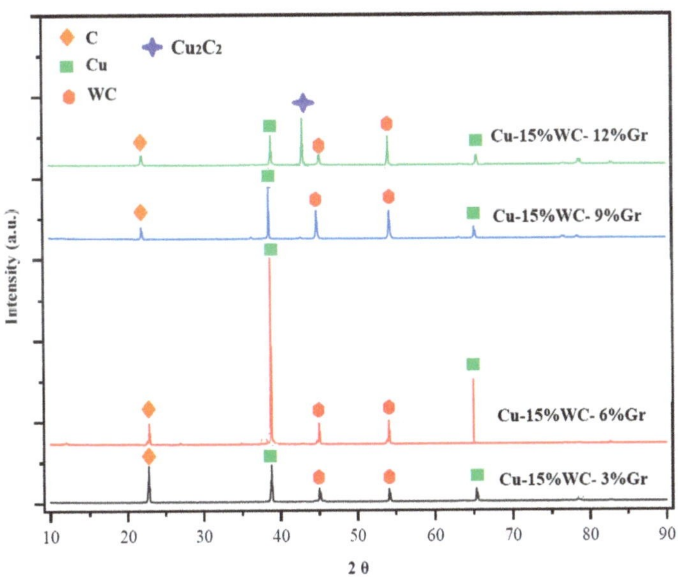

Figure 7. XRD pattern for the Cu-15%WC-X%Gr (X = 3, 6, 9, 12.) hybrid microwave sintered composite.

https://doi.org/10.3390/lubricants14020054

6.2. Mechanical Properties of the Synthesised Cu-WC-Gr Composite

6.2.1. Hardness Result for the Cu-WC-Gr Composite

Figure 8 illustrates the variation in the Vickers hardness with the Gr wt%. The hardness increases from 180 HV for the Cu-15%WC to 239 HV for the Cu-15%WC-12%Gr composite, corresponding to an enhancement of 32.7% for the 12 wt% Gr addition compared with the unreinforced graphite (Cu-15%WC). The hardness improvement was due to the wetting property of the reinforced Gr particles among the copper matrix and reinforcement phases. The enhanced wettability accelerates the efficient load transfer among the interfaces of the matrix and reinforcement phases, thereby resulting in the higher Vickers values [34,35]. In addition to this, the incorporation of the Graphite phases into the ductile copper matrix, along with their uniform dispersion, significantly improves the resistance to the penetration of the indenter of the hardness tester. The attained hardness values are 14.32% and 20% higher than the hardness values attained by Manikandan et al. [36] and Abhijit Bhowmik et al. [37]. The improvement in hardness was due to the addition of the graphite particles, which establishes the strong interfacial and intermetallic bonding with the adjacent reinforcements. Additionally, the uniform sintering during the MAS process at short durations ensures a rapid heat dispersion and promotes the generation of clean and robust interfaces, which further improves the hardness of the microwave sintering Cu-WC-Gr composite [15,38].

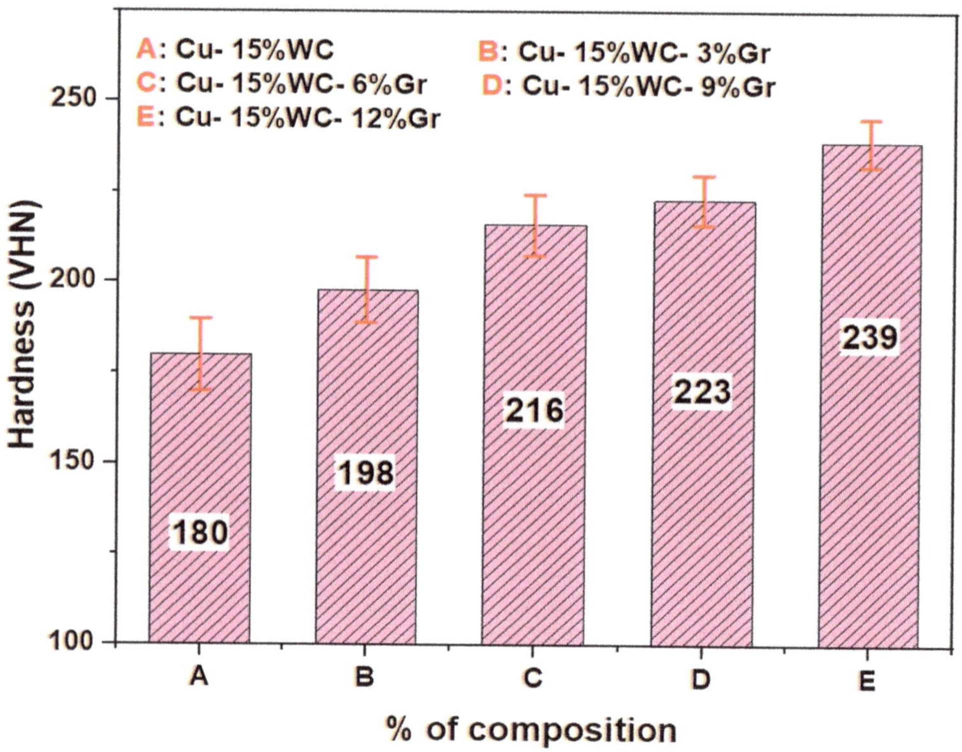

Figure 8. Variation in the hardness of the Cu-WC-Gr composite with the Gr content.

6.2.2. Compression Strength Analysis of the Synthesised Cu-WC-Gr Composite

The variation in the compression strength of the Cu-WC-Gr composite with the wt% of Gr particles is illustrated in Figure 9. It was observed that the compression strength

https://doi.org/10.3390/lubricants14020054

of the Cu-15%WC-9%Gr composite exhibited a remarkable increase of 38.59% over the Cu-15%WC composite. The optimum compression strength of 395 MPa was attained for the Cu-15%WC-9%Gr composite. The improvement in the compression strength was due to the incorporation of the Gr particles, which refine and strengthen the grain boundaries at the matrix and reinforcement interfaces [39]. These strengthened grain boundaries act as effective barriers to microcrack initiation and propagation under compressive loading, thereby enhancing the composite compression strength up to the 9 wt% graphite addition. Furthermore, the incorporation of graphite as an impurity restricts the dislocation movement near the interfaces, leading to dislocation pile-up at the grain boundaries and consequently increasing the strength of the composite through the Orowan strengthening mechanism [40]. However, when the graphite particles exceeded more than 9 wt%, a reduction in the compression strength was observed. This decline is attributed to the formation of the intermetallic at the interfaces, as shown in Figure 6c. These agglomerations develop additional slip planes, facilitating atomic movement even under the lower magnitude of the compressive loads [41]. Hence, based on the compression strength results, the composite corresponding to the 9 wt% graphite reinforcement was selected for the tribological investigations in the present study.

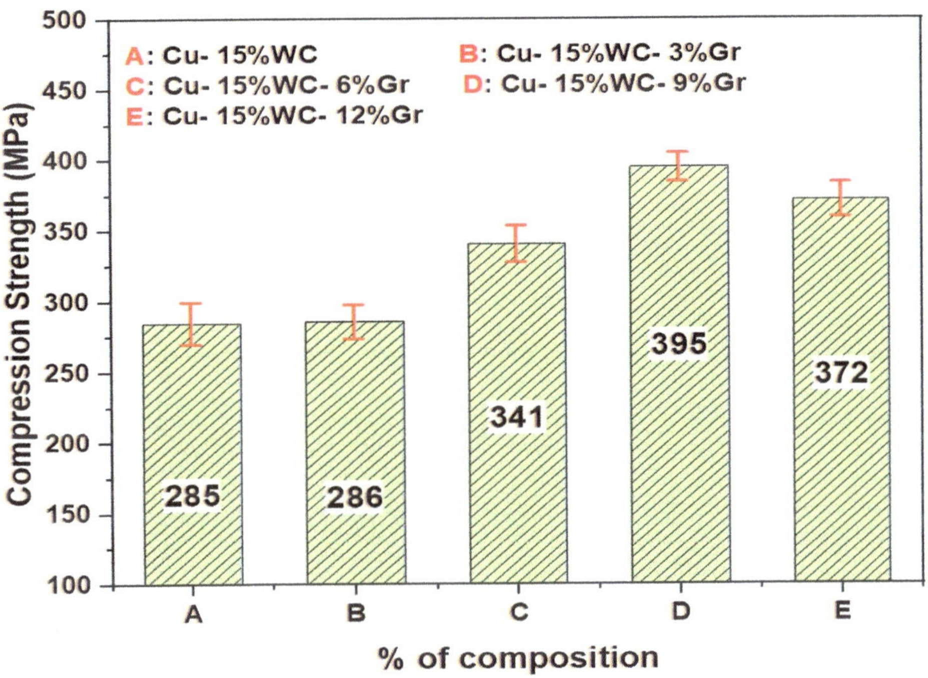

Figure 9. Variation in the compression strength of the Cu-WC-Gr composite with the Gr content.

6.3. Tribological Analysis

6.3.1. Wear Rate

The variation in the wear rate of the Cu-15%WC-9 wt%Gr composite under different lubrication conditions as a function of the sliding time is shown in Figure 10. The results indicate that the presence of the nanofluids at the interface of the pin and disc remarkably declines the wear rate of the composite compared to dry and oil lubrication. At an applied load of 40 N and a sliding duration of 30 min, the wear rate under 2 wt% SiC nanofluid lubrication decreased by 53.18% and 58.7% relative to oil-lubricated and dry sliding con-

https://doi.org/10.3390/lubricants14020054

ditions, respectively. The lowest wear rate of 0.0235 mm^3/m observed under 2 wt% SiC nanofluid conditions is due to the generation of a stable tribofilm at the interface. This tribofilm serves as a protective film, preventing the direct asperity contact between the mating surfaces and partially lifting the pin form the disc surface, resulting in an apparent reduction in the wear rate of the Cu-15%WC-9%Gr composite.

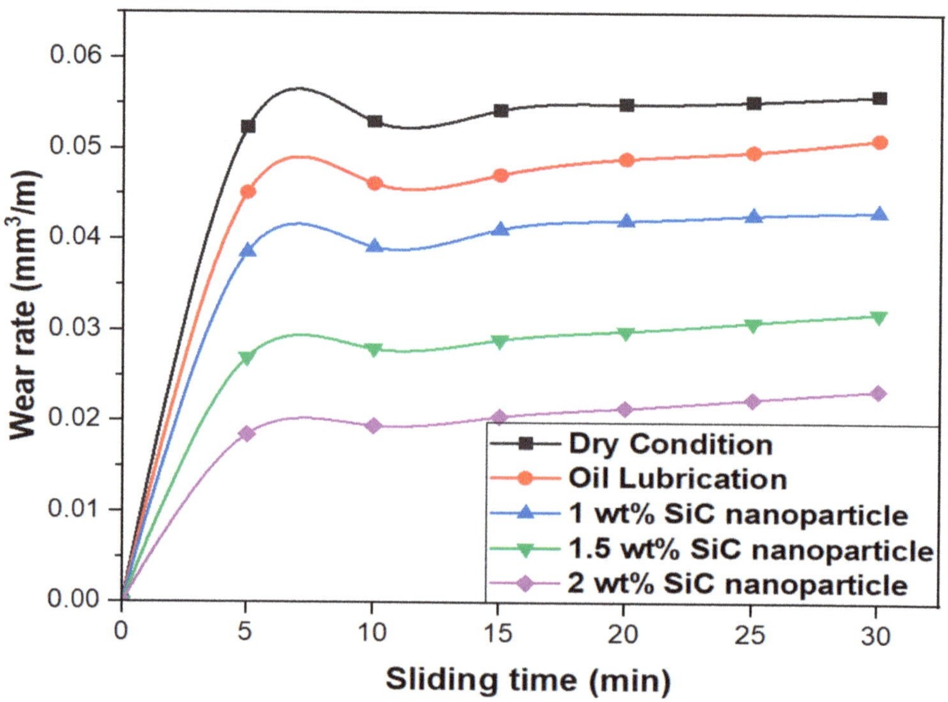

Figure 10. Effect of lubrication regimes on the wear of the HMMC pin with time.

Despite this, when the sliding time exceeds 30 min, the heat generated at the interface leads to the breakdown of the tribofilm, thereby increasing the wear rate of the composite pin against the steel disc. The sustained presence of the nanofluids containing SiC particles at the interfaces of the composite pin and counter disc minimises the surface contact and leads to an improved lubrication performance [42]. Apart from this, the SiC nanofluids present at the interfacial regions are capable of absorbing and dissipating the generated heat, which declines the plastic deformation of the Cu-WC-Gr composite and surpasses the wear of the composite pin.

From the figure, it is also clear that the rate of wear also increases with prolonging the sliding time for all lubricating regimes. However, the magnitude of the wear rate is strongly influenced by the type of lubricant used between the contact surfaces. After 30 min of sliding time, the wear rate values recorded under dry, oil, and nanofluid lubrication with 1 wt%, 1.5 wt%, and 2 wt% were approximately 0.0563 mm^3/m, 0.0502 mm^3/m, 0.0432 mm^3/m, 0.0320 mm^3/m, and 0.0235 mm^3/m, respectively. Notably, the presence of the 2 wt% SiC nanofluid resulted in a substantial decline in the wear rate. These results clearly highlight the superior lubricating efficiency of nanofluids, particularly at higher nanoparticle concentrations.

The enhancement in the wear resistance when increasing the SiC nanofluid content is attributed to the synergistic effect of the improved thermal conductivity, homogeneous

https://doi.org/10.3390/lubricants14020054

dispersion of the nanoparticles, and development of the protective tribofilm at the contact interface. The uniform dispersion of the SiC nanoparticles induces rolling and polishing mechanisms that substantially reduce abrasive contact among the disc and composite pin surfaces. Apart from this, these nanoparticles exhibit a mending effect by filling surface microcracks and defects, thereby limiting material loss. Overall, the findings demonstrate that nanofluids containing 2 wt% SiC deliver a superior tribological performance, clearly outperforming both conventional oil lubrication and dry sliding conditions.

6.3.2. Coefficient of Friction

The C.O.F. for the microwave sintered Cu-15%WC-9%Gr composite is depicted in Figure 11. The maximum C.O.F. of approximately 0.38 was observed under dry sliding conditions, primarily due to the strong adhesive interactions between unlubricated contact surfaces, which result in elevated frictional forces. In contrast, the lowest C.O.F. of 0.28 was recorded under 2 wt% SiC nanofluid conditions. This significant reduction in the friction is partly attributed to the presence of the SiC nanoparticles at the interfaces, which leads to the formation of a tribofilm, thereby preventing direct metal-to-metal contact [43]. Apart from this, the presence of graphite phases in the composite, which form the wear debris during the sliding, further contributes to lowering of the C.O.F. In contrast, the highest value of the C.O.F., 0.38, under dry conditions was attained due to the pure metal-to-metal contact during the sliding. Notably, nanofluid lubrication demonstrated the most effective friction reduction, with the C.O.F. decreasing progressively as the SiC nanoparticle concentration increased.

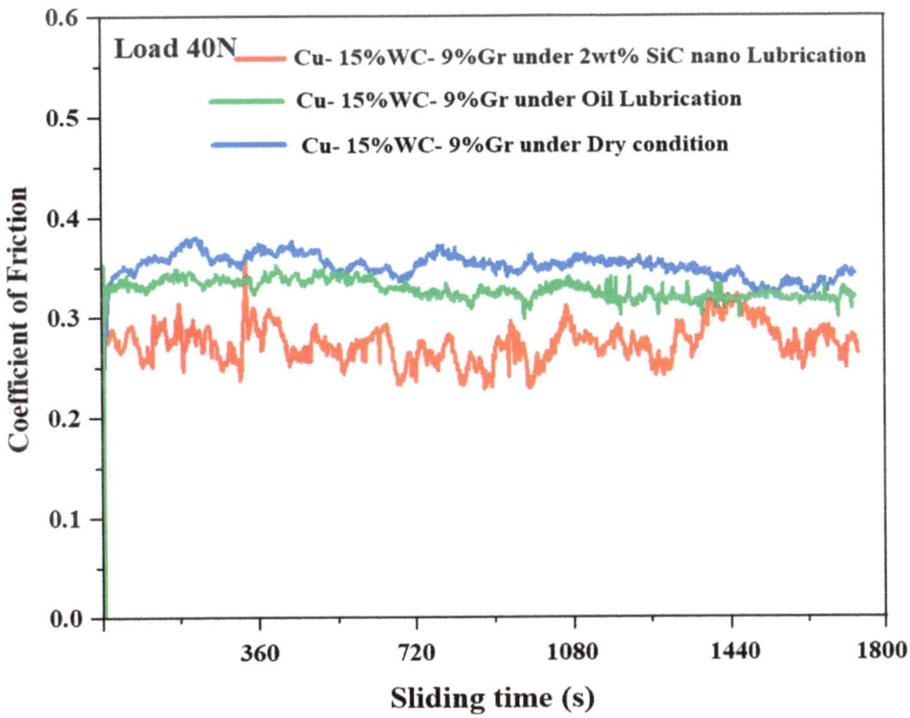

Figure 11. Effect of the lubrication regimes on the C.O.F. of the HMMC pin with time.

The C.O.F. findings demonstrate that nanofluids provide a superior friction reduction compared to conventional oil, with their effectiveness improving as the nanoparticle con-

https://doi.org/10.3390/lubricants14020054

centration increases. The enhanced lubricating performance is attributed to the generation of a protective tribofilm, improved heat dissipation, and the rolling action of uniformly dispersed nanoparticles, which collectively minimise direct surface contact [44]. Among the tested conditions, the nanofluid containing 2 wt% SiC exhibited the highest efficiency, resulting in a substantial reduction in friction and a marked improvement in the overall tribological performance of the composite.

6.4. Worn Surface Morphology

Figure 12 represents the SEM micrographs for the worn surfaces of the Cu-15 wt% WC-9 wt% Gr composite under dry, oil, and 2 wt% SiC nanofluid lubricant conditions for a 30 min time interval sliding time. Under all lubrication regimes, continuous grooves aligned with the sliding direction are evident, arising from the ploughing action of the hard WC particles against the EN31 counter steel disc.

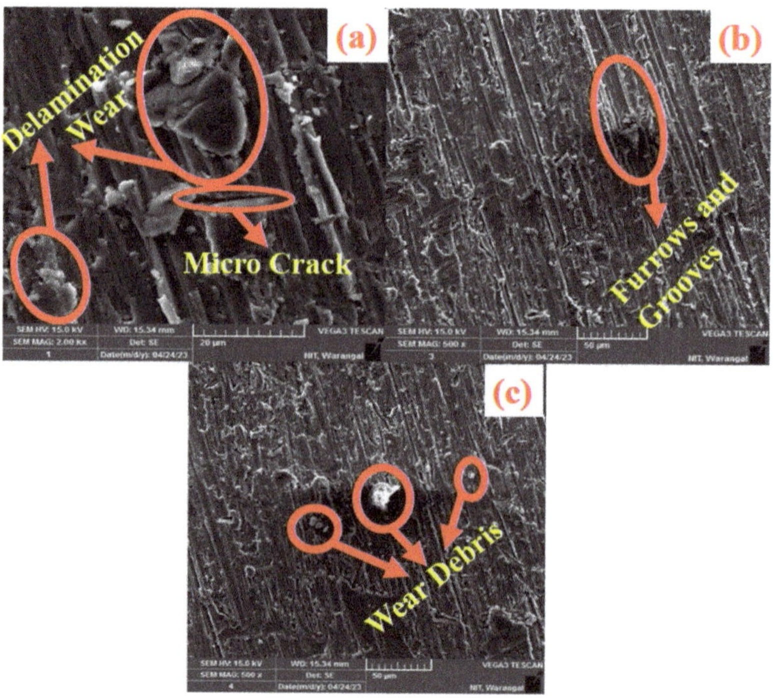

Figure 12. Wear track morphology of the HMMC under (**a**) dry, (**b**) oil, and (**c**) SiC nanofluid lubrication conditions.

Under dry sliding conditions, Figure 12a, the direct interaction between the Cu-WC-Gr composite pin and counter surface results in significant frictional heat, leading to thermal softening and severe damage to the composite surface [45]. Moreover, the hard WC particles tend to debond from the matrix and fracture into fine abrasive debris. The fractured reinforcements from the surface create the preferential sites for the initiation of microcracks, ultimately promoting the delamination wear mechanism. The presence of surface microcracks and the delamination wear under dry sliding conditions was depicted in Figure 12a. The extent of surface damage is further influenced by localised stress concentrations at asperity contacts, which intensify with increasing applied loads.

Figure 12b depicts the worn surface morphology under soluble oil lubrication. Although pronounced furrows and grooves are still observed, the overall worn surface

https://doi.org/10.3390/lubricants14020054

appears comparatively smoother than that formed under dry conditions. The reduced wear rate in oil-lubricated conditions is primarily attributed to the diminished direct asperity contact due to the presence of the lubricating film, which facilitates the transition from severe to mild wear regimes [46]. Furthermore, the existence of the oil film surpasses the temperature rise at the interfaces, leading to lower friction. The lubricant also aids in removing the wear debris from the contact zone, thereby minimising third-body abrasion and oxidation wear compared to dry conditions.

In contrast, under a 2 wt% SiC nanofluid lubricating condition, the presence of SiC nanofluid particles transforms the dominant sliding contact into a combined rolling–sliding mechanism. Uniformly dispersed SiC nanoparticles promote controlled micro-abrasion, producing a polishing effect at the contact regions, as depicted in Figure 12c. The combined rolling and polishing actions significantly reduce the C.O.F. at an optimal concentration of 2 wt% SiC. Beyond this level, nanoparticle agglomeration diminishes these beneficial effects. Additionally, the nanoparticles exhibit a mending effect by filling surface defects and microcracks, thereby limiting material loss. As a result, the synergistic rolling, polishing, and mending mechanisms account for the superior anti-wear and anti-friction performance under 2 wt% SiC nanofluid lubrication.

Although a direct chemical characterization of the tribofilm was not performed, its existence is strongly supported by indirect tribological evidence. The marked reductions in the wear rate and C.O.F. under a 2 wt% SiC nanofluid lubrication, along with the smoother worn surfaces and the absence of severe delamination, indicate the formation of a stable protective tribofilm. This tribofilm is proposed to comprise compacted SiC nanoparticles, graphite-rich wear debris from the Cu–WC–Gr composite, and tribo-oxidised copper species. The uniform dispersion of SiC nanoparticles promotes rolling, mending, and load-bearing effects, while graphite debris provides solid lubrication and shear accommodation. In addition, the high thermal conductivity and stability of the SiC nanofluid enhance heat dissipation, suppressing thermal softening and tribofilm degradation. The synergistic action of these mechanisms results in a mechanically stable and continuously replenished tribofilm, accounting for the improved wear resistance observed. The significant reduction and stabilization of the C.O.F, along with the transition from severe ploughing to mild polishing wear observed in the worn surface morphology, strongly indicate the operation of a third-body lubrication mechanism.

7. Conclusions

The key conclusions drawn from the present study are briefly presented below.

(1) Cu–15%WC–X%Gr composites fabricated through microwave sintering showed a uniform reinforcement dispersion up to 9 wt% Gr, while 12 wt% Gr caused Cu_2C_2 agglomeration and interfacial defects.

(2) The hardness increased from 180 HV (Cu–15%WC) to 239 HV (Cu–15%WC–12%Gr), achieving a 32.7% improvement due to enhanced interfacial bonding and load transfer.

(3) A maximum compression strength of 395 MPa was obtained for Cu–15%WC–9%Gr, representing a 38.59% increase over Cu–15%WC, attributed to grain boundary strengthening and dislocation blocking.

(4) The tribological performance improved significantly under 2 wt% SiC nanofluid lubrication, yielding the lowest wear rate of 0.0235 mm^3/m, with reductions of 53.18% and 58.7% compared to oil and dry conditions, respectively, while simultaneously reducing the C.O.F. from ~0.38 (dry) to ~0.28 due to tribofilm formation, nanoparticle rolling, and the lubricating action of graphite.

https://doi.org/10.3390/lubricants14020054

(5) The Cu–15%WC–9%Gr composite exhibited a superior tribological performance when combined with 2 wt% SiC nanofluid lubrication, demonstrating a clear synergistic effect between the optimised material system and the lubrication condition.
(6) The presence of graphite within the composite and uniformly dispersed SiC nanoparticles in the nanofluid promoted stable tribofilm formation, enhanced heat dissipation, and reduced direct asperity contact, resulting in the lowest wear rate of 0.0235 mm^3 m^{-1} and a reduced C.O.F. of 0.28.
(7) Worn surface analysis revealed a severe delamination under dry sliding, moderate abrasion with oil, and smooth, crack-free surfaces under nanofluid lubrication due to rolling, polishing, and mending effects.

Author Contributions: Conceptualization, V.S.S.V.; Methodology, V.S.S.V.; Validation, B.S.; Formal analysis, B.S.; Investigation, P.R.V.; Data curation, M.M.M.; Writing—original draft, V.S.S.V. and M.M.M.; Writing—review & editing, P.R.V. and M.M.M.; Supervision, B.S.; Project administration, P.R.V. All authors have read and agreed to the published version of the manuscript.

Funding: This research received no external funding.

Data Availability Statement: The original contributions presented in this study are included in the article. Further inquiries can be directed to the corresponding author.

Conflicts of Interest: The authors declare no conflict of interest.

References

1. Sakib-Uz-Zaman, C.; Khondoker, M.A.H. A Review on Extrusion Additive Manufacturing of Pure Copper. *Metals* **2023**, *13*, 859. [CrossRef]
2. Kocich, R.; Opěla, P.; Marek, M. Influence of Structure Development on Performance of Copper Composites Processed via Intensive Plastic Deformation. *Materials* **2023**, *16*, 4780. [CrossRef]
3. Zhai, Z.; Dong, H.; Li, D.; Wang, Z.; Sun, C.; Chen, C. Effect of TiC Particles on the Properties of Copper Matrix Composites. *Inorganics* **2024**, *12*, 120. [CrossRef]
4. Kumar, N.; Bharti, A.; Dixit, M.; Nigam, A. Effect of Powder Metallurgy Process and its Parameters on the Mechanical and Electrical Properties of Copper-Based Materials: Literature Review. *Powder Metall. Met. Ceram.* **2020**, *59*, 401–410. [CrossRef]
5. Yan, Y.F.; Kou, S.Q.; Yang, H.Y.; Shu, S.L.; Qiu, F.; Jiang, Q.C.; Zhang, L.C. Ceramic particles reinforced copper matrix composites manufactured by advanced powder metallurgy: Preparation, performance, and mechanisms. *Int. J. Extreme Manuf.* **2023**, *5*, 032006. [CrossRef]
6. Long, X.; Chong, K.; Su, Y.; Du, L.; Zhang, G. Connecting the macroscopic and mesoscopic properties of sintered silver nanoparticles by crystal plasticity finite element method. *Eng. Fract. Mech.* **2023**, *281*, 109137. [CrossRef]
7. Liu, K.; Shi, X.; Wang, D.; Feng, Y.; Jian, Y.; Li, W. A method for the dynamic characteristic analysis of a rotor-rolling bearing system influenced by elastohydrodynamic lubrication. *J. Sound Vib.* **2025**, *608*, 119075. [CrossRef]
8. Bahador, A.; Umeda, J.; Hamzah, E.; Yusof, F.; Li, X.; Kondoh, K. Synergistic strengthening mechanisms of copper matrix composites with TiO_2 nanoparticles. *Mater. Sci. Eng. A* **2020**, *772*, 138797. [CrossRef]
9. El-Zaidia, M.M.; Zaki, M.Z.; Abomostafa, H.M.; Taha, M.A. Comprehensive studies for evaluating promising properties of Cu/graphene/fly ash nanocomposites. *Sci. Rep.* **2024**, *14*, 2236. [CrossRef]
10. Samal, C.P.; Parihar, J.S.; Chaira, D. The effect of milling and sintering techniques on mechanical properties of Cu-graphite metal matrix composite prepared by powder metallurgy route. *J. Alloys Compd.* **2013**, *569*, 95–101. [CrossRef]
11. Akbarpour, M.R.; Mousa Mirabad, H.; Alipour, S. Microstructural and mechanical characteristics of hybrid SiC/Cu composites with nano- and micro-sized SiC particles. *Ceram. Int.* **2019**, *45*, 3276–3283. [CrossRef]
12. Manohar, G.; Pandey, K.M.; Maity, S.R. Effect of sintering mechanisms on mechanical properties of AA7075/B4C composite fabricated by powder metallurgy techniques. *Ceram. Int.* **2021**, *47*, 15147–15154. [CrossRef]
13. Yang, Z.; Liu, J.; Liu, J.; Chen, X.; Yan, T.; Chen, Q. Investigation on physicochemical properties of graphene oxide/nano-hydroxyapatite composites and its biomedical applications. *J. Aust. Ceram. Soc.* **2021**, *57*, 625–633. [CrossRef]
14. Wang, C.; Zhang, Z.; Jing, X.; Yang, Z.; Xu, W. Optimization of multistage femtosecond laser drilling process using machine learning coupled with molecular dynamics. *Opt. Laser Technol.* **2022**, *156*, 108442. [CrossRef]
15. Ashwath, P.; Anthony Xavior, M. The effect of ball milling & reinforcement percentage on sintered samples of aluminium alloy metal matrix composites. *Procedia Eng.* **2014**, *97*, 1027–1032. [CrossRef]

https://doi.org/10.3390/lubricants14020054

16. Venkatesh, V.S.S.; Deoghare, A.B. Effect of Sintering Mechanisms on the Mechanical Behaviour of SiC and Kaoline Reinforced Hybrid Aluminium Metal Matrix Composite Fabricated through Powder Metallurgy Technique. *Silicon* **2021**, *14*, 5481–5493. [CrossRef]
17. Dey, D.; Bhowmik, A.; Biswas, A. Effect of SiC Content on Mechanical and Tribological Properties of Al2024-SiC Composites. *Silicon* **2020**, *11*, 1 11. [CrossRef]
18. Liu, J.; Xiang, S.; Zhou, X.; Lin, S.; Dong, K.; Liu, Y.; He, D.; Fan, Y.; Liu, Y.; Xiong, B.; et al. Lubrication Performance Promotion of GTL Base Oil by BN Nanosheets via Cascade Centrifugation-Assisted Liquid-Phase Exfoliation. *Lubricants* **2025**, *13*, 281. [CrossRef]
19. Poria, S.; Sahoo, P.; Sutradhar, G. Tribological Characterization of Stir-cast Aluminium-TiB2 Metal Matrix Composites. *Silicon* **2016**, *8*, 591–599. [CrossRef]
20. Venkatesh, V.S.S.; Vundavilli, P.R. Mechanical Properties and Tribological Study of Bottom Pouring Stir-Cast A356 Alloy Reinforced with Graphite Solid Lubricant Extracted from Corn Stover. *Lubricants* **2024**, *12*, 341. [CrossRef]
21. Kumar, A.; Patnaik, A.; Bhat, I.K. Tribology Analysis of Cobalt Particulate Filled Al 7075 Alloy for Gear Materials: A Comparative Study. *Silicon* **2019**, *11*, 1295–1311. [CrossRef]
22. Yan, X.; Hu, J.; Zhang, X.; Xu, W. Obtaining superior low-temperature wear resistance in Q&P-processed medium Mn steel with a low initial hardness. *Tribol. Int.* **2022**, *175*, 107803. [CrossRef]
23. Zhang, Z.; Liu, S.; Zhang, Y.; Wang, C.; Zhang, S.; Yang, Z.; Xu, W. Optimization of low-power femtosecond laser trepan drilling by machine learning and a high-throughput multi-objective genetic algorithm. *Opt. Laser Technol.* **2022**, *148*, 107688. [CrossRef]
24. Zou, Y.; Tang, S.; Guo, S.; Song, X. Tool wear analysis in turning inconel-657 using various tool materials. *Mater. Manuf. Process.* **2024**, *39*, 1363–1368. [CrossRef]
25. Zheng, K.; Min, Z.; Zhang, F.; Ren, Z.; Lin, Y. High Heat-fade Resistance, Metal-free Resin-based Brake Pads: A Step towards Replacing Copper by Using Andalusite. *Chin. J. Mech. Eng.* **2025**, *38*, 153. [CrossRef]
26. Ramadhan, A.I.; Diniardi, E.; Basri, H.; Almanda, D.; Azmi, W.H. Thermal-Properties Correlation of Al_2O_3-TiO_2-SiO_2 Nanofluids in Water-Ethylene Glycol Mixture for Application Radiator Cooling System. *Semarak Int. J. Mech. Precis. Eng.* **2025**, *1*, 51–61. [CrossRef]
27. Zeng, L.; Deng, X.; Li, F.; Dong, C.; Wang, S.; Yang, H.; Tang, C.; Li, Y. Study on dynamic wear evolution of modified gear rack considering the real-time variation of contact characteristics. *Wear* **2025**, *571*, 205845. [CrossRef]
28. Ezekwem, C.; Dare, A. Thermal and electrical conductivity of silicon carbide nanofluids. *Energy Sources Part A Recover. Util. Environ. Eff.* **2020**, *46*, 11320–11338. [CrossRef]
29. Choudhary, R.; Khurana, D.; Kumar, A.; Subudhi, S. Stability analysis of Al_2O_3/water nanofluids. *J. Exp. Nanosci.* **2017**, *12*, 140–151. [CrossRef]
30. Wang, S.; Zhang, Z.; Sun, C.; Gong, L.; Zhang, X.; Gao, S.; Zhang, C.; Han, Q.; Yan, S. Multifunctional Tribovoltaic Coating for Self-Powered In Situ Sensing with Exceptional Tribological Robustness and Charge Transport. *Adv. Funct. Mater.* **2025**, 14190. [CrossRef]
31. Xiao, C.; Peng, J.; Jiao, Y.; Shen, Q.; Zhao, Y.; Zhao, F.; Li, H.; Song, Q. Strong and Tough Multilayer Heterogeneous Pyrocarbon Based Composites. *Adv. Funct. Mater.* **2024**, *34*, 2409881. [CrossRef]
32. Manohar, G.; Pandey, K.M.; Maity, S.R. Effect of Variations in Microwave Processing Temperatures on Microstructural and Mechanical Properties of AA7075/SiC/Graphite Hybrid Composite Fabricated by Powder Metallurgy Techniques. *Silicon* **2022**, *14*, 7831–7847. [CrossRef]
33. Yuan, F.; Han, Y.; Xiao, K.; Xiang, G.; Wang, Z.; Li, L.; Wang, C.; Chen, X. Spatial modification optimization methods for harmonic drives using a 3D non-uniform line-contact elastohydrodynamic lubrication model. *Tribol. Int.* **2026**, *214*, 111288. [CrossRef]
34. Guo, J.; Han, Y.; Xiao, K. Evaluation of linear and nonlinear methods for predicting the dynamic behavior of plain and textured water-lubricated bearings. *Phys. Fluids* **2025**, *37*, 083621. [CrossRef]
35. Zhou, S.; Lin, Z.; Tao, X.; Li, W.; Chen, B.; Li, F.; Li, Y.; Du, Y.; Jiang, Y.; Deng, Y. Sintering of Chang'e-5 high-fidelity lunar soil simulant for providing high-strength materials for lunar base construction. *Sci. China Technol. Sci.* **2025**, *68*, 1820602. [CrossRef]
36. Manikandan, R.; Arjunan, T.V.; Akhil, A.R. Studies on micro structural characteristics, mechanical and tribological behaviours of boron carbide and cow dung ash reinforced aluminium (Al 7075) hybrid metal matrix composite. *Compos. Part B Eng.* **2020**, *183*, 107668. [CrossRef]
37. Bhowmik, A.; Dey, D.; Biswas, A. Comparative Study of Microstructure, Physical and Mechanical Characterization of SiC/TiB2 Reinforced Aluminium Matrix Composite. *Silicon* **2020**, *13*, 2003–2010. [CrossRef]
38. Wang, H.; Zhou, S.; Zhang, X.; Zhou, Q.; Jiang, Y.; Deng, Y.; Liu, J.; Lin, Z.; Li, F.; Zhang, C.; et al. Particle Morphology Controls the Bulk Mechanical Behavior of Far-Side Lunar Regolith from Chang ' e-6 Samples and Deep Learning. *Research* **2026**, *9*, 1064. [CrossRef]
39. Prasad Reddy, A.; Vamsi Krishna, P.; Rao, R.N. Tribological Behaviour of Al6061–2SiC-xGr Hybrid Metal Matrix Nanocomposites Fabricated through Ultrasonically Assisted Stir Casting Technique. *Silicon* **2019**, *11*, 2853–2871. [CrossRef]

https://doi.org/10.3390/lubricants14020054

40. Hu, Y.; Chen, J.; Zhen, Y.; Han, Y.Z.; Yu, C.; Xiong, K.; Zhang, S. Study on the wettability, intermetallic compound growth, voids formation and mechanical properties of Cu/Sn joints with changes in substrate roughness for electronic packaging. *J. Mater. Sci Mater. Electron.* **2025**, *36*, 851. [CrossRef]
41. Xu, H.; Liu, M.; Hu, J.; Wang, Y. Enhanced electromagnetic wave absorption through in-situ mineralization constructed robust Fe_3O_4/PDA/$Ti_3C_2T_x$ MXene composites with brick-and-mortar microstructure. *Ceram. Int.* **2026**. [CrossRef]
42. Wu, Y.Y.; Tsui, W.C.; Liu, T.C. Experimental analysis of tribological properties of lubricating oils with nanoparticle additives. *Wear* **2007**, *262*, 819–825. [CrossRef]
43. Irfan Ul Haq, M.; Raina, A.; Anand, A.; Sharma, S.M.; Kumar, R. Elucidating the Effect of MoS_2 on the Mechanical and Tribological Behavior of AA7075/Si3N4 Composite. *J. Mater. Eng. Perform.* **2020**, *29*, 7445–7455. [CrossRef]
44. Siva, V.S.B.; Ganguly, R.I.; Srinivasa Rao, G.; Sahoo, K.L. Wear behaviour of novel Al based composite reinforced with ceramic composite (Al_2O_3-SiC-C) developed from colliery shale material. *Tribol. Mater. Surfaces Interfaces* **2014**, *8*, 117–124. [CrossRef]
45. Venkatesh, V.S.S.; Deoghare, A.B. Modelling and Optimisation of Wear Parameters for Spark Plasma Sintered Al—SiC—Kaoline Hybrid Composite. *Adv. Mater. Process. Technol.* **2021**, *8*, 1286–1304. [CrossRef]
46. Mukhopadhyay, A.; Duari, S.; Barman, T.K.; Sahoo, P. Tribological Performance Optimization of Electroless Ni–B Coating under Lubricated Condition using Hybrid Grey Fuzzy Logic. *J. Inst. Eng. Ser. D* **2016**, *97*, 215–231. [CrossRef]

https://doi.org/10.3390/lubricants14020054

lubricants

MDPI

Article

On the Microstructural and Tribological Investigations of WC-12Co/NiCrFeSiAlBC HVOF Cermet Coatings: Effects of WC-12Co Fraction

Fida Harabi [1], Basma Ben Difallah [1,*], Faten Nasri [1], Clisia Aversa [2], Mohamed Kharrat [1], Massimiliano Barletta [2] and Antonio Pereira [3,4,*]

1 Laboratory of Electromechanical Systems, National School of Engineers of Sfax, University of Sfax, Sfax 3038, Tunisia; fida.harabi@enis.tn (F.H.); nasrifaten85@gmail.com (F.N.); mohamed.kharrat@ipeis.usf.tn (M.K.)
2 Mechanical Department, University of Roma Tre, 00146 Rome, Italy; massimiliano.barletta@uniroma3.it (M.B.)
3 Department of Mechanical Engineering, Centre for Mechanical Technology and Automation (TEMA), University of Aveiro, 3810-193 Aveiro, Portugal
4 LASI—Intelligent Systems Associate Laboratory, 4800-058 Guimarães, Portugal
* Correspondence: basma.bendhaifallah@ipeis.usf.tn (B.B.D.); abastos@ua.pt (A.P.)

Abstract

Previous research indicates that WC-12Co contents above 60 wt.% in feedstock powders for cermet coatings impair adhesion and wear resistance. This study characterizes NiCrFeSiAlBC coatings—unreinforced or reinforced with 65 wt.% or 85 wt.% WC-12Co—applied via high-velocity oxy-fuel (HVOF) spraying onto stainless steel substrates under controlled parameters. It quantifies the influence of high carbide volume fractions within the NiCrFeSiAlBC matrix on microstructure and tribomechanical performance. Microstructural analysis revealed uniformly distributed cermet layers featuring dissolved reinforcements and WC hard phase formation, with minimal W_2C crystallization. Elevated WC-12Co incorporation promoted densification and reduced porosity. Vickers microhardness tests (HV 0.3) demonstrated increased hardness upon WC-12Co addition, attributable to finer particle sizes, lower porosity, and the presence of WC phases alongside crystallographic refinements. Under dry reciprocating sliding conditions, friction coefficients and wear volumes decreased markedly. Consequently, the coating with 85 wt.% WC exhibited the best mechanical and tribological properties.

Keywords: HVOF process; NiCrFeSiAlBC; WC-12Co; cermet coatings; microstructure; microhardness; friction; wear

1. Introduction

Modern automotive and powertrain components, such as piston rings, cylinder liners, valve seats, and camshafts, operate under severe conditions that combine high contact stresses, sliding motion and thermal gradients. These harsh operating conditions accelerate surface degradation mechanisms, including abrasive and adhesive wear, oxidation, and tribo-corrosion, which often govern the lifetime of critical components [1,2]. Surface engineering via advanced coatings is crucial to enhance durability, curtail servicing intervals, and offer operational efficacy while supporting light-weighting strategies through superficial densification that preserves the resilience and plastic deformability of low-density substrates.

High-velocity oxy-fuel (HVOF) thermal spraying has emerged as an effective technique for producing protective coatings with superior adhesion, low porosity, and excellent mechanical properties [3]. In fact, during HVOF spraying, the elevated particle velocities and thermal energy strongly affect splat formation, carbide retention, and interfacial bonding, thereby substantially altering the coatings' microstructure [4,5] to attain the desired properties. HVOF coatings offer a sustainable alternative to hard chromium plating, which is increasingly restricted due to the health and environmental hazards associated with hexavalent chromium compounds [2]. The ability to deposit dense, wear-resistant coatings makes HVOF particularly suitable for engine and powertrain applications where components are subjected to repeated sliding and high thermal loads [6].

Nickel-based self-fluxing alloys are widely used as protective coatings in industry due to their excellent corrosion and wear resistance [7]. Consequently, extensive research has focused on their deposition processes, microstructural features, and mechanical performance. According to the literature, deposition technologies significantly affect coating's properties, including adhesion, porosity, and tribological properties [8]. For instance, laser cladding studies under varying conditions have yielded diverse morphologies, planar cellular structures [9] or flower-like patterns [10]. Similarly, Huang et al. reported that atmospheric plasma-sprayed (APS) Ni-based coatings presented lamellar structures [7]. The variety in the structure is accompanied by a broad range of phases formed during deposition. Notably, Cr, B, Si, Fe and C in Ni-based powders enhance self-fluxing behavior, promote hard phase formation within the Ni matrix, and improve wear and corrosion resistance [11–13]. This composition enables crystallization of a wide variety of crystalline phases influenced by powder chemistry and deposition parameters [9,14], and resulting in borides and carbides embedded in the microstructure of NiCrBSiFeC alloy [9,14,15]. Such phases boost their resistance to abrasion and wear [16]. Moreover, Ni-based coatings exhibit high hardness, strong metallurgical adhesion to substrates, and superior high-temperature corrosion and wear performance [15,16]. NiCrBSiFeC coatings are widely employed as matrix materials in HVOF coatings due to their toughness, high-temperature stability, and corrosion resistance [17]. However, metallic matrices alone often lack sufficient resistance to severe wear in high-load sliding applications. To overcome this limitation, hard ceramic reinforcements, particularly tungsten carbide bonded with cobalt (WC-12Co), are incorporated to form cermet coatings. The key benefit of WC-12Co lies in its composite structure: the carbide provides exceptional hardness and wear resistance, while the Co binder matrix imparts the necessary toughness to the system [18]. Nevertheless, sprayed WC-12Co is brittle and susceptible to cracking [19,20]. Moreover, it forms highly brittle phases (W_2C, Co-W-C) during decomposition and decarburization above 450 °C [21,22], which degrades fracture toughness [22]. NiCrBSiFeC is employed to support and bond WC particles, ensuring effective wetting and adhesion of the reinforcing material [9]. The literature highlights significant interest in optimizing WC-12Co content within metallic matrices. Several studies, using different processes (laser-cladding, HVOF), have shown that in NiCrBSi coatings reinforced with WC/12Co, the progressive addition of ceramic particles (10–50 wt.%) modified the microstructure and the tribological properties of cermet coatings [9,15,23] significantly. In contrast, excessive hard phase levels (60–80 wt.%) hinder particle wetting, introduce defects, and adversely affect bonding, cohesion and wear mechanisms [15,23]. Thus, balancing the matrix–reinforcement ratio is essential for optimizing hardness, toughness and wear resistance while maintaining performance under elevated stress and temperature conditions [6]. Coatings are particularly suitable for engine and powertrain components, where coatings must endure high contact stresses, repeated sliding motion, and thermal cycling while preserving consistent tribological performance [1,6,24,25].

https://doi.org/10.3390/lubricants14030100

In the present study, industrial NiCrFeSiAlBC powder, incorporating aluminum (Al) as a novel alloying element, was used to extend prior findings. NiCrFeSiAlBC-based cermet coatings were subsequently developed, both unreinforced and reinforced with WC/12Co at 65 wt.% and 85 wt.%. Optimized HVOF parameters ensured consistent microstructural development, even at these high hard-phase contents. The resulting coatings underwent comprehensive characterization of their microstructure, porosity, mechanical properties, and tribological behavior. Quantitative phase composition data were also obtained.

2. Materials and Methods

2.1. Materials

2.1.1. Substrate

Coatings were deposited on AISI 316L stainless steel substrates, selected for their excellent corrosion resistance, thermal stability, and mechanical properties, which are essential for demanding industrial applications. Prior to deposition, the substrates underwent a standardized preparation protocol, including degreasing with acetone to remove surface contaminants, abrasive blasting to achieve a controlled arithmetic surface roughness of about 4–7 μm for mechanical interlocking, and final cleaning to ensure a defect-free surface. This procedure enhances coating adhesion and ensures reproducibility. The substrate dimensions are 30 mm × 20 mm × 2 mm.

2.1.2. Feedstock Powders

The feedstock material employed in this study was a NiCrFeSiAlBC alloy powder supplied by Praxair Surface Technologies. Table 1 presents the detailed chemical composition of the multi-component alloy. Nickel represents the primary constituent, ensuring adequate ductility and toughness of the coating matrix, while chromium enhances oxidation and corrosion resistance. The presence of silicon, aluminum, boron, and carbon contributes to solid-solution strengthening and the formation of hard phases, which collectively improve hardness and wear resistance. Owing to this combination of mechanical robustness and chemical stability, NiCrFeSiAlBC was selected as a suitable matrix material for high-performance cermet coatings produced by high-velocity oxy-fuel (HVOF) spraying. As we can see in Figure 1, the NiCrFeSiAlBC powder morphology is predominantly spherical particles with a few slightly elongated semispherical particles; a predominance of particles with sizes between 30 and 45 microns is seen.

Table 1. Chemical composition of the NiCrFeSiAlBC powder.

Element	Ni	Cr	Fe	Si	Al	B	C
Wt. (%)	67–85	6–18	1–5	2–13	0–2	0–4	0–2

2.1.3. Powder Preparation

Composite powder mixtures were prepared by blending the NiCrFeSiAlBC master alloy with tungsten carbide–cobalt (WC-Co) reinforcement at two mass fractions, namely 65 wt.% and 85 wt.%, using ball milling. The powders were introduced into a cylindrical jar mounted on a roller-driven milling system to ensure homogeneous mixing. To enhance the dispersion of the ceramic phase within the metallic matrix and promote inter-particle adhesion, steel balls were added to the container, thereby converting the mixing operation into a ball milling process.

Ball milling was used as a blending technique to achieve a homogeneous dispersion of ceramic particles within the metallic master alloy matrix while maintaining particle characteristics suitable for both additive manufacturing and HVOF spraying [26]. Ball milling and related mechanical alloying processes are well documented in the powder

https://doi.org/10.3390/lubricants14030100

metallurgy literature as effective methods to control particle morphology, reinforcement dispersion, and internal strain through process variables such as milling speed, ball-to-powder ratio, and milling time. Several experimental studies have demonstrated that these milling parameters directly influence composite powder morphology and properties; for example, variations in milling speed, ball-to-powder ratio, and milling duration significantly affect powder particle size, homogeneity, and densification behavior [27,28].

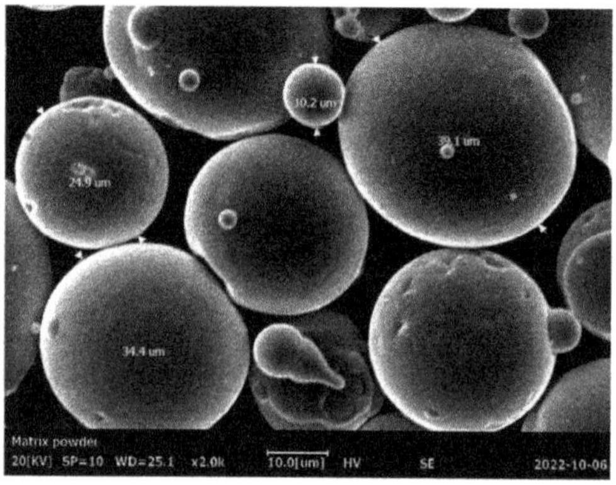

Figure 1. SEM image of NiCrFeSiAlBC powder showing predominantly spherical morphology.

The powders were mixed in a rotating steel jar containing steel balls, resulting in a low-energy ball milling process that enhances the mechanical adhesion of ceramic particles onto metal powder surfaces. The mixing efficiency is influenced by parameters such as ball-to-powder weight ratio, rotation speed, and drum filling level, which govern the motion of the grinding media and the resulting powder homogeneity [26,27]. To prevent oxidation and preserve powder chemistry, milling was performed under an inert argon atmosphere. The selected milling parameters (Table 2) were chosen based on the established literature guidelines to ensure homogeneous blending while preserving powder flowability [26].

Table 2. Ball milling parameters used for powder blending.

Parameter	Value
Milling device	Roller/drum mill
Jar material	Steel
Ball material	Steel
Ball-to-powder weight ratio (BPR)	2:1
Mixing time	1 h
Atmosphere	Argon
Rotation speed	60 rpm

The prepared powders were subsequently used for microstructural characterization and coating deposition. Figure 2 presents SEM images showing the morphology of composite formulations of NiCrFeSiAlBC reinforced with WC-12Co.

https://doi.org/10.3390/lubricants14030100

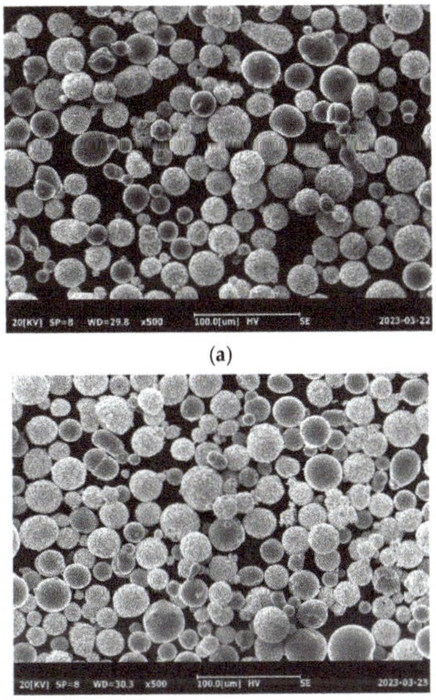

(**a**) (**b**)

Figure 2. (**a**) NiCrFeSiAlBC + 65% WC/12Co morphology and (**b**) NiCrFeSiAlBC + 85% WC/12Co morphology obtained by SEM.

2.2. Methods

2.2.1. HVOF Spraying Process

Coatings were deposited using a JP-5000 HVOF system (TAFA/OerlikonMetco, Concord, NH, USA) at RINA Consulting S.p.A., which enabled industrial-scale precision. The system combusts kerosene with high-purity oxygen in a high-pressure chamber to generate a supersonic jet, accelerating powder particles to velocities exceeding 700–800 m/s. This high kinetic energy enables particle deformation and bonding with the substrate while limiting thermal degradation. The deposition process utilized a robotic arm to control gun trajectory, stand-off distance (330–370 mm), and traverse speed (500 mm/s), with a digitally controlled powder feeder ensuring consistent flow.

Experimental methodology for the different investigated coatings has been summarized in Figure 3.

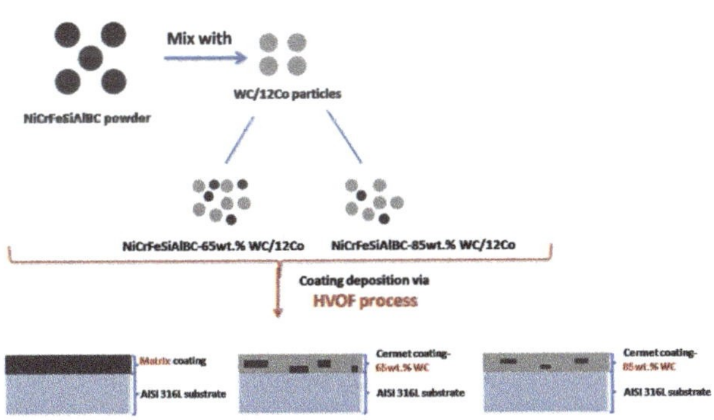

Figure 3. Experimental methodology for the different investigated coatings.

https://doi.org/10.3390/lubricants14030100

2.2.2. Characterization Techniques

Following deposition, the coatings were subjected to comprehensive characterization to assess their microstructural, mechanical, and tribological characteristics. SEM imaging was performed using a COXEM CX-200Plus microscope (Daejeon, Republic of Korea) to assess particle distribution, porosity, and interfacial quality. Samples were cut to appropriate sizes, embedded in thermosetting resin, and subjected to progressive grinding and polishing with fine abrasives and diamond paste to achieve sub-micron surface finishes. A thin gold layer was sputter-coated on the samples using a Cressington Sputter Coater 108 Auto (Cressington Scientific Instruments Ltd., Watford, UK) to enhance conductivity and prevent surface charging. SEM images enabled analysis of WC-12Co particle dispersion, matrix-ceramic interfaces, and potential microstructural defects.

X-ray diffraction (XRD) analysis was performed using a RigakuSmartLab diffractometer (Rigaku Corporation, Tokyo, Japan) to identify crystalline phases, detect phase transformations, and confirm the presence of metallic and ceramic phases. Scans were conducted under controlled conditions with Cu-Kα radiation, providing accurate phase identification. The resulting diffraction patterns were analyzed with X'Pert High Score Plus software (3.0d (3.0.4)) to evaluate existing phases and their mass fraction in the coatings using Rietveld refinement.

Microhardness measurements were performed using a Leitz RZD-DO Vickers microdurometer (Leitz, Wetzlar, Germany) under a 3 N load with a dwell time of 10 s. Multiple indentations per sample provided statistically significant results and allowed evaluation of the influence of WC/12Co content on coating hardness and mechanical integrity.

Tribological performance was evaluated using a CSM Instruments Tribometer Plus (CSM Instruments, Peseaux, Switzerland) in a ball-on-flat configuration and reciprocating linear motion. Al_2O_3 balls (6 mm diameter) were used as the counterbody under a normal load of 10 N, a stroke length of 5 mm, a frequency of 30 Hz, and 5000 cycles in dry ambient conditions (25 °C). Friction forces were continuously monitored, and wear profiles were analyzed using profilometry. Wear volume loss was quantified by measuring the cross-sectional area S (in mm^2) of the wear tracks with a Surtronic® S-100 profilometer (AMETEK Taylor Hobson, Leicester, UK). For each specimen, measurements were taken at the track endpoints and midpoint; the mean S value was then used to calculate volume via $V = S \times d$, where stroke length d = 5 mm. Wear mechanisms were qualitatively assessed through optical microscopy of the tracks.

3. Results

3.1. Microstructural, Crystallographic and Morphological Analyses

3.1.1. Microstructural Analysis

The morphological aspects of the surface and cross-sectional microstructures of the different sprayed coatings, examined by SEM, are presented in Figure 4. A preliminary examination of the coating surfaces reveals that all the layers are continuous and relatively homogeneous. However, the NiCrFeSiAlBC alloy matrix exhibits noticeable pores and microcracks (Figure 4a). In contrast, the incorporation of WC-12Co particles into the NiCrFeSiAlBC alloy (Figure 4c,e) results in denser coatings with significantly reduced porosity. The micrographs of the latter coatings reveal the presence of lamellar splats and loosely bonded unburnt NiCrFeSiAlBC/WC-12Co particles. The lamellar features observed in the mixed coatings indicate the formation of an anisotropic microstructure [29]. Furthermore, melted or partially melted WC-12Co particles predominantly occupy the coating surface (light gray regions in Figure 4c), whereas the melted matrix appears darker. This dominance of WC/12Co increases with higher WC/12Co content in the matrix, as

https://doi.org/10.3390/lubricants14030100

observed in Figure 4e. These microstructural features can be attributed to the thermal and kinetic conditions imposed during the spraying process [4,5].

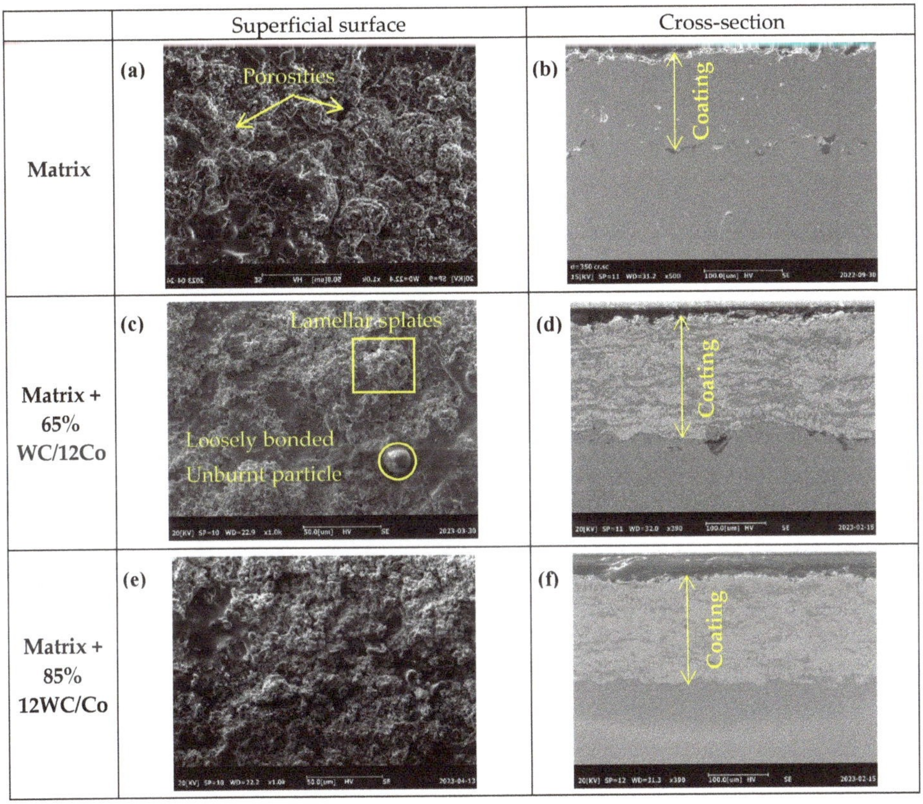

Figure 4. SEM micrographs of: (**a**) NiCrFeSiAlBC coating (surface); (**b**) NiCrFeSiAlBC coating (cross-section); (**c**,**d**) NiCrFeSiAlBC + 65 wt.% WC–12Co cermet coating (surface and cross-section, respectively); (**e**,**f**) NiCrFeSiAlBC + 85 wt.% WC–12Co cermet coating (surface and cross-section, respectively).

The cross-sectional microstructures of the different coatings (Figure 4b,d,f) show compact and thick layers of about 150 μm thickness. Microscopic examination confirms the SEM surface observations previously discussed (Figure 4a,c,e). Indeed, the alloy matrix coating (Figure 4b) exhibits an undulated surface morphology with internal defects such as pores and microcracks. In contrast, the mixed coatings (Figure 4d,f) reveal a lamellar structure composed of alternating strata: light-gray WC agglomerates and dark-gray undulated layers of NiCrFeSiAlBC, illustrating the discrete mixing characteristic of successive thermal spray deposits [30]. The micrographs highlight a dense material, free of pores and voids, both within the coating and at the coating–substrate interface. The coated surface and the interface are particularly distinguished by the formation of this lamellar structure. The arithmetic surface roughness values (Ra) are summarized in Table 3. The results show a systematic increase in roughness with the incorporation of WC–12Co powder, since the two feedstock materials are comparable in terms of particle size and shape. This increase is, therefore, mainly attributed to the intrinsic thermo-physical properties of the WC–Co particles during HVOF spraying. Due to their high melting temperature and limited plastic deformation compared to the NiCrFeSiAlBC alloy, WC–Co particles are frequently deposited in a partially molten state, resulting in incomplete splat spreading

https://doi.org/10.3390/lubricants14030100

upon impact. The presence of hard WC-rich agglomerates protruding from the coating surface enhances surface heterogeneity and promotes the formation of micro-asperities. As the WC–12Co content increases, this effect becomes more pronounced, leading to higher arithmetic surface roughness values.

Table 3. Comparison of arithmetic surface roughness values (Ra).

Coating	NiCrFeSiAlBC	65 wt.% WC/12Co	85 wt.% WC/12Co
Ra (μm)	2.43	4.12	4.29

3.1.2. Crystallographic Analysis

Figure 5 shows the XRD diagrams obtained during Matrix (Figure 5a), Matrix-65% WC-12Co (Figure 5b) and Matrix-85% wt.% WC-12Co (Figure 5c) analysis. First general observation of the diffractograms indicates that Matrix and mixed coatings form a crystalline structure. The resulting structure seems to be well-crystallized in the case of mixed layers, showing well-defined and intense peaks.

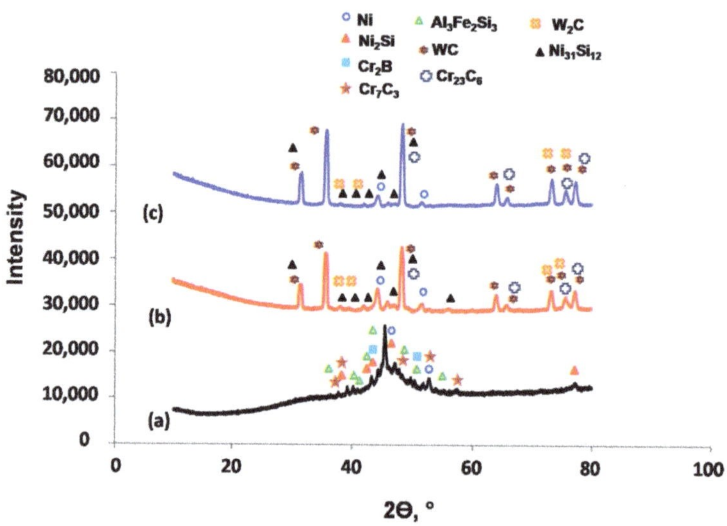

Figure 5. X-ray diffraction patterns of (a) Matrix, (b) Matrix-65% WC/12Co and (c) Matrix-85% WC/12Co.

Regarding NiCrFeSiAlBC coatings (Figure 5a), the major detected phases were: γ-Ni (34.9 wt.%) and the intermetallic phase δ-Ni2Si (35 wt.%). Additionally, Cr7C3 (22.1 wt.%) phase was identified, and Cr2B was present in a smaller fraction (1.2 wt.%), indicating the formation of hard phases: mainly carbides and borides. Moreover, the phase $Al_3Fe_2Si_3$ was identified with a mass fraction of 6.3 wt.% in the coating, demonstrating that Al is incorporated into the matrix structure and forms intermetallic phases with Fe and Si. The intermetallic phases are hard, which enhances wear resistance but may induce brittleness when excessively abundant, or if they are unevenly distributed or poorly formed [31,32]. Then, mixing with 65 wt.%-WC/12Co (Figure 5b) causes many phases to disappear while new phases—from mixing matrix elements and WC/12Co—to appear, as though the presence of WC causes a change in matrix structure. In fact, the γ-Ni phase with a smaller fraction (16.4 wt.%) was still identified after mixing. In addition, the intermetallic phases of Ni and Si were detected from matrix elements, but with a new crystalline structure, Ni31Si12 (34.1 wt.%). WC patterns were detected as well, confirming the WC phase's existence, and its mass fraction was in the order of 28.8 wt.%. Moreover, a new carbide

https://doi.org/10.3390/lubricants14030100

phase was detected, Cr23C6 phase (20.6 wt.%), along with traces of W2C (ε wt.%), which were produced by decarburizing the WC phase [15]. The increase in the hard phase content to 85 wt.% in the matrix caused the WC patterns' intensity to rise (Figure 5c) while the matrix patterns' intensity decreased. Indeed, the WC phase's crystallization seemed to be greater than the matrix phases with 85 wt.% WC/12Co. This interpretation was confirmed by Rietveld refinement: γ-Ni (8.4 wt.%), Ni31Si12 (15.5 wt.%), WC (59.4 wt.%) and Cr23C6 (16.6 wt.%). Since the percentage of un-decarburized WC increased, the decrease in Cr23C6 mass fraction confirms that the appearance of this phase results from the WC decomposition. Although peaks corresponding to W2C were detected, its mass fraction was still negligible even when the WC content increased in the initial mixed powder and the coatings.

The appearance of these latter phases means that diffusion and precipitation of the elements occur in the molten bath [33]. In fact, WC dissolves at the edges under the action of heat, and the dissolved W and C elements diffuse into the molten bath and combine with Cr to form Cr23C6. However, the density of WC is greater than that of the molten matrix, and the melting point is higher. Therefore, the WC will not be completely dissolved. Furthermore, WC coatings grown by HVOF processes do not show completely decarburized phases [15]. The undissolved WC presents the hardest phase, and as the WC content increases, the size of these particles decreases [9], which confirms the microscopic observation of loosely bonded unburnt particles of large size for 65%wt.% WC/12Co mixed coatings, which was not the case for 85%wt.% layers. Indeed, optimized HVOF parameters allow developing cermet coatings without substantial crystallization of highly brittle phases (W2C, Co-W-C) during decomposition and decarburization of WC/12Co.

3.1.3. Porosity Analysis

The porosity levels in the coatings were quantified through image analysis of cross-sectional backscattered electron (BSE) images (Figure 6), using ImageJ software (version 1.53t, National Institutes of Health, Bethesda, MD, USA). For each coating, three representative images from different regions were analyzed. The images were converted to 8-bit grayscale, and a threshold was applied to distinguish pores from solid material, generating binary images (pores = black, solid = white). The area fraction of pores was measured using the "Analyze Particles" function, and the reported values represent the average $\pm$ standard deviation of the three measurements. The mean porosities of the different coatings are represented in Figure 7. NiCrFeSiAlBC coatings exhibited the highest porosity (0.35 $\pm$ 0.02 vol.%), while 65 wt.% WC/12Co and 85 wt.% WC/12Co coatings showed lower porosities of 0.15 $\pm$ 0.01 vol.% and 0.10 $\pm$ 0.01 vol.%, respectively. The lower porosity in WC/12Co coatings, particularly 85 wt.% WC/12Co, is consistent with SEM and XRD results discussed earlier, which demonstrated good wetting of WC grains by NiCrFeSiAlBC and Co binders, more extensive particle melting, and enhanced densification during deposition. Prior XRD analyses also revealed higher carbon retention in 85 wt.% WC/12Co coatings through prominent WC peaks, indicating complete binder melting, which drastically reduces porosity [22,34]. Additionally, retaining solid WC particles likely increases kinetic energy during deposition, promoting tighter inter-particle bonding and improved coating cohesion [21].

3.2. *Micromechanical and Tribological Investigations*

3.2.1. Micro-Hardness Results

Figure 8 displays the average hardness values recorded on the surface of the NiCrFeSiAlBC coating and the two NiCrFeSiAlBC + WC/12Co cermet coatings. The NiCrFeSiAlBC+ 85%WC/12Co coating exhibits the highest average hardness at 1540 HV, roughly 150% greater than that of the NiCrFeSiAlBC coating (960 HV0.3). Notably, the intermediate

https://doi.org/10.3390/lubricants14030100

NiCrFeSiAlBC+ 65%WC/12Co coating shows an average hardness about 9% lower than the 85%WC/12Co variant.

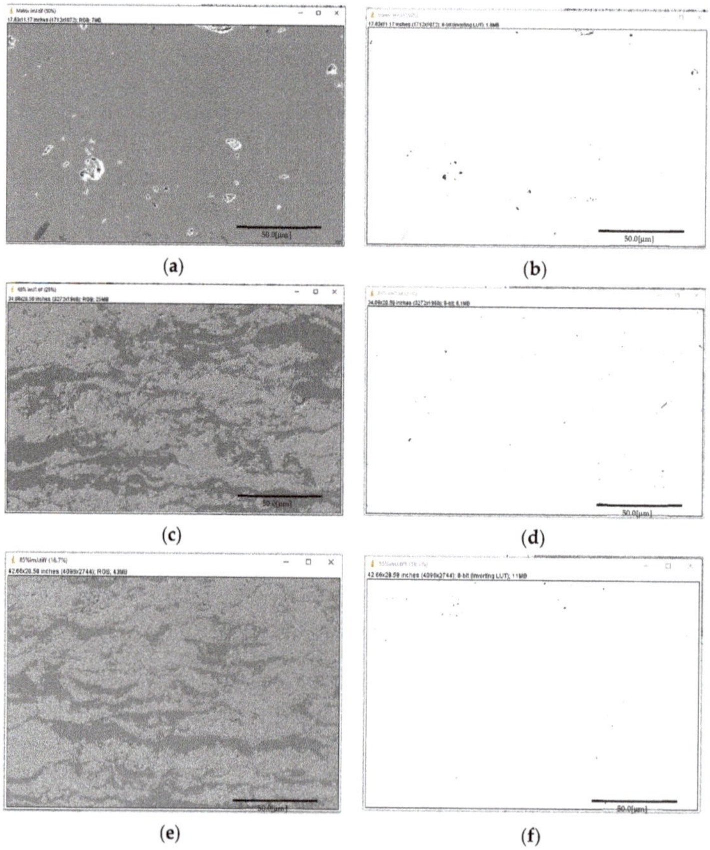

Figure 6. Scheme of porosity assessment in HVOF sprayed coatings for (**a**,**b**) Matrix, (**c**,**d**) Matrix + 65%WC/12Co, (**e**,**f**) Matrix + 85%WC/12Co.

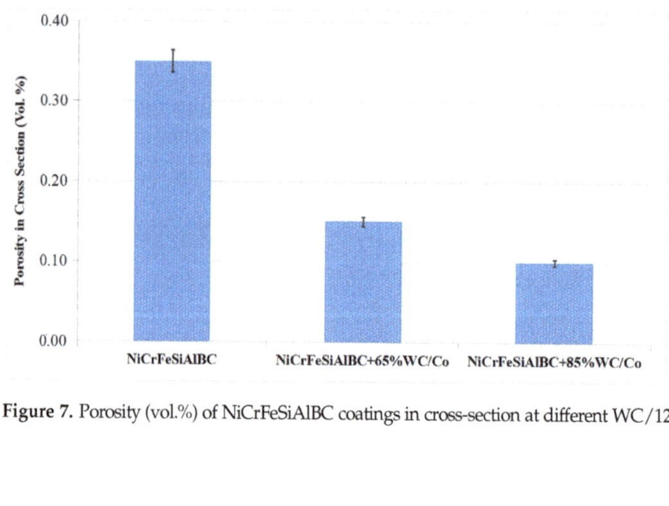

Figure 7. Porosity (vol.%) of NiCrFeSiAlBC coatings in cross-section at different WC/12Co weight fractions.

https://doi.org/10.3390/lubricants14030100

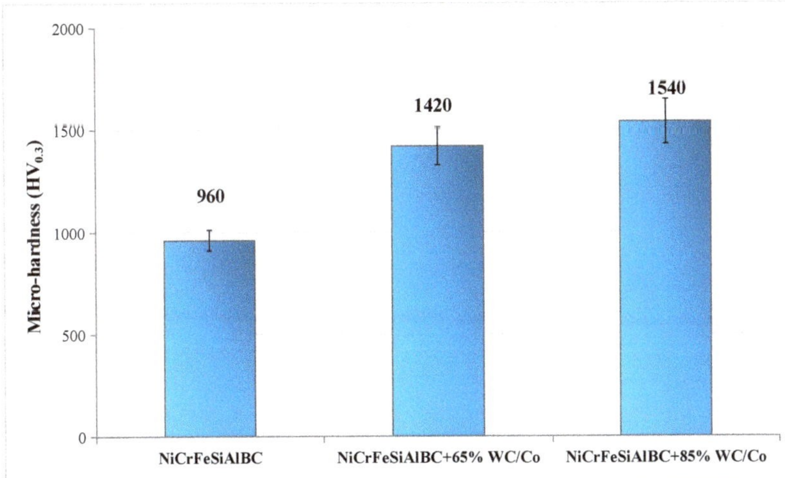

Figure 8. Microhardness (HV0.3) of NiCrFeSiAlBC coatings at different WC/12Co weight fractions.

This trend aligns with findings from Çelik and Benli, who showed that WC-12Co coatings applied via HVOF can exceed 1300 HV0.1 in hardness, depending on feedstock powders and spraying parameters [35]. The substantial variation in hardness stemmed from differences in the HVOF spraying equipment used. The setup in this investigation delivered greater power, accelerating particle velocities during flight and yielding a more compact structure with a harder microstructure [15]. This approach also minimized or effectively suppressed WC decomposition and decarburization. With identical matrix composition and no novel phases identified in the XRD patterns, the improved hardness could be directly linked to the elevated WC fraction. The wide error bars for the NiCrFeSiAlBC + WC/12Co coatings suggest no statistically significant differences in their hardness levels. Despite its lower inherent hardness, the incorporation of NiCrFeSiAlBC did not cause any substantial drop in overall hardness, likely due to enhanced densification from the molten NiCrFeSiAlBC covering the WC cemented phase. These findings, both in terms of mean values and variability, align well with prior research on similar coating systems [30,36].

3.2.2. Tribological Investigations

Figure 9 illustrates typical friction coefficient trends as a function of sliding cycles at varying tungsten carbide mass fractions. The curve profiles clearly reflect the decline in the friction coefficient with increasing WC/12Co weight percentages. After 5000 cycles, the average friction coefficient drops from 0.342 to 0.261 with the incorporation of 85% WC/12Co.

The obtained results prove satisfactory regarding both friction and wear performance, evidenced by the progressive reduction in worn volume following the addition of hard WC particles, as shown in Figure 10. The unreinforced nickel matrix, notwithstanding its high density and adequate fracture toughness, manifests the lowest wear resistance, registering a worn volume of 9.32×10^{-2} mm^3. This underscores its pronounced susceptibility to plastic deformation and adhesive wear mechanisms under sliding contact regimes. Incorporation of 65 wt.% WC/12Co shows a marked diminution in worn volume to 5.47×10^{-2} mm^3, due to the dispersion of hard carbide particulates that enhance load distribution under stress. Superior wear resistance is evidenced in the 85 wt.% WC/12Co reinforced coating, which attains a minimal worn volume of 4.98×10^{-2} mm^3. This enhancement correlates directly with elevated microhardness measurements from Figure 10, wherein increased hardness inversely scales with wear rate. Analogous hardness–weight loss correlations have been

https://doi.org/10.3390/lubricants14030100

documented in prior investigations of sliding wear in WC/Co cermet coatings [21,37]. These observations also corroborate findings by Yao et al., who established that NiCrBSi-WC(Co) cermet coatings exhibit a pronounced decrease in the wear rates with WC(Co) additions up to 50 wt.%, beyond which further increases to 80 wt.% yield only marginal improvements [35].

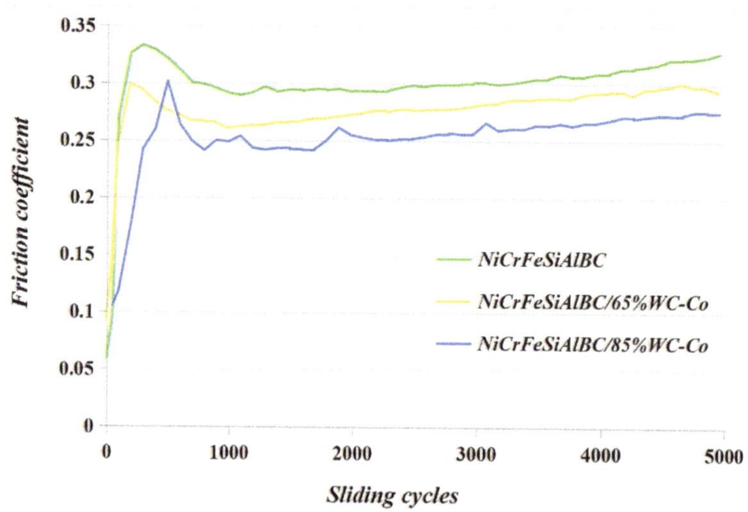

Figure 9. Friction coefficient variation versus number of sliding cycles at different WC/12Co weight fractions.

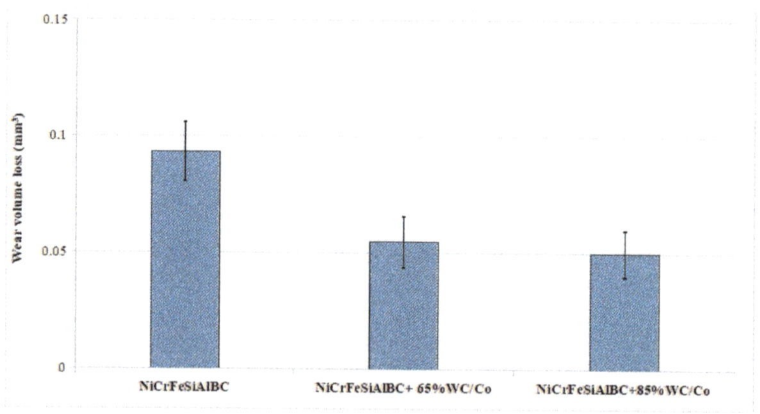

Figure 10. Wear volume after 5000 sliding cycles of NiCrFeSiAlBC coatings at different WC/12Co weight fractions.

Figure 11a–c display optical micrographs of wear tracks formed on NiCrFeSiAlBC matrix coatings reinforced with graded WC/12Co weight fractions after 5000 cycles of reciprocating sliding, and Figure 11d shows typical transversal wear profiles of the different coatings. Progressive WC/12Co incorporation yielded a marked diminution in track width and depth (Figure 11d). This observation correlates directly with quantified wear volume reductions, exhibiting an approximate 40% decrease upon integration of 85 wt.% WC/12Co (Figure 11). Friction and wear constitute system-inherent properties governed by the tribological pair and test environment. Among these factors, antagonist material selection exerts a significant

https://doi.org/10.3390/lubricants14030100

influence on frictional response, particularly when promoting tribochemical interactions with the hard metal surface or inducing transferred particle accumulation on the coating. Previous extensive research on HVOF-sprayed NiCrBSi–WC(Co) coatings (0–80 wt% WC(Co)) using reciprocating wear testing against AISI 52100 steel ball demonstrates a wear mechanism transition: low WC content shows matrix-dominated adhesive wear via inter/intragranular cracking, while high WC fractions exhibit carbide-controlled loading with tribo-oxidation (WxOy) and particle fracture [15]. Other tribological evaluations of WC/Co cermet coatings sliding against Si3N4 ball reveal that initial sliding contact induces severe plastic deformation of the ductile Co matrix, followed by WC particle microcracking, pullout, and debris generation due to loss of intergranular support [37]. For conventional WC/Co coatings tested against alumina counterparts, wear rates correlate primarily with carbide grain size rather than coating hardness or fracture toughness [38]. Submicron WC reinforcements thus critically govern wear performance by refining particle size distributions. In WC cermet coatings with ductile matrices, the superior hardness of WC phases fundamentally drives wear rate reductions. The involved mechanism acts by transitioning from matrix severe adhesive wear at low carbide fractions to carbide-dominated load-bearing regimes at elevated WC contents. The optimal performances are achieved through critical inter-particle spacing that maximizes hardness while preserving matrix continuity.

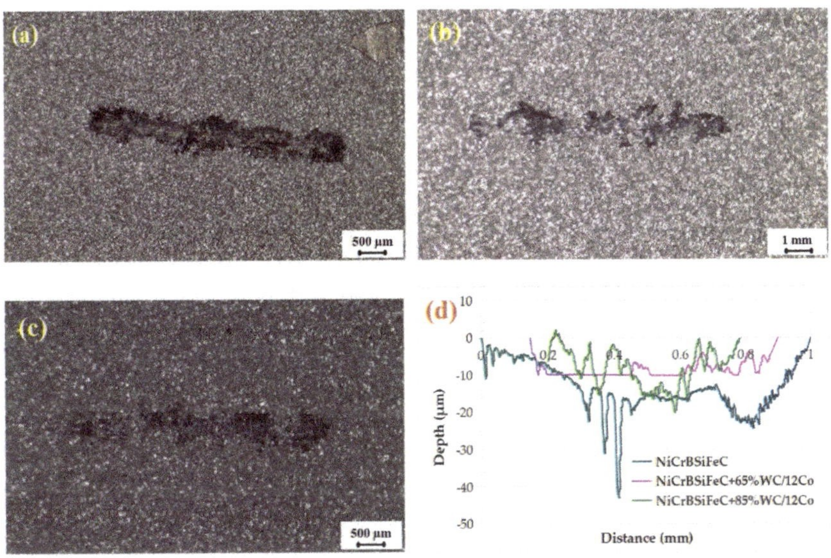

Figure 11. Optical micrographs of wear tracks after 5000 cycles: (**a**) NiCrFeSiAlBC, (**b**) NiCrFeSiAlB + 65% WC/12Co, (**c**) NiCrFeSiAlBC + 85% WC/12Co and (**d**) typical wear profiles of the different coatings.

4. Conclusions

WC/12Co-reinforced NiCrFeSiAlBC cermet coatings were produced by HVOF spraying, and the influence of carbide content on microstructure, hardness, and tribological response was comprehensively assessed. HVOF processing successfully resulted in well-adhered Ni-based coatings and WC/12Co cermet coatings with characteristic lamellar architectures. The introduction of a high content of WC/12Co improved coating compactness and reduced internal defects, even with the WC content increased up to 85 wt.%.

XRD analysis revealed crystalline structures across all coatings. Matrix coatings primarily contained carbides and borides as hard phases, with dominance of pure Ni and

https://doi.org/10.3390/lubricants14030100

intermetallic phases containing Si. The addition of aluminum also enabled the formation of intermetallic hard phases. WC addition promoted dominant WC crystallization, with the detection of $Cr_{23}C_6$, and traces of W_2C from decarburization of WC. In fact, HVOF optimized parameters limited the formation of brittle phases, such as W_2C.

Porosity levels decreased with the increase in ceramic phase content. Mixed coatings with high WC contents in the matrix (on the order of 65 wt.%) showed dense, cohesive and adhesive layers deposited on stainless steel substrates.

Microhardness values increased significantly with carbide fraction, with the highest hardness measured for the coating containing 85 wt.% WC/12Co, reflecting the strong reinforcing effect of the carbide phase and improved densification. The formation of hard phases and particle size refinement likely contributed to enhanced microhardness.

Tribological testing revealed a clear improvement in wear resistance and a reduction in friction coefficient as WC content increased. The highest carbide fraction promoted a transition toward carbide-controlled load support, resulting in minimal wear loss under dry sliding conditions.

Overall, the results demonstrate that carbide content plays a decisive role in governing the mechanical and tribological behavior of NiCrFeSiAlBC-based cermet coatings. Moreover, well controlled parameters of HVOF enable production of high-content WC/12Co cermet coatings that exhibit strong potential for use in severe sliding environments, particularly in engine and powertrain components while offering a sustainable alternative to conventional hard chromium surface treatments.

Author Contributions: Conceptualization, B.B.D.; Methodology, B.B.D.; Validation, B.B.D., M.K. and M.B.; Investigation, B.B.D., F.N., C.A. and A.P.; Resources, A.P.; Writing—original draft, F.H. and F.N.; Writing—review and editing, B.B.D. and M.K.; Supervision, B.B.D., C.A., M.K. and M.B.; Project administration, M.K. All authors have read and agreed to the published version of the manuscript.

Funding: This work was funded by the project AM2R—Agenda Mobilizadora para a inovação empresarial do setor das Duas Rodas [C644866475-00000012—project n. 15], financed by PRR—Recovery and Resilience Plan under the Next Generation EU from the European Union, and had laboratory support of the Centre for Mechanical Technology and Automation (TEMA).

Data Availability Statement: The original contributions presented in this study are included in the article. Further inquiries can be directed to the corresponding authors.

Acknowledgments: The kind assistance provided by RINA (Italy) members in developing the HVOF coatings is gratefully acknowledged.

Conflicts of Interest: The authors declare no conflicts of interest.

References

1. Liu, T.; Li, J.; Li, H.; Liu, J.; Huang, Y.; Wang, Q.; Ramachandran, C.S. Optimization of Preparation Process Parameters for HVOF-Sprayed WC-10Co-4Cr Coatings and Study of Abrasive and Corrosion Performances. *Lubricants* **2025**, *13*, 516. [CrossRef]
2. Ben Difallah, B.; Mebdoua, Y.; Serdani, C.; Kharrat, M.; Dammak, M. Advanced HVOF-Sprayed Carbide Cermet Coatings as Environmentally Friendly Solutions for Tribological Applications: Research Progress and Current Limitations. *Technologies* **2025**, *13*, 281. [CrossRef]
3. Vidaković, I.; Šimunović, K.; Heffer, G.; Špada, V. Microstructural Analysis of Nickel-Based Self-Fluxing Alloy Coatings. *Weld. World* **2024**, *68*, 2819–2836. [CrossRef]
4. Brioua, S.; Belmokre, K.; Debout, V.; Jacquot, B.; Conforto, E.; Touzain, S.; Creus, J. Influence of Spray Parameters on the Metallurgical and Functional Properties of HVOF WC Based Cermets Deposited onto Low Alloy Steel. *J. Mater. Sci. Technol. Res.* **2022**, *9*, 375. [CrossRef]
5. Shuecamlue, S.; Taman, A.; Khamnantha, P.; Banjongprasert, C. Influences of flame remelting and WC-Co addition on microstructure, mechanical properties and corrosion behavior of NiCrBSi coatings manufactured via HVOF process. *Surf. Interfaces* **2024**, *48*, 104135. [CrossRef]
6. Richert, M. Thermally Sprayed Coatings for the Protection of Industrial Fan Blades. *Materials* **2024**, *17*, 3903. [CrossRef]

https://doi.org/10.3390/lubricants14030100

7. Huang, S.; Zhou, J.; Sun, K.; Yang, H.; Cai, W.; Liu, Y.; Zhou, P.; Wu, S.; Li, H. Microstructural Charactistics of Plasma Sprayed NiCrBSi Coatings and Their Wear and Corrosion Behaviors. *Coatings* **2021**, *11*, 170. [CrossRef]
8. González, R.; Cadenas, M.; Fernández, R.; Cortizo, J.L.; Rodríguez, E. Wear behaviour of flame sprayed NiCrBSi coating remelted by flame or by laser. *Wear* **2007**, *262*, 301–307. [CrossRef]
9. Bergant, Z.; Batic, B.S.; Felde, I.; Storm, R.; Sedlacek, M. Tribological Properties of Solid Solution Strengthened Laser Cladded NiCrBSi/WC-12Co Metal Matrix Composite Coatings. *Materials* **2022**, *15*, 342. [CrossRef]
10. Zhao, L.; Yu, H.; Wang, Y.; Zhao, Z.; Song, W.; Chen, C. NiCrBSi Coatings Fabricated on 45 Steel Using Large Spot Laser Cladding. *Materials* **2022**, *15*, 6246. [CrossRef]
11. Shieh, Y.-H.; Wang, J.-T.; Shih, H.C.; Wu, S.-T. Allowing and post-heat treatment of thermal sprayed coatings of self-fluxing alloys. *Surf. Coat. Technol.* **1993**, *58*, 73–78. [CrossRef]
12. Cha, S.C.; Gudenau, H.W.; Bayer, G.T. Comparison of corrosion behaviour of thermal sprayed and diffusion-coated materials. *Mater. Corros.* **2002**, *53*, 195–205. [CrossRef]
13. Mahesh, C. Multiscale abinitio simulation of Ni-based alloys: Real-space distribution of atoms in γ+γ' phase. *Comput. Mater. Sci.* **2015**, *108*, 192–204. [CrossRef]
14. Zhu, L.; Wang, J.; Deng, X.; Dong, Y.; Yang, Y.; Li, D. Comparison of Microstructure and Properties of In-Situ TiN- and WC-Reinforced NiCrBSi Composite Coatings Prepared by Plasma Spraying. *Materials* **2018**, *11*, 2182. [CrossRef]
15. Yao, S.H. Tribological behaviour of NiCrBSi–WC (Co) coatings. *Mater. Res. Innov.* **2014**, *18*, S2-332–S2-337. [CrossRef]
16. Kazamer, N.; Kossman, S.; Baranyi, I.; Chicot, D.; Serban, V.; Rajnai, Z.; Voda, M. Influence of TiB_2 addition on the mechanical and tribological properties of NiCrBSi deposited by flame spraying. *Mater. Tech.* **2018**, *106*, 202. [CrossRef]
17. Singh, G.; Kaur, M. High-Temperature Wear Behaviour of HVOF-Sprayed NiCrSiFeBC–WC–Co Composite Coating. *Surf. Eng.* **2020**, *36*, 1139–1155. [CrossRef]
18. Zhao, L.; Maurer, M.; Fischer, F.; Dicks, R.; Lugscheider, E. Influence of Spray Parameters on the In-Flight Particle Properties and the Properties of HVOF Coating of WC-CoCr. *Wear* **2004**, *257*, 41–46.
19. Richard, C.S.; Béranger, G.; Lu, J.; Flavenot, J.F.; Grégoire, T. Four-point bending tests of thermally produced WC-Co coatings. *Surf. Coat. Technol.* **1996**, *78*, 284–294. [CrossRef]
20. Tahir, A.; Li, G.-R.; Liu, M.-G.; Yang, G.-J.; Li, C.-X.; Wang, Y.-Y.; Li, C.-J. Improving WC-Co coating adhesive strength on rough substrate: Finite element modeling and experiment. *J. Mater. Sci. Technol.* **2020**, *37*, 1–8. [CrossRef]
21. Chakradhar, R.P.S.; Prasad, G.; Venkateswarlu, K.; Srivastava, M. An Investigation on the Wear and Corrosion Behavior of HVOF-Sprayed WC-12Co-Al_2O_3 Cermet Coating. *J. Mater. Eng. Perform.* **2018**, *27*, 1241–1248. [CrossRef]
22. Nahvi, S.M.; Jafari, M. Microstructural and mechanical properties of advanced HVOF-sprayed WC-based cermet coatings. *Surf. Coat. Technol.* **2016**, *286*, 95–102. [CrossRef]
23. Shen, X.; Peng, H.; Xue, Y.; Wang, B.; Su, G.; Zhu, J.; Li, A. Microstructure and Properties of WC/Ni-Based Laser-Clad Coatings with Different WC Content Values. *Materials* **2022**, *15*, 6309. [CrossRef]
24. Berger, L.M. Application of hard metals as thermal spray coatings. *Int. J. Refract. Met. Hard Mater.* **2015**, *49*, 350–364. [CrossRef]
25. Ibarra, J.d.J.; González, M.A.; Rodríguez, E.; Vásquez, G.I.; Medina, A.; Bernal, J.; Aguilar, C.; Velez, E.E. Enhanced Fracture Toughness of WC-CoCr Thermally Sprayed Coatings by the Addition of NiCrFeSiBC and Mo and its Influence on Sliding Wear Behavior. *Coatings* **2024**, *14*, 1207. [CrossRef]
26. Suryanarayana, C. Mechanical alloying and milling. *Prog. Mater. Sci.* **2001**, *46*, 1–184. [CrossRef]
27. Zhuang, Z.; Li, Z.; Xu, L.; Feng, Y.; Wang, R.; Peng, C. Influence of ball milling conditions on powder characteristics and sintering densification of Mo–Cu alloy. *Int. J. Refract. Met. Hard Mater.* **2024**, *125*, 106914. [CrossRef]
28. Ni, M.; Zhang, X.; Liu, Y.; Zhou, H. Effect of milling time on powder characteristics and mechanical performance of Ti–4 wt.% Al alloy. *Powder Technol.* **2019**, *342*, 275–287. [CrossRef]
29. Singh, J.; Kumar, S.; Mohapatra, S.K. Erosion wear performance of Ni-Cr-O and NiCrBSiFe-WC(Co) composite coatings deposited by HVOF technique. *Ind. Lubr. Tribol.* **2019**, *71*, 610–619. [CrossRef]
30. Ramesh, M.R.; Prakash, S.; Nath, S.K.; Sapra, P.K.; Venkataraman, B. Solid particle erosion of HVOF sprayed WC-Co/NiCrFeSiB coatings. *Wear* **2010**, *269*, 197–205. [CrossRef]
31. Mathew, J.; Remy, G.; Williams, M.A.; Tang, F.; Srirangam, P. Effect of Fe intermetallics on microstructure and properties of Al-7Si alloys. *JOM* **2019**, *71*, 4362–4369. [CrossRef]
32. Li, L.; Li, X.; Shi, L.; Huang, S.; Xu, C.; Lu, G.; Guan, S. CALPHAD-Assisted Analysis of Fe-Rich Intermetallics and Their Effect on the Mechanical Properties of Al-Fe-Si Sheets via Continuous Casting and Direct Rolling. *Metals* **2025**, *15*, 578. [CrossRef]
33. Feng, L.; Li, C.; Han, X.; Luo, F.; San, H. Experimental Study on Preparation of Tungsten-Carbide- Particle-Reinforced Nickel-Based Alloy Plasma Surfacing Layer on Descaling Roller Surface. *Coatings* **2024**, *14*, 729. [CrossRef]
34. Liu, J.; Bai, X.Z.; Chen, T.; Yuan, C. Effects of Cobalt Content on the Microstructure, Mechanical Properties and Cavitation Erosion Resistance of HVOF Sprayed Coatings. *Coatings* **2019**, *9*, 534. [CrossRef]

https://doi.org/10.3390/lubricants14030100

35. Çelik, İ.; Benli, B. The effect of WC-12Co and CrC-NiCr hard coatings applied by HVOF method on the microstructure, mechanical and surface properties of steel. *Metall. Res. Technol.* **2024**, *121*, 318. [CrossRef]
36. Khuengpukheiw, R.; Wisitsoraat, A.; Saikaew, C. Wear behaviors of HVOF-sprayed NiSiCrFeB, WC-Co/NiSiCrFeB and WC-Co coatings evaluated using a pin-on-disc tester with C45 steel pins. *Wear* **2021**, *484–485*, 203699. [CrossRef]
37. Yuan, J.; Ma, C.; Yang, S.; Yu, Z.; Li, H. Improving the wear resistance of HVOF sprayed WC-Co coatings by adding submicron-sized WC particles at the splats' interfaces. *Surf. Coat. Technol.* **2016**, *285*, 17–23. [CrossRef]
38. Yang, Q.; Senda, T.; Ohmori, A. Effect of carbide grain size on microstructure and sliding wear behavior of HVOF-sprayed WC–12%Co coatings. *Wear* **2003**, *254*, 23–34. [CrossRef]

https://doi.org/10.3390/lubricants14030100

MDPI AG
Grosspeteranlage 5
4052 Basel
Switzerland
Tel.: +41 61 683 77 34

Lubricants Editorial Office
E-mail: lubricants@mdpi.com
www.mdpi.com/journal/lubricants

www.ingramcontent.com/pod-product-compliance
Lightning Source LLC
Chambersburg PA
CBHW061546210926
51850CB00009B/2013
* 9 7 8 3 7 2 5 8 7 3 7 2 2 *